S0-APO-398

Introduction to Computer Engineering

Franco P. Preparata
University of Illinois at Urbana—Champaign

HARPER & ROW, PUBLISHERS, New York
Cambridge, Philadelphia, San Francisco,
London, Mexico City, São Paulo, Singapore, Sydney

Sponsoring Editor: Peter Richardson
Project Editor: David Nickol
Text Art: Reproduction Drawings, Ltd.
Production: Delia Tedoff
Compositor: Science Press
Printer and Binder: The Maple Press

Introduction to Computer Engineering

Library of Congress Cataloging in Publication Data

Preparata, Franco P.
Introduction to computer engineering.

Bibliography: p.
Includes index.
1. Computer engineering. I. Title.
TK7885.P624 1985 621.3819'58 84-15666
ISBN 0-06-045271-4

84 85 86 87 9 8 7 6 5 4 3 2 1

Contents

PREFACE vii

Part One Overview

Chapter 1. Representation of Information 3

1.1. Introduction 3
1.2. Representation of Information (Letters and Numbers) 5
1.3. Base Conversions 6
1.4. Introduction to Binary Arithmetic: Addition and Subtraction 12
1.5. Hexadecimal and Octal Notations 13
1.6. Other Important Codes 14
1.7. Codes for Error Control 17
NOTES AND REFERENCES 19
PROBLEMS 20

Chapter 2. Introduction to Computer Organization and Design 23

2.1. A smooth transition 23
2.2. A Simplistic Computer 25
2.3. The Instruction Repertoire of SEC (Stripped-Down Version) 28
2.4. Some Simple Programs in Machine Language 30
2.5. Applications of Digital Computers 34
2.6. Levels of Digital System Design 35
NOTES AND REFERENCES 40
PROBLEMS 41

Part Two General Techniques

Chapter 3. Combinational Networks and Switching Algebra 47

3.1. Introduction to Binary Functions of Binary Variables 47
3.2. Switching Functions of One and Two Variables 49
3.3. Networks and Expressions 51
3.4. Switching Algebra 54
3.5. Boolean Expressions: Normal and Canonical Forms 58
3.6. Other Important Boolean Connectives 63
3.7. Review Example 67
NOTES AND REFERENCES 68
PROBLEMS 69

Chapter 4. Elements of Logical Design, Combinational Networks and Modules 72

4.1. Introduction 72
4.2. Analysis and Design of Combinational Networks 73
4.3. Minimization Techniques 78
4.4. Minimization Technique Based on the Karnaugh Map 79
4.5. Switching Functions with "Don't Care" Conditions 88
*4.6. A Tabular Minimization Technique (Quine-McCluskey) 92
*4.7. Design of All-NAND (or All-NOR) Combinational Networks 97
4.8. Combinational Modules (MSI Modules) 99
NOTES AND REFERENCES 112
PROBLEMS 113

Chapter 5. Sequential Networks 117

5.1. Introduction 117
5.2. Timing Diagrams 117
5.3. Feedback and Memory (Latch) 118
*5.4. Asynchronous Sequential Networks: A Brief Discussion 121
5.5. Gated Latches and Flip-Flops (Master-Slave and Edge Triggered) 130
5.6. Other Types of Clocked Flip-Flops 134
5.7. Parallel Registers 138
5.8. Synchronous Sequential Networks 138
5.9. Analysis of Sequential Networks 143
5.10. Synthesis of Sequential Networks 147
5.11. A General Design Technique Based on ROMs and PLAs 157
5.12. Other Important Sequential Modules 160
NOTES AND REFERENCES 164
PROBLEMS 165

Part Three System Organization

Chapter 6. Binary Arithmetic and the Arithmetic-Logic Unit (ALU) 173

6.1. Addition, Subtraction, and the Representation of Negative Numbers 173

6.2. Addition and Subtraction of Integers in the Two's Complement Notation 178
*6.3. The One's Complement Notation 183
6.4. Adder Cell (Full Adder) 184
6.5. Parallel Adder Subsystems 186
6.6. Arithmetic-Logic Unit (ALU) 195
NOTES AND REFERENCES 202
PROBLEMS 203

Chapter 7. Computer Organization: CPU and Memory 207
7.1. Register Transfer Language 207
7.2. Microsequence Implementation 212
7.3. Basic Organization of the SEC Processor 215
7.4. FETCH and EXECUTE Microsequences 221
7.5. Index Registers 223
7.6. Subroutines 228
7.7. Nested Subroutines—Stacks 233
*7.8. Paged Memory and Indirect Addressing 236
7.9. A Review of the Addressing Modes of the SEC 239
NOTES AND REFERENCES 240
PROBLEMS 241

Chapter 8. Input/Output 244
8.1. General Considerations 244
8.2. Programmed Input/Output 249
8.3. Interrupt Input/Output 256
*8.4. Direct Memory Access (Data Break) 264
*8.5. The "Cold Start" of the System 267
NOTES AND REFERENCES 269
PROBLEMS 270

Chapter 9. Control Unit: Microprogramming 272
9.1. Introduction 272
9.2. Microstep Implementation and Control Signals 272
9.3. The Function of the Control Unit 274
9.4. Microprogrammed Control Unit 279
NOTES AND REFERENCES 286
PROBLEMS 287

Bibliography 288

Appendix—Microprocessors 291
A.1. Introduction 291
A.2. Designer's Architecture and Programmer's Architecture 293
A.3. Control Instructions 299
A.4. External Bus Control and I/O Processes 301
A.5. Software Interrupts—Trap Instructions 307

Index 309

Preface

This book is intended as a text for a one-semester course (possibly, two one-quarter courses) in computer engineering, electrical engineering, or computer science. This course is intended as a first course in computer systems, and, as such, it has no mandatory prerequisite, except college standing. Indeed, some cursory references are made to electricity and high-level languages, only because it is presumed that most students will have some familiarity with those disciplines and thus will find some comfort in relating to them. However, the student will benefit immensely from having attended an introductory course in computer programming. This not because of explicit references to a specific high-level language he or she may have learned (such as FORTRAN or Pascal), but because the prior exposure to the notions of problem, algorithm, and program will greatly motivate the exploration and study of the machines that execute those programs. For this reason, the present course is envisioned as a sophomore/junior level course, where the general maturity stemming from one or two years of higher education can only create a more favorable intellectual climate.

The objective of the present course is to provide a basic knowledge of the organization and operation of computing systems. Superficially one may say, according to commonly held distinction, that this course plays with regard to hardware the role played by an introductory programming course with respect to software. However, the pedagogical objective is more mature and more ambitious; indeed it aims at closing the gap between these two complementary facets of computing, by showing that the hardware–software trade-off is an all-encompassing aspect of computing experience and is dictated by economical criteria. For these reasons, the course is placed centrally in both computer science and computer engineering, and provides the conceptual background for further (possibly diverging) studies in computer architecture, logical design, switching theory, data structures, and system software.

Two main features of the text deserve some illustration and discussion. The first concerns the sequence of presentation of the various topics, the other a choice of pedagogical nature.

The order of presentation is a judicious application of the top-down viewpoint. As is obvious, the top-down approach—starting from the global and introducing successive analytic refinements all the way down to the emergence of fine details—has the advantage of building at each stage of study the appropriate motivational framework for the next stage, although it suffers from some inevitable vagueness at the top levels of analysis. At the other extreme, the bottom-up approach—starting from the level of finest details, and synthesizing increasingly complex subsystems using previously defined objects—maintains at each stage a complete explanation of the objects being studied, but lacks the very powerful educational guide represented by a vision of the global objective of the study.

In concrete terms, in a rigorous top-down approach, starting from a simple algorithm, one expresses it as a simple program to be executed by a machine, which consists of black boxes that reflect the classical von Neumann architecture. One could continue on this path by revealing the internal structure of each black box as an interconnection of simpler black boxes (registers, etc.). However, the interconnection rules and the behavior of data paths would appear rather artificial, when completely divorced from actual implementations by means of logical devices.

Therefore, I have chosen what I hope is a suitable and effective compromise between the two methodologies. The computer is first presented as a system described at the functional unit level which is capable of executing programs expressed in machine language; the rest of the text is devoted to the design of a system that supports this language. In other words, after providing the essential motivation for the structure of a computer, we resort to the bottom-up approach to develop the techniques needed for the design of digital networks, both combinational and sequential. Once these techniques have been absorbed in sufficient detail, we revert to the original plan, by providing the detailed register structure of the various functional units. On this detailed structure, machine language instructions are realized as sequences of data transfers and other operations; finally, the sequencing is realized by a microprogrammed control unit.

But perhaps more important than the presentational stategy—the aim of which is just to improve teacher–student communication—is the choice of the pedagogical character of the text, that is, the subject matter being presented. The basic tenet is that there are fundamental notions in computing systems that have appeared in the past 30 or so years, such as stored program, instruction fetch and execute, command-address partition of an instruction word, and so on, which transcend the innumerable embodiments of the von Neumann machine. On the other hand, many of the currently standard features of existing processors, such as index registers, general registers, stacks, subroutine-call commands, sophisticated I/O controls, and so on, are devices aimed primarily at improving (sometimes dramatically) the efficiency of computing, and are the results of years of application of the hardware–software trade-off principle. With these ideas in mind, when I had to choose the processor to be illustrated in the text, after a long, agonizing reflection, I decided in favor of an "educational machine." I am fully aware of the objections raised against an artificial machine, which still keeps the students one step removed from the reality in which they will have to operate soon after the completion of their college experience. To respond to these objections, I have "tied" the educational machine to the real world in the last chapter of this text (called Appendix), where the main features of the powerful Motorola MC68000 microprocessor are described and contrasted with those of the much simpler educational computer.

There are two main reasons that helped me arrive at my decision in favor of an artificial machine, one philosophical and one very practical.

The philosophical reason is that existing processors are engineering products, and as such their design is optimized to the ultimate. To give just one example, there is nothing fundamental in variable length instructions (as common in today's micropocessors), except the wish to minimize memory occupancy and execution times. On the other hand, the complications in describing the structure and the behavior of a control unit designed to manage such instructions, are an unnecessary burden which tends to blur together two separate classes of issues: functionality and optimization. There is certainly merit in an educational approach where engineering optimization follows the treatment of the underlying "basic science." Therefore, the possibility of isolating the "fundamental" from the "contingent" provides additional weight in favor of an educational machine (called SEC).

The practical reason, as pointed out by other authors of related works, is that manufacturers do not disclose the ultimate details of their microprocessors, whereas the student of computer engineering or computer science should be exposed to all aspects of computer design, not just those that are visible to the user. Indeed, what is available in the literature on existing machines is basically the "user's architecture," to be contrasted with the "designer's architecture" where the system is described at the gate level. In addition, the computing principles illustrated in this text have retained their validity for years and are likely to do so well beyond the useful lifetime of today's commercially available microprocessors.

Once this difficult choice was made, another extremely attractive possibility emerged. Indeed, it was possible to conceive SEC as a "modular" machine, built around a basic unit, consisting of a RAM and an accumulator and endowed with the most rudimentary instruction reportoire. With this plan, one could fully see the hardware–software trade-off principle at work. Indeed, only through the frustrations of managing programming loops by updating pointers and counters in main memory, is one forced to "invent" the index registers. The awkwardness of managing subroutines through the conventional repertoire provides the motivation for the introduction of special instructions. The demands on the programmer represented by polling tests in "programmed I/O" are alleviated with the introduction of "interrupt I/O." And one may continue with additional examples. In other words, the choice of a pedagogical machine affords an interesting dividend: It enables us to present the computer as an "evolving system," in which—starting from a bare-bone structure—new features are introduced one by one in response to explicit needs of functionality and performance. In some sense, this step-by-step building of the computer is nothing but a retracing of the history of the von Neumann machine.

Of course, there are many omissions. For example, many important techniques of logical design are not even mentioned; the treatment of asynchronous sequential networks is barely an introduction; important features in the computer organization, as general registers, are omitted; the topic of system software—which is an integral part of a computing system—is not even approached, except for the illustration of a simple bootstrap loader.

The limited time for a semester of instruction and the wish not to inflate the text much beyond that required for a one-semester course are constraints on the selection and extension of topics. But more importantly, the aim of the projected course is the study of the "fundamentals" of computer systems rather than the immediate attainment of a "professional" state-of-the-art competence. And this has a twofold reason.

First, it would be unrealistically ambitious to expect to attain levels of truly professional competence in an introductory course, which for many readers may just break the ground of this fascinating field. Therefore, although binary arithmetic, logical design, computer architecture, machine language programming, and microprocessors are the topics of the text, this cannot be an in-depth course in any of these topics. In other words, rather than attempt to replace a number of existing courses with a bare exposure to the above topics, in a well-structured curriculum in computer science and engineering, this introductory course forms the background to many specialized, professional level courses. Indeed, it should be emphasized that, while a course based on this text may be adequate for the engineer whose professional interest is not the field of computers, a good curriculum should follow up with in-depth treatments of each of the above topics for the student who intends to become a professional in computer science and engineering.

The second reason is the belief—supported by decades of experience—that a large fraction of the presented material is the basis for lifelong learning in computing systems, beyond ephemeral features dictated by contingent technologies. In particular, I believe that the detailed study of the complex microprocessors available today will be greatly facilitated by the general background provided by a course like the one for which this book is intended.

And now I wish to have a final word on the "detail" of the presentation. The editorial objective has been uniformity of coverage, within a general framework of conciseness. In other words, from my long experience of classroom usage, I estimate that about 6 to 8 pages of printed text correspond to the material covered in 50 minutes of lecture, uniformly throughout the book.

I like to conclude by acknowledging the invaluable suggestions made by many of my colleagues here and elsewhere, who tested preliminary versions of this material in the classroom. Particularly I would like to thank for their advice, A. Apostolico, D. J. Brown, R. M. Brown, M. C. Loui, F. Luccio, and D. L. Waltz. I am also very grateful to G. M. Masson, J. Pugsley, and E. R. Robbins, who acted as reviewers for the manuscript and whose extremely constructive criticism has been incorporated in this final version.

FRANCO P. PREPARATA

PART ONE

OVERVIEW

Chapter 1

Representation of Information

1.1 INTRODUCTION

The first question to be addressed in information processing is how we choose to represent "information." Without delving into the philosophical aspects of what is information, we may just reflect a moment about some common experiences. For example, when we want to know the present temperature, we call our "friendly" telephone number, and we obtain an answer: 76°F. Information consists of receiving *one* message (in this case, 76°) out of *a set of possible messages* (for example, the temperatures in Fahrenheit degrees expressed as integers between −30 and 110).

In this course we shall only consider cases in which the "messages" are in finite number (for example, the integers between 0 and 1,000,000). It is also a familiar experience that our messages, or *objects,* are represented by means of symbols: *numbers* are represented by means of the *numerals* 0, 1, . . . , 9, *words* are represented by means of the *letters* A, B, . . . , Z. In these examples numbers and words are the objects, while numerals and letters are the symbols. Notice that the sets of numerals and of letters are both finite (and small).

In order to convey information without error (or with as little error as possible), the symbols used for representation must be *easily distinguishable* from each other; also we require that they be *reliable,* that is, that they do not alter themselves during the time in which they are to be used.

Thus we have the following definition:

Information is represented using a finite number of symbols that are reliable and easily distinguishable.

The symbols used may take on a variety of forms. They could be marks on a sheet of paper (such as numerals or letters), as is common in printed communication; or, they could be different voltage levels on a communication line, a hole (or its absence) in a punched card, and so on. Inside digital equipment information will be represented by voltage levels on wires, or by the magnetization of appropriate devices, and the like. The requirements of reliability and easy distinguishability call for as small a set of symbols as possible; and since a set of symbols for representing information cannot have fewer than two elements (with *one* symbol, there is nothing to distinguish it from), this is why, almost universally, information inside digital equipment is represented by two symbols.

The set of two symbols is normally referred to as the *binary alphabet* and the two symbols themselves are 0 and 1. In general, we shall represent our objects by means of *strings,* or *vectors,* of k binary symbols (*bits*). Since each such vector can be used to represent a distinct object, a natural question is "How many are the distinct binary vectors with k components?" The answer is simply 2^k. In fact, with one component we have two distinct vectors: 0 and 1. Assume (inductively) that there are 2^{k-1} vectors with $(k - 1)$ components and imagine appending to each such vector on the left an additional component. Since the latter can be either 0 or 1, *each* vector generates *two* new vectors, thereby obtaining $2 \times 2^{k-1} = 2^k$ vectors with k components, as claimed.

Up to this point, we have explicitly restricted ourselves to the problem of representing the elements of a finite set of messages. A typical instance of this situation is, for example, the set of letters of the alphabet. But there may be cases in which the *finite* set of messages arises from the measurement of a continuous physical entity (such as a pressure, a temperature, a distance), which has been "granulated," or *quantized* or *made discrete,* according to the precision of the measuring instrument (for example, if temperatures are given in integer degrees, the degree is the granule, or *quantum,* of our measurement). In such a case one may select an alternative way of transmitting and processing information; specifically, we may transform the original continuous entity into another continuous entity proportional to it (typically an electrical voltage), process it in this form, and possibly put in discrete form the result of processing. This type of information processing is called *analog,* as opposed to *digital;* analog, because the object of processing is an entity proportional—hence, analogous—to the original one. A typical case where analog and digital processing are ideally contrasted is in the transmission of speech over a communication channel. Through a microphone, speech is initially transformed into a continuous electrical waveform; next, (1) in analog communication, this waveform is directly transmitted; (2) in digital communication, the waveform value is measured at evenly spaced instants, the measurements are quantized (analog-to-digital conversion), that is, each of them is transformed into a string of symbols, and these symbols are finally transmitted over the channel.

This digression on analog information processing has been made for the purpose of completeness; in this course we shall restrict ourselves to *digital information processing.*

1.2 REPRESENTATION OF INFORMATION (LETTERS AND NUMBERS)

As we have seen, representation of information (or *coding,* for short) is the assignment of a distinct string of symbols to each of the distinct objects we want to represent. Obviously, in digital computers, numbers are fundamental objects to be dealt with, so we will consider the problem of representing them.

We begin with the problem of representing the 10 numerals 0, 1, . . . , 9 by means of fixed-length strings of bits. As we shall see, the representation may vary depending on the nature of the application. We may want to code for the following objectives:

1. *For mere distinguishability.* Because $2^3 = 8 < 10 < 16 = 2^4$, 3 bits are not enough, but 4 are. So 4 bits are required. Since we only need to assign a different 4-bit string to each numeral, the number of possible codings is enormous. In fact, we begin by choosing a representation for numeral 0: this can be done in 16 ways. Next we choose a representation for numeral 1: this can be done in 15 ways (one string has been used for 0). Similarly, 2 can be coded in 14 ways, and so on. Thus, the number of different codings for the 10 numerals is

$$16 \times 15 \times 14 \ldots \times 7 = \frac{16!}{6!} \cong 2.905 \times 10^{10}$$

Typically, mere distinguishability is all we want for letters of the alphabet and punctuation symbols; for numbers we may want more, as we shall see.

2. *For ease in applications.* Such applications are, for example, display and analog-to-digital conversion. We shall discuss the latter in Sec. 1.6, and consider here a reasonable code for display of numerals in the familiar 7-segment pattern [Figure 1.1(a)]. Specifically, we may use 1 bit to denote the status of each segment (1 for lighted; 0 for not lighted). The code is shown in Figure 1.1(b); for example, segments S_0, S_2, S_3, S_5, and S_6 will be lighted to display numeral 5.

3. *For use in arithmetic.* We would like to adopt a number representation which simplifies the execution of arithmetic operations on numbers. In this

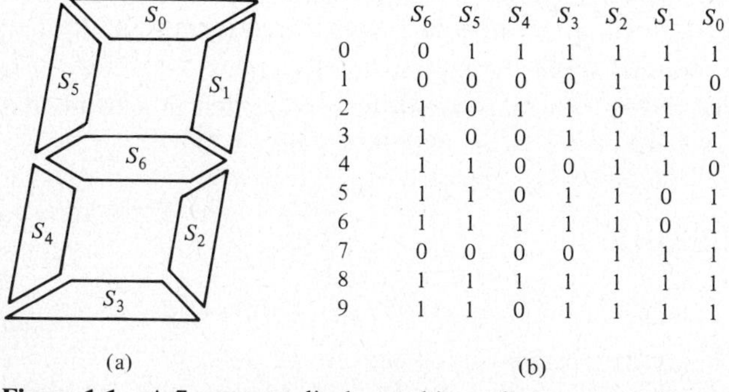

	S_6	S_5	S_4	S_3	S_2	S_1	S_0
0	0	1	1	1	1	1	1
1	0	0	0	0	1	1	0
2	1	0	1	1	0	1	1
3	1	0	0	1	1	1	1
4	1	1	0	0	1	1	0
5	1	1	0	1	1	0	1
6	1	1	1	1	1	0	1
7	0	0	0	0	1	1	1
8	1	1	1	1	1	1	1
9	1	1	0	1	1	1	1

(b)

Figure 1.1 A 7-segment display and its coding.

context we no longer focus on the numerals 0, . . . , 9, but consider the set of nonnegative integers 0, 1, 2, ...

The familiar representation of numbers, known as *arabic notation* (or *decimal notation*), is a special case of the more general class of *positional notations*. In these notations, a number is represented by a string of symbols, and each symbol has a different *weight* depending on the position where it occurs. Specifically, in the number 752, 7 has weight 100, 5 has weight 10, and 2 has weight 1, so that the *value* of 752 is $7 \times 10^2 + 5 \times 10^1 + 2 \times 10^0$. The number 10, whose powers give the weights of the positions, is called the *radix* or *base* of the notation.

The positional notation can be used also to represent mixed numbers, that is, integer + fraction. For example, the decimal number 23.56 has the value $2 \times 10^1 + 3 \times 10^0 + 5 \times 10^{-1} + 6 \times 10^{-2}$.

There is nothing magic about the base 10; in fact any integer base $b \geq 2$ can be used, together with a set of integers $\{0, 1, \ldots, b - 1\}$ used as numerals. In our context, we shall normally use the base $b = 2$, so the integers $\{0, 1\}$ will be our numerals.

In general, given a number N, a string of binary digits $a_n, \ldots, a_{-m}$, as shown below, is used for its base b representation,

Integer — Point — Fraction

$$N \equiv (a_n a_{n-1} \ldots a_0 \,\vdots\, a_{-1} a_{-2} \ldots a_{-m})_b$$

Base[1]

Positional Representation

while the value of N is given by the formula

Number — Digit — radix, base

$$N = \sum_{i=-m}^{n} a_i b^i$$

Value

The fundamental importance of the positional notation is the great ease with which arithmetic operations are performed on numbers represented positionally. Indeed, as we shall see later (Chapter 6), the addition (or subtraction) of two numbers can be carried out by processing one position at a time (see Sec. 1.4).

4. *For error control*. Since information is represented by values of physical entities (typically, electric voltages) and is processed or transmitted through physical devices, ever present disturbances or malfunctions may alter the intended value, cause an *error*. Information can be represented in a form that allows either to detect an error or even to correct it (see Sec. 1.7).

1.3 BASE CONVERSIONS

In connection with the binary notation, two problems are important:

1. To obtain the binary representation of a given value N.
2. The converse of **1.** to obtain the value of a given binary representation.

[1]Whenever confusion might be possible, a base b representation should have a subscript b.

These two problems are usually referred to as base conversions, because normally a value N is given in the decimal notation and we want to express it in binary, and conversely. Of course, one may also want to convert a representation in base b_1 to its equivalent in base b_2; however, for simplicity, we shall concern ourselves here with conversions between decimal and binary notations.

Before discussing a coherent and general framework for base conversions, it is appropriate to discuss some rather crude but popular methods. These methods, which are not very attractive from the point of view of efficiency of operation (especially if carried out by an automatic device, such as a digital computer), are nevertheless quite useful for human users, since we normally prefer the comfort of habit to efficient, but hard to remember, methods.

The most natural brute force method is "substitution," which is quite handy in obtaining the decimal value of a number represented in binary. For example, if $N \equiv 1\,0\,1\,1\,0\,1\,0_2$, we know that the value of N is $N = 1 \times 2^6 + 0 \times 2^5 + 1 \times 2^4 + 1 \times 2^3 + 0 \times 2^2 + 1 \times 2^1 + 0 \times 2^0 = 64 + 16 + 8 + 2 = 90$. What we have done is the *substitution* of each of the nonzero *binary* digits of N by its weight represented in *decimal,* and then tallied the resulting terms. This method is quite comfortable in converting from binary to decimal, since each binary digit is either 0 or 1. It is much less comfortable, however, in the opposite conversion, since we have little or no familiarity with the binary representations of powers of 10, and even less familiarity with binary arithmetic.

Therefore, in converting an integer N from decimal to binary, it is easier to determine first the largest power 2^k of 2 contained in N, and then to obtain the new integer $N' = N - 2^k < 2^k$, which is in turn to be expressed in binary. For example, given $N = 211$, we have

	$2^7 = 128$ is largest power of 2 contained in 211
$211 - 128 = 83$	$2^6 = 64$ is largest power of 2 contained in 83
$83 - 64 = 19$	$2^4 = 16$ is largest power of 2 contained in 19
$19 - 16 = 3$	$2^1 = 2$ is largest power of 2 contained in 3
$3 - 2 = 1$	$2^0 = 1$ is largest power of 2 contained in 1

In this manner we obtain $N = 1\,1\,0\,1\,0\,0\,1\,1$.

After this brief digression on the naive conversion methods, we are ready to consider a more general framework. For convenience we shall consider separately two cases, integers and fractions.

1.3.1 Problem 1. Conversion of Integers

Let N be an integer. Recall that N is given by

$$N = a_n 2^n + a_{n-1} 2^{n-1} + \cdot \cdot \cdot + a_0 \tag{1.1}$$

which can be reexpressed as

$$N = a_0 + 2(a_1 + 2(a_2 + \cdot \cdot \cdot + 2(a_{n+1} + 2a_n) \ldots)) \tag{1.2}$$

Now, we define the following numbers $S_n, S_{n-1}, S_{n-2}, \ldots$:

$$S_n = a_n$$
$$S_{n-1} = a_{n-1} + 2S_n$$
$$\vdots$$
$$S_i = a_i + 2S_{i+1} \tag{1.3}$$
$$\vdots$$

From Eqs. (1.2) and (1.3), it is clear that $S_0 = N$. This yields the following simple conversion procedures:

1. Binary → decimal (integer), that is, given $a_n a_{n-1} \ldots a_0$ find its value N.

   a. Set $S_n = a_n$.
   b. For $i = n - 1, n - 2, \ldots, 0$ compute $S_i = a_i + 2 \times S_{i+1}$.
   c. Set $N = S_0$.

Notice that in this procedure we scan the sequence $a_n, \ldots, a_0$ from left to right, that is, from the *m*ost *s*ignificant *d*igit (MSD) to the *l*east *s*ignificant *d*igit (LSD), and at each step we obtain S_i by *multiplying* S_{i+1} by 2 and adding a_i.

EXAMPLE 1.1

Find the value of 1101001 (in this case $n = 6$).

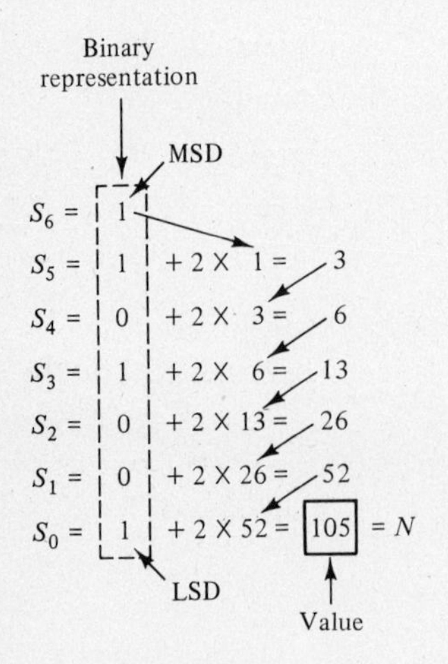

2. Decimal → binary (integer), that is, given an integer N find its binary representation. We shall use again the sequence $S_n, S_{n-1}, \ldots$, but in reverse order, that is, we start from $S_0 = N$ and successively compute $S_1, S_2, \ldots$, at each step obtaining a new digit in the binary representation of N. The following

simple procedure will obtain $a_n a_{n-1} \dots a_0$:

a. Set $S_0 = N$.
b. For $i = 0, 1, \dots, n$, compute a_i and S_{i+1} as the remainder and quotient of the division of S_i by 2, that is, $S_i = 2 \cdot S_{i+1} + a_i$.

Notice that in this procedure we obtain $a_n \dots a_0$ from the LSD to the MSD, and at each step we obtain a_i by *dividing* S_i by 2. Notice also that the process can be stopped as soon as the quotient is 0 (continuing it only generates a string of zeros to the left of the most significant 1).

EXAMPLE 1.2

Find the binary representation of the integer 105.

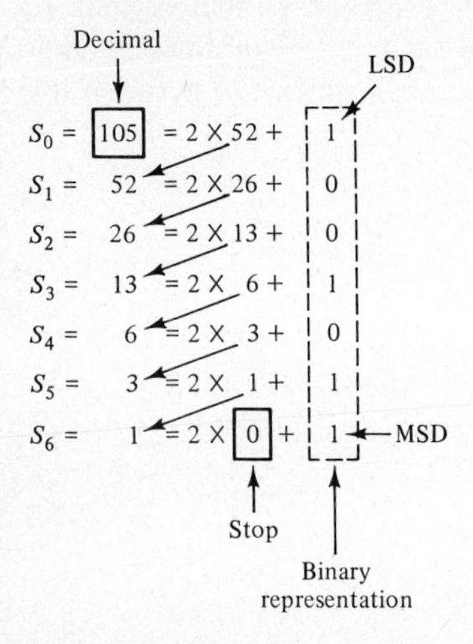

1.3.2 Problem 2. Conversion of Fractions

Let $F < 1$ be a proper fraction. The binary representation of F will be a sequence of bits $a_{-1}a_{-2}a_{-3} \dots$, which is related to F by the formula

$$F = a_{-1}2^{-1} + a_{-2}2^{-2} + a_{-3}2^{-3} + \dots \tag{1.4}$$

If we multiply both sides of Eq. (1.4) by 2, we obtain

$$2F = a_{-1} + (a_{-2}2^{-1} + a_{-3}2^{-2} + \dots)$$

and we recognize that the term within parentheses is itself a proper fraction. Thus we can define a sequence of fractions $F_0, F_{-1}, F_{-2}, \dots$ as follows

$$F = F_0$$
$$2F_0 = a_{-1} + F_{-1}$$
$$2F_{-1} = a_{-2} + F_{-2}$$
$$\vdots$$
$$2F_{-i} = a_{-(i+1)} + F_{-(i+1)}$$

that is, $a_{-(i+1)}$ and $F_{-(i+1)}$ are, respectively, the integer and the fractional part of $2F_{-i}$. (Note also that $2^{-i} \cdot F_{-i}$ is the contribution to the fraction F due to the bits to the right of a_{-i}.)

Remark. An interesting question arises at this point: When do we stop the conversion process? We know that any fraction (a rational number) can be represented in any base by a finite string of digits followed by a periodic finite string (which repeats indefinitely). So, suppose, for example, we are given a fraction F represented in decimal. If we wish to represent F *exactly* in binary, we have to carry out the conversion process until we reach the periodic portion. However, a different criterion is normally adopted, based on the fact that the two representations (in decimal and in binary, respectively) should have comparable "accuracies." Specifically, consider a decimal fraction F represented with a string of m digits to the right of the point (see illustration below). Suppose now to truncate this string to n decimal digits (with $n < m$) and to disregard the digits to the right of the truncation: it is easy to realize that the difference between the

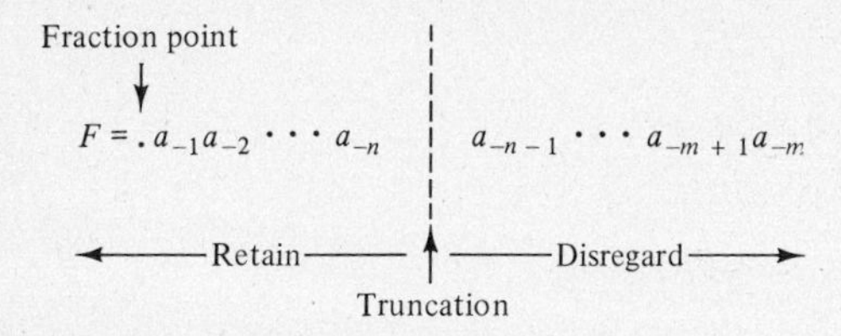

value of the original sequence and that of the truncated sequence is $< 10^{-n}$. To attain a comparable accuracy, how many bits should the binary representation of F have? Denoting by n' this number of bits, by repeating the previous reasoning with respect to base 2, we obtain

$$2^{-n'} \leq 10^{-n}$$

and, by taking binary logarithms of both sides, we have:

$$-n' \leq -n \log_2 10$$

or (changing signs and using the fact that $\log_2 10 = 3.322 \ldots$)

$$n' \geq n \cdot 3.32$$

From this argument we derive the conclusion that the binary representation of a fraction should be normally expressed with about three times as many digits as its decimal counterpart, if comparable accuracies are desired.

We now examine the conversion procedures:

1. Decimal → binary (fraction) (with s bits, i.e., as $a_{-1}a_{-2} \ldots a_{-s}$)

a. Set $F_0 = F$.

b. For $i = 1, 2, \ldots, s$ compute a_{-i} and F_{-i} as the integer and the fractional parts, respectively, of the product $2 \times F_{-(i-1)}$.

Notice that we compute the representation from its MSD to its LSD, and at each step we obtain a_{-i} after *multiplying* $F_{-(i-1)}$ by 2. (The conversion process can be stopped earlier than the target number of bits if we reach the periodic part of the representation, in particular, when the period is the single bit 0.)

EXAMPLE 1.3

Find the binary representation of the fraction 0.67578125.

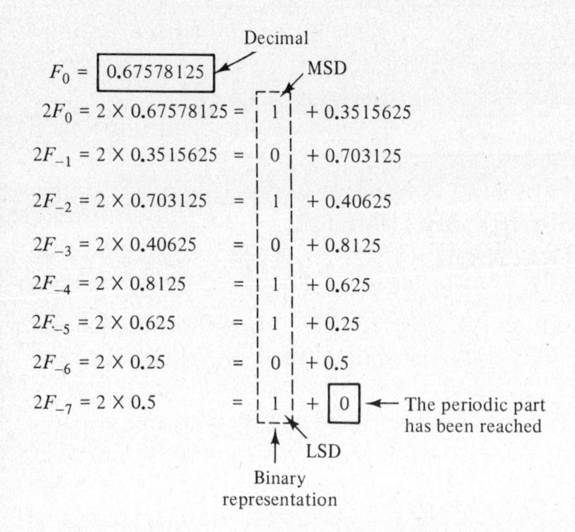

2. Binary ⟶ decimal (fraction). Given the binary sequence $a_{-1}a_{-2} \ldots a_{-s}$ we use $F_0, F_{-1}, F_{-2}, \ldots$ and the following simple procedure (notice that $2^{-s} \cdot F_{-s}$, the contribution to F of the bits to the right of a_{-s}, is 0):

a. Set $F_{-(s-1)} = a_{-s}/2$.
b. For $i = s - 2, s - 3, \ldots, 0$ compute $F_i = (a_{-(i+1)} + F_{-(i+1)})/2$.
c. Set $F = F_0$.

Here we compute the value scanning the representation from its LSD to its MSD. Also, F_i is obtained by *dividing* by 2 the sum $(a_{-(i+1)} + F_{-(i+1)})$.

EXAMPLE 1.4

Find the value of 0.10101101 (in this case $s = 8$).

LSD ↓

$F_{-7} = 1/2 = 0.5$

$F_{-6} = (0 + 0.5)/2 = 0.25$

$F_{-5} = (1 + 0.25)/2 = 0.625$

$F_{-4} = (1 + 0.625)/2 = 0.8125$

$F_{-3} = (0 + 0.8125)/2 = 0.40625$

$F_{-2} = (1 + 0.40625)/2 = 0.703125$

$F_{-1} = (0 + 0.703125)/2 = 0.3515625$

$F_0 = (1 + 0.3515625)/2 =$ 0.67578125 ← Decimal

↑ MSD — Binary representation

We can now briefly summarize the procedures we have previously studied in detail. We will assume that N is a mixed number, with an integer part and a fractional part. The features of the methods to obtain the decimal from the binary representation, and vice versa, are concisely given below:

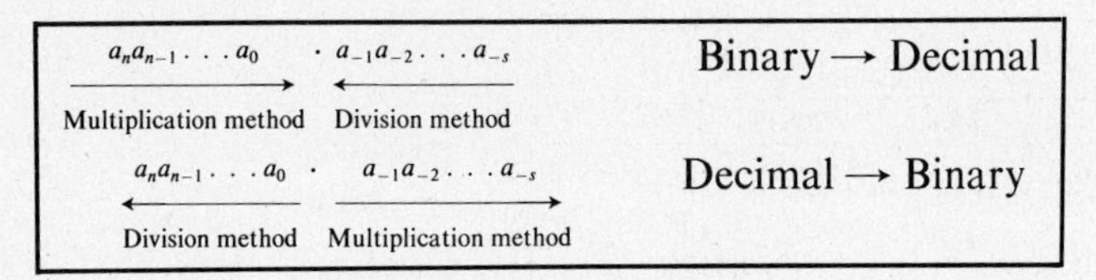

1.4 INTRODUCTION TO BINARY ARITHMETIC: ADDITION AND SUBTRACTION

After discussing the binary notation, we consider how to perform arithmetic operations on numbers expressed in that notation (and we shall convince ourselves of the advantages of the notation).

We shall consider the operations of addition and subtraction, and, exactly as we would in the decimal case, we construct the respective operation tables with single-digit operands (see Figure 1.2). Clearly, the entry 2 in the addition table and the entry -1 in the subtraction table are not represented in our binary notation. Specifically, we will represent 2 in binary as $a_1 a_0 = 10_2$, which means a *sum* of 0 and a *carry of 1* to the position of immediately higher significance. The representation of the negative entry -1 is somewhat more subtle, since we have not yet dealt with negative numbers. Note, however, that $-1 = (-1) \times 2^1 + (1) \times 2^0$. Therefore, -1 may be represented in base 2 as $a_1 a_0 = (-1)1_2$,[2] where $a_0 = 1$ means a *difference* of 1 and $a_1 = (-1)$ means a *borrow* of 1 from the position of immediately higher significance. Thus we find again the notions of carry and borrow that are familiar from arithmetic with the decimal notation.

Let us now consider in detail the operation of addition of two n-bit binary numbers. Let $A \equiv a_{n-1}a_{n-2} \ldots a_0$ and $B \equiv b_{n-1}b_{n-2} \ldots b_0$ be the two addends. In the position of weight 2^i we have bits a_i of A and b_i of B; let bit c_i be the carry resulting from the addition in the position of weight 2^{i-1}. The three bits a_i, b_i, and c_i must be added together to yield the bit s_i (of weight 2^i) of the sum

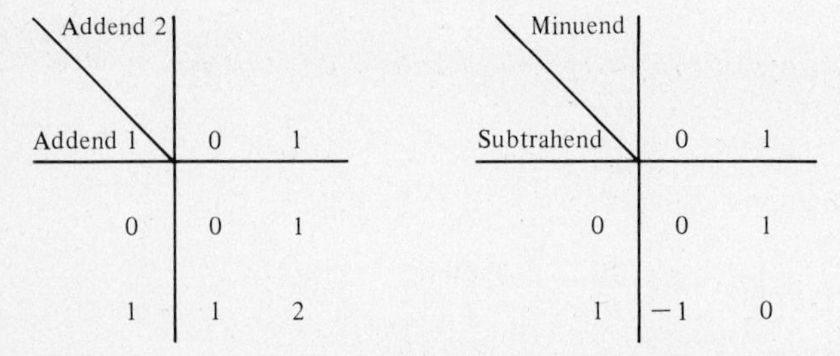

Figure 1.2 Addition (left) and subtraction (right) tables for single-digit operands.

[2]Notice that we are exceptionally using the digit (-1)!

and the bit c_{i+1} (of weight 2^{i+1}) which is the carry into the next position. It is easily realized that the triple (a_i, b_i, c_i) can have *eight* possible configurations, as shown in Figure 1.3. For each of these configurations we compute the value of

a_i	b_i	c_i	Value of $a_i + b_i + c_i$	c_{i+1}	s_i
0	0	0	0	0	0
0	0	1	1	0	1
0	1	0	1	0	1
0	1	1	2	1	0
1	0	0	1	0	1
1	0	1	2	1	0
1	1	0	2	1	0
1	1	1	3	1	1

Figure 1.3 Binary adder table.

the sum $(a_i + b_i + c_i)$ which equals the number of 1s in the configuration. When we represent the value of the sum by a 2-bit binary number, the leftmost bit is c_{i+1} while the other is s_i. Therefore, the table in Figure 1.3 completely illustrates how to compute c_{i+1} and s_i from a_i, b_i, c_i. This table can now be used in a straightforward manner to implement the addition of two binary numbers. The carry c_0 is conventionally set to 0.

EXAMPLE 1.5

Add $A \equiv 10011$ and $B \equiv 00110$.

```
Carries   0   1   1   0
A ≡       1   0   0   1   1
B ≡       0   0   1   1   0
          -----------------
          1   1   0   0   1
```

The operation of subtraction is susceptible of analogous treatment and is left as an exercise.

1.5 HEXADECIMAL AND OCTAL NOTATIONS

When the base 16 is chosen, we obtain the so-called *hexadecimal notation.* There is a very simple correspondence between the binary and the hexadecimal notations. Indeed, assume that the number n of bits in the binary representation is a multiple of 4, that is, $n = 4s$. Then we have

$$\sum_{j=0}^{n-1} a_j 2^j = \sum_{j=0}^{4s-1} a_j 2^j = \sum_{r=0}^{s-1} (a_{4r+3} \cdot 2^3 + a_{4r+2} \cdot 2^2 + a_{4r+1} \cdot 2 + a_{4r}) 2^{4r}$$

(we have just grouped the bits of $a_{n-1}a_{n-2} \ldots a_0$ in quadruplets starting from the right). We now let $u_r \equiv a_{4r+3}a_{4r+2}a_{4r+1}a_{4r}$, and notice that the value of u_r satifies $0 \leq u_r \leq 15$. We obtain

$$\sum_{r=0}^{s-1} (a_{4r+3} \cdot 2^3 + a_{4r+2} \cdot 2^2 + a_{4r+1} \cdot 2 + a_{4r}) 2^{4r} = \sum_{r=0}^{s-1} u_r 16^r$$

that is, $(u_{s-1}u_{s-2} \ldots u_0)_{16}$ is the hexadecimal representation of the number represented in binary by $(a_{n-1}a_{n-2} \ldots a_0)_2$. Therefore, to obtain the hexadecimal representation from the binary representation, starting from the LSD we subdivide the binary string into consecutive substrings of 4 bits each and consider each such substring as the binary coding of a hexadecimal integer. As we have seen, the latter is an integer between 0 and 15; since it is desirable to represent each such digit with just one symbol, the adopted convention is to choose the first six letters of the alphabet to represent, in the order, the integers 10, 11, 12, 13, 14, 15. Therefore, the *sequence* of the hexadecimal digits is

0, 1, 2, 3, 4, 5, 6, 7, 8, 9, A, B, C, D, E, F

Another frequently used representation is in base 8, and is called the *octal notation.* The only difference from the hexadecimal notation is that the bits of $a_{n-1}a_{n-2} \ldots a_0$ are now grouped in triplets rather than in quadruplets, so that the digits in the octal notation are the numerals 0, 1, . . . , 7.

EXAMPLE 1.6

Convert 010111010100_2 in hexadecimal and in octal.

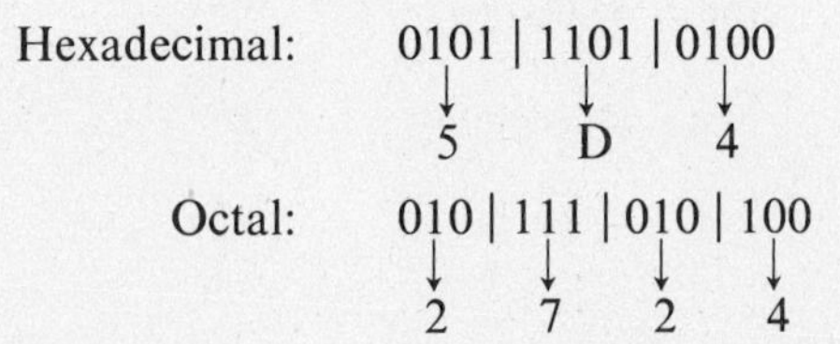

so that $5D4_{16}$ and 2724_8 are, respectively, the hexadecimal and octal for the given string. (Recall that the subscripts 2, 8, and 16 denote the bases.)

What is the usefulness of the octal and hexadecimal notations? Not a profound one, except that they are very compact (and much less prone to transcription errors) ways to express long binary sequences.

1.6 OTHER IMPORTANT CODES

We shall briefly mention some other binary codes which are relevant to, or find application in, digital systems.

1.6.1 Binary Coded Decimals (BCD)

It is a scheme whereby the numerals 0, 1, . . . , 9 are each coded by means of its 4-bit binary correspondent. Thus, the integer 359_{10} is coded in BCD as 0011 0101 1001, whereas it would be coded in binary as 101100111_2.

1.6.2 Gray Code

It is a binary coding scheme of positive integers with the property that the representations of two consecutive integers differ in exactly 1 bit. There are several applications where this property is highly desirable. A typical one is the analog-to-digital conversion of the position of a rotating disk or shaft. The 360° angle is normally subdivided into 2^k angular sectors of identical width (see Figure 1.4, where $k = 3$), each of which can be coded by a distinct configuration of k bits. The digital reading of the position can be done either by mechanical or by optical means. For example, each of the k bits of the code is assigned to a distinct circular track on the disk. Each track is partially coated with insulating material, so that a brush placed on the track can selectively establish electric contact with the disk (a situation corresponding to binary 1) or be disconnected from it (a situation corresponding to a binary 0).

Of course, if distinguishability is the only issue, we have a lot of flexibility in the choice of the code. For example, we may choose the binary positional code; such realization is shown in Figure 1.4(b). Note, however, that the operation may not be reliable. Indeed in changing from 001 to 010, for example, if the brushes are not perfectly radially aligned, the sequence of position readings may be 001 → 000 → 010, with a spurious 000 [this is the situation illustrated in Figure 1.4(b)]. This difficulty is completely avoided by the adoption of a Gray code, as illustrated in Figure 1.4(c), since the codings of two adjacent sectors differ in just one digit.

We shall now consider the relation between the binary and the Gray code

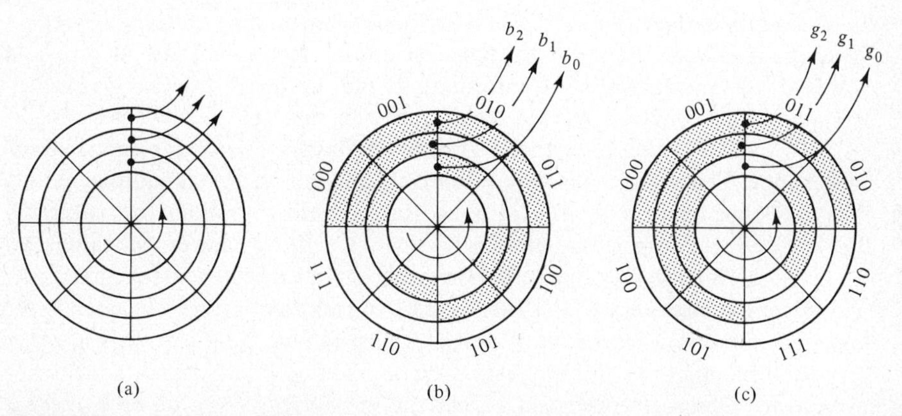

Figure 1.4 A shaft position encoder: (a) track and brush arrangement; (b) binary encoding; (c) Gray encoding.

representations of an integer N. Letting $b_{n-1}b_{n-2} \ldots b_0$ and $g_{n-1}g_{n-2} \ldots g_0$ be, respectively, these binary and Gray code representations, we have[3]

$$\begin{cases} g_{n-1} = b_{n-1} \\ g_{n-2} = (b_{n-2} + b_{n-1}) \bmod 2 \\ \vdots \\ g_i = (b_i + b_{i+1}) \bmod 2 \\ \vdots \\ g_0 = (b_0 + b_1) \bmod 2 \end{cases} \tag{1.5}$$

Thus Eq. (1.5) can be used directly to convert from binary to Gray; for the reverse, set $b_{n-1} = g_{n-1}$ and $b_i = (g_i + b_{i+1}) \bmod 2$, for $i = n - 2, n - 3, \ldots, 0$.

EXAMPLE 1.7

Convert 10011 from binary to Gray and 11101 from Gray to binary.

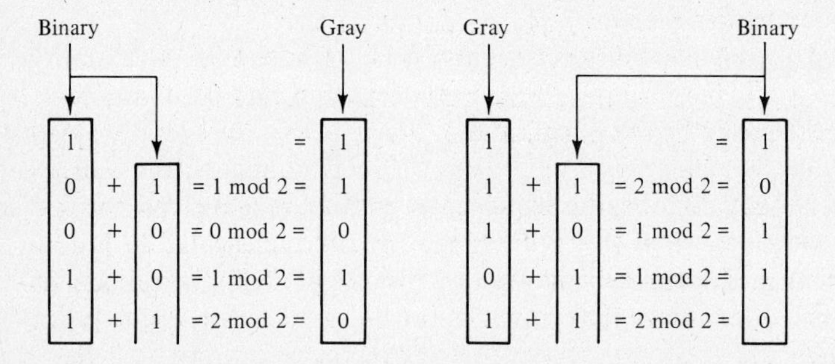

1.6.3 Alphanumeric Codes

As mentioned earlier, the coding of alphabetic characters is entirely arbitrary. However, it is very desirable to agree on one or few standards, in order to facilitate the transmission of information and to avoid useless and time-consuming code conversions. One such standard is the ASCII code (pronounced "askee"), which is the acronym of *A*merican *S*tandard *C*ode for *I*nformation *I*nterchange. This standard is very extensively adopted in digital equipment and deserves being illustrated. ASCII is a 7-bit code for encoding the numerals {0, . . . , 9}, the upper case alphabet {A, . . . , Z}, the lower case alphabet {a, . . . , z}, a large set of punctuation symbols, and special control commands used in communication. ASCII is referred to as an *alphanumeric* code (alphabetic and numeric). The table of ASCII, given in Figure 1.5, reveals a lot of structure. In other words, alphanumeric symbols are not coded arbitrarily; rather, the codings are assigned to suit the most natural use of each type of

[3]The notation a mod 2 means "the remainder of the division of a by 2," that is, a mod 2 is 0 if a is even and is 1 if a is odd.

blank	010 0000	A	100 0001	a	110 0001
.	010 1110	B	100 0010	b	110 0010
(	010 1000	C	100 0011	c	110 0011
+	010 1011	D	100 0100	d	110 0100
$	010 0100	E	100 0101	e	110 0101
*	010 1010	F	100 0110	f	110 0110
)	010 1001	G	100 0111	g	110 0111
—	010 1101	H	100 1000	h	110 1000
/	010 1111	I	100 1001	i	110 1001
,	010 1100	J	100 1010	j	110 1010
'	010 0111	K	100 1011	k	110 1011
=	011 1101	L	100 1100	l	110 1100
		M	100 1101	m	110 1101
0	011 0000	N	100 1110	n	110 1110
1	011 0001	O	100 1111	o	110 1111
2	011 0010	P	101 0000	p	110 0000
3	011 0011	Q	101 0001	q	110 0001
4	011 0100	R	101 0010	r	110 0010
5	011 0101	S	101 0011	s	111 0011
6	011 0110	T	101 0100	t	111 0100
7	011 0111	U	101 0101	u	111 0101
8	011 1000	V	101 0110	v	111 0110
9	011 1001	W	101 0111	w	111 0111
		X	101 1000	x	111 1000
		Y	101 1001	y	111 1001
		Z	101 1010	z	111 1010

Figure 1.5 Table of the ASCII alphanumeric code.

symbol, whenever applicable. The codings are strings of 7 bits, which are in turn broken down into two substrings, of 3 and 4 bits, respectively. The leftmost 3-bit string is by and large used to classify the type of symbols (011 for numerals, 100 and 101 for upper case letters, 110 and 111 for lower case letters, 010 for punctuation and special symbols). Moreover, within each type, the rightmost 4-bit string has a precise rationale. For example, the numerals are in BCD form. The letters are coded so that, if we list them alphabetically, their codes—viewed as binary numbers—are in natural order; this is particularly convenient if one wishes to obtain alphabetic sorting of nonnumerical information.

1.7 CODES FOR ERROR CONTROL

Either disturbances in transmission or malfunctions in the equipment may alter some of the digits used in the representation of messages; such alterations are usually referred to as *errors*. Errors may have very undesirable consequences; think, for example, of a transmission error on the balance of a bank account. The following question is therefore very natural: "Is it possible to code the messages of a set—for example, the letters of the alphabet in binary—so that, if an error occurs, either we become aware of it (error detection) or, even better, we recover the original message (error correction)?"

To correctly approach the key idea in error detection and error correction (briefly, *error control*), consider the case in which the number of possible strings over the code alphabets is equal to the number of messages to be encoded. For example, the set of messages consists of the 26 letters of the English alphabet, of the symbols for the comma, semicolon, period, colon, and question mark, and of the special symbol "space" (for a total of 32 messages), and we choose to code this set with strings of 5 bits. Since the latter are $2^5 = 32$ in number, each string will be used in the code. If now a disturbance alters one or more bits of the string being transmitted the resulting string still represents a *legal* message; therefore, an equivocation occurs, and there is absolutely no way we may become aware of it.

It is crucial to error control that the string resulting from an error be not a legal message, that is, that there be many more strings available for encoding than there are messages: this is, in essence, the notion of *redundancy*. Obviously, the larger the redundancy, the larger—in principle—the error protection (and the cost).

In this section, we shall briefly mention some elementary codes for error control. It is assumed throughout that the binary alphabet $\{0, 1\}$ is used and that the disturbance may alter at most 1 bit (single error). We shall assume throughout that only strings of a *fixed* length be used for encoding (*block codes*).

1.7.1 Parity Check Codes

Suppose that we have 2^n messages to be coded. Initially we code each of them with a string of n bits, that is, in a nonredundant fashion. Next we take any such n-bit string w and we extend it with an additional bit b_{n+1} as follows: if the number of 1s of w is odd, then $b_{n+1} = 1$, otherwise $b_{n+1} = 0$; b_{n+1} is called the *parity check bit*. This simple error-control scheme is frequently applied to the ASCII code, thereby obtaining an 8-bit code.

EXAMPLE 1.8

For $n = 5$, $w_1 = 01101$ is extended to 011011 and $w_2 = 11000$ is extended to 110000 (the parity check bit is the rightmost one). Notice that the string extended with the parity check bit always has an even number of 1s; thus, if just 1 bit is altered (i.e., either 1 becomes 0 or vice versa), the resulting string will have an odd number of 1s, that is, a single error can be detected.

Assume now that the number n of information bits is of the form $n = r \times s$, where r and s are both integers. We arrange our n information bits into a matrix of r rows and s columns (see Figure 1.6), and append an even parity check bit for each row and each column. The resulting code is technically called a *product code* and has $n + r + s$ bits. Suppose now that a single error occurs in the bit position at the intersection of row i and column j. This alters the parity of the 1s in both row i and column j, so that (in the hypothesis that no other error has occurred!) the error position can be identified and a correction performed. The

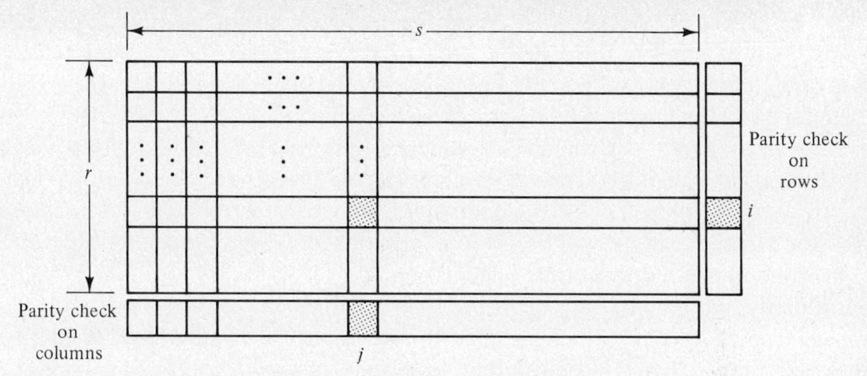

Figure 1.6 Illustration of a single-error correcting product code.

reader should convince himself that if two arbitrary positions are in error, the situation can always be detected.

The study of parity check codes is the fascinating subject of coding theory and the interested reader is referred to specialized textbooks.

1.7.2 2-out-of-5 Codes

The message set consists of the numerals 0, 1, . . . , 9. We use strings of 5 bits of which exactly two are 1s, that is, 00011, 00101, 01001, 10001, 00110, 01010, 10010, 01100, 10100, 11000. Since there are 10 such strings, each one of them can be chosen to code a decimal digit. Notice if that a single bit is altered, the resulting string will have either 1 or 3 bits equal to 1, that is, it no longer represents a digit. Here again, we have a single-error detecting code.

NOTES AND REFERENCES

The positional notation to represent integers and the notion of zero, which are crucial factors in the development of arithmetics, were unknown to the western classical civilizations. The Romans and the Greeks used cumbersome notations, with a rationale for representation, but with poor arithmetic capabilities. Those two important elements—positional notation and zero—were known to the Mayas of Central America already in the first century A.D.; curiously enough, they used a base 20 notation. This discovery, however, had no influence on the development of modern mathematics, since the Mayan civilization blossomed in isolation and mysteriously died. The same discovery was made in the Indo-Persian region in about the same period, and was brought by the Arabs to the Western World. The "arabic number system" made its appearance in Europe around 1000 A.D. Thus, the positional notation, which we take for granted and consider almost natural, is a relatively recent discovery of mankind. A most astonishing fact is that the classical languages, as well as many of the modern western languages, denote numbers with phrases that in some sense reflect the decimal notation.

The binary number system, which is today universally used in digital computers, was first proposed by the great German philosopher-mathematician Gottfried Leibnitz (1646–1716), but was adopted in practice only when the technology of calculating machinery changed from mechanical to electrical.

In this chapter we have also taken a glimpse at binary arithmetics. Needless to say, this is a very well-developed field. The reader is advised at this point to content himself with the rudimentary concepts presented here; references for further reading will be given in Chapter 6, where binary arithmetics will be studied in a more detailed fashion.

Finally, we have barely opened our eyes on the topic of codes for error control. Coding theory, which deals with this topic, is by now an extremely developed and sophisticated discipline. Prompted by the revolutionary work of C. E. Shannon on information theory (Shannon, 1949), coding theory was pioneered by R. W. Hamming, who discovered the first nontrivial class of error-correcting codes (Hamming, 1950). The study of coding theory requires a solid background in modern algebra. The reader who feels comfortable with modern algebra, may refer to the advanced texts of McWilliams and Sloane (1978) and Berlekamp (1968); otherwise, the first three chapters of the text by S. Lin (1970) require only an elementary knowledge of linear algebra.

PROBLEMS

1.1. What is the minimum number of bits required to encode a typewriter character set consisting of upper *and* lower case letters A–Z, the numbers 0–9, and the plus and minus signs and decimal point?

1.2. How many binary digits are required to encode a typewriter character set consisting of the letters A–Z (uppercase), the numerals 0–9, and the punctuation symbols . , : ; ′ ! ? ( ) ?

1.3. How many binary places would be required to represent positive numbers less than 100_{10} whose fractional parts are either zero or a multiple of 0.125_{10}?

1.4. Convert the following to 8-bit binary integers: 23_{10}, 34_5, 63_{10}, 63_8. (For 34_5, convert first to decimal notation.)

1.5. Convert the following decimal numbers to binary:
(a) 320 **(b)** 12.125 **(c)** 24.250 **(d)** 0.7

1.6. Convert the following decimal numbers to their binary equivalents:
(a) 42 **(b)** 95 **(c)** 0.6875 **(d)** 39.725

1.7. Convert the following binary numbers to their decimal equivalents:
(a) 11010 **(b)** 111011 **(c)** 0.1101 **(d)** 1101.00111

1.8. Convert the following binary numbers to decimal:
(a) 10101 **(b)** 10000 **(c)** 110.01 **(d)** 0.0101

1.9. Find the bases b (positive integers) for which the following equalities hold:
(a) $21_b + 131_b = 24_b + 120_b + 4_b$
(b) $(23_b)^2 - 240_b = (14_b)^2 + 60_b$
(c) $40_b + 251_b + 12_b = 131_b + 31_b + 141_b$

1.10. Find the integer r (bases) for which the following statements are correct:
(a) $23_r + 21_r + 10_r + 2_r = 111_r$
(b) $\sqrt{232_r} = 14_r$
(c) $3_r x^2 - 12_r x - 231_r = 0$ has roots -3_{10} and 5_{10}.

1.11. Find the base b (a positive integer) for which the following equalities hold:
(a) $14_b + 52_b + 3_b = 113_b$
(b) $(13_b)^2 = 202_b$

1.12. Carry out the following operations:

(a)
```
1 0 0 1 1 1 0 1 +
1 0 1 0 1 1 1 0
```
(b)
```
1 0 0 1 1 +
    1 1 0 1
```
(c) 1 1 0 0 1. 1 1 + 1 0 1 1. 0 1
(d) 1 0 1 1 1. 1 0 1 + 1 0 1 0. 0 1

1.13. Execute the following operations between binary numbers:

(a)
```
1 0 1 1 0 1 +
  1 0 1 0 1
```
(b)
```
1 1 1 1 +
1 1 1 1
```
(c)
```
1 0 1 1. 0 0 1 +
    1 0. 1 0 1
```
(d)
```
1 0 1 0 1 -
  1 1 0 0
```
(e)
```
1 0 0 0 -
      1
```

1.14. Extend your knowledge of binary addition, carrying out the following additions among three operands:

(a)
```
1 0 1 1 0 +
  1 1 0 1 +
    1 1 1
```
(b)
```
1 1 1 0 1 +
1 1 1 0 1 +
1 1 1 1 0
```

1.15. Longhand binary multiplication is analogous, in a straightforward manner, to the familiar longhand decimal multiplication. Compute the product

$$P = 0\ 1\ 0\ 1_2 * 1\ 0\ 1\ 1_2$$

and check your calculation in decimal.

1.16. Write the binary equivalent of the BCD integer

0 1 1 0   1 0 0 1   0 1 0 1.

1.17. Write the sequence of the integers 0 through 9 in Gray code using 4 bits.

1.18. Perform the following code conversions:
(a) From Gray to binary:

1 1 0 0 1,   0 1 0 0 1,   0 0 1 0 1

(b) From binary to Gray:

1 0 1 1 1,   1 0 0 0 1,   0 1 0 0 1

1.19. A (2, 4, 2, 1) BCD code is a scheme whereby the bit string $a_3a_2a_1a_0$ is used to represent the decimal numeral $(0 \leq N \leq 9)$

$$N = a_3 2 + a_2 4 + a_1 2 + a_0$$

with the convention that $a_3 = 0$ for $0 \leq N \leq 4$ and $a_3 = 1$ for $5 \leq N \leq 9$. Write the 10 decimal numerals in the (2, 4, 2, 1) code.

1.20. A (5, 4, 2, 1) BCD code (also called biquinary) is a scheme whereby the bit string $a_3 a_2a_1a_0$ is used to represent the decimal numeral N as

$$N = a_3 5 + a_2 4 + a_1 2 + a_0$$

with the convention that

$$a_3 = \begin{cases} 0 & \text{for } 0 \leq N \leq 4 \\ 1 & \text{for } 5 \leq N \leq 9 \end{cases}$$

Write the 10 decimal numerals in the (5, 4, 2, 1) code.

1.21. How many different nonnegative integers can be encoded in BCD format when one has 16 binary places (bits) available for the encoding?

Chapter 2

Introduction to Computer Organization and Design

2.1 A SMOOTH TRANSITION

Consider a simple and familiar problem, solving a system of linear equations

$$\begin{cases} a_{11}x_1 + a_{12}x_2 = b_1 \\ a_{21}x_1 + a_{22}x_2 = b_2 \end{cases} \tag{2.1}$$

in the two variables x_1 and x_2, where a_{11}, a_{12}, a_{21}, a_{22}, b_1, and b_2 are given numbers. If the determinant $a_{11}a_{22} - a_{21}a_{12} \neq 0$, the solutions are given by the well-known relations

$$x_1 = \frac{b_1a_{22} - b_2a_{12}}{a_{11}a_{22} - a_{21}a_{12}}, \qquad x_2 = -\frac{b_1a_{21} - b_2a_{11}}{a_{11}a_{22} - a_{21}a_{12}} \tag{2.2}$$

Otherwise the system is either inconsistent or underconditioned. The first case occurs when $a_{11}/a_{21} \neq b_1/b_2$; otherwise, the system has infinite solutions, that is, we can express one variable in terms of the other. So if $a_{11} \neq 0$, we set $x_1 = (b_1 - a_{12}x_2)/a_{11}$, otherwise we set $x_2 = b_1/a_{12}$. On the basis of these formulae we can now develop a simple procedure for calculating the solutions x_1 and x_2. This procedure, which is also diagrammatically shown in Figure 2.1 by means of a flowchart, can be expressed as follows:

Step 1. Compute the quantity $\Delta = a_{11}a_{22} - a_{21}a_{12}$.

Step 2. If $\Delta = 0$ and $a_{11}/a_{21} \neq b_1/b_2$, stop, for the system is inconsistent; if $a_{11}/a_{21} = b_1/b_2$ and $a_{11} \neq 0$, set $x_1 = b_1/a_{11} - x_2(a_{12}/a_{11})$, or (if $a_{11} = 0$) set $x_2 = b_1/a_{12}$ and stop.

Step 3. If $\Delta \neq 0$, set $x_1 = (b_1a_{22} - b_2a_{12})/\Delta$ and $x_2 = -(b_1a_{21} - b_2 \cdot a_{11})/\Delta$ and stop.

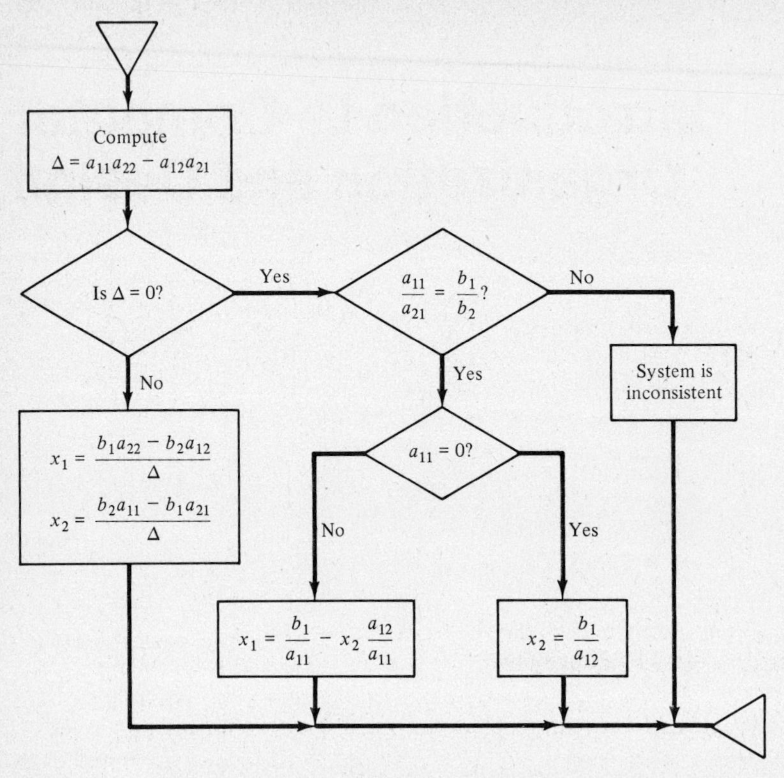

Figure 2.1 A flowchart of the procedure for solving a system of two linear equations.

The preceding procedure consists of precisely defined actions (steps 1–3), and we may think of supplying it to an unimaginative but very diligent worker whose task is the solution of linear systems. Let us picture the work desk of our collaborator (Figure 2.2): we would see a sheet of paper with the procedure printed on it, a few sheets of scratch paper for writing intermediate results, and a calculator. We would then hand the worker a sheet of paper containing the values of $a_{11}, \ldots, a_{22}, b_1, b_2$ and ask him to solve the equation. He would begin by reading step 1 of the procedure and performing the action specified by it, he would then proceed to the subsequent step 2, and so on. In general his activity would be a sequence of the following pairs of actions:

1. Read a step in the procedure sheet.
2. Carry out the step, possibly using the calculator and the scratch paper.

After completing his work, he would write his results on a sheet of paper to be handed out.

We now see that all the actions required to solve a simple problem are so well specified that there is hardly the need of a human operator to carry them out. In fact, we might think of replacing him by some "automatic" device: of course the other items on our "computing desk," such as procedure sheet, scratch

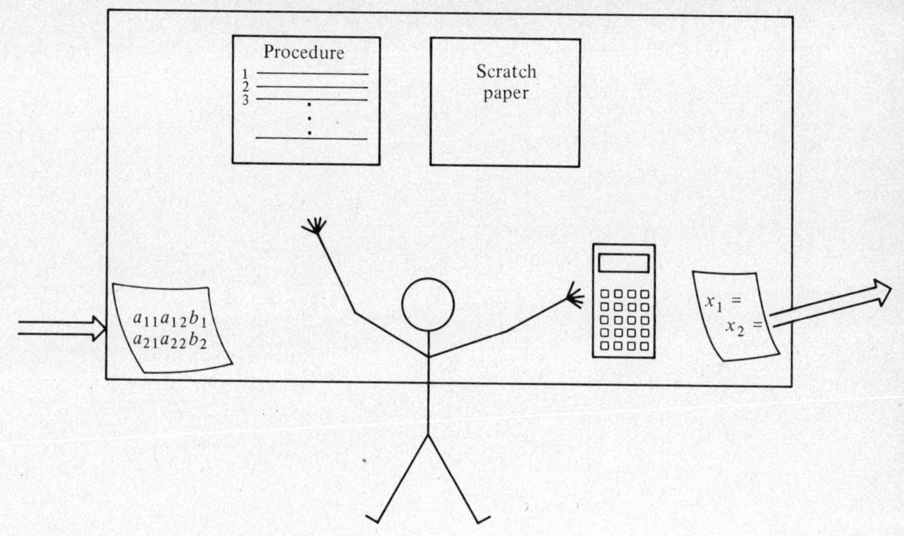

Figure 2.2 A human calculator.

paper, data sheet, and result sheet, would have to be provided in forms suited to the automatic operator. Independently of how we realize those items in practice, conceptually we identify the following basic components:

1. *Storage* (or memory): a facility to store the procedure, the initial data, the intermediate results, the final results.
2. *Arithmetic function:* the capability to perform arithmetic (and other operations) on data.
3. *Input/output:* means to receive data from and deliver results to the exterior.
4. *Control:* to execute the steps of procedures by coordinating the flow of data between the various components introduced above.

This is precisely the basic philosophy of digital computers. In Figure 2.3 we illustrate how the components, also called *units,* are identified in our previous model of human computation. The corresponding model of automatic computation is known as the *von Neumann computer,* after one of its inventors. For well-known reasons, electronics is the technology used in the realization of these machines.

2.2 A SIMPLISTIC COMPUTER

We shall now begin the study of the structure and operation of a digital computer. The computer we shall study is not one of those that are commercially available; for the latter, the sound engineering outlook to optimize the cost-effectiveness of the system leads to specific solutions which are quite complicated

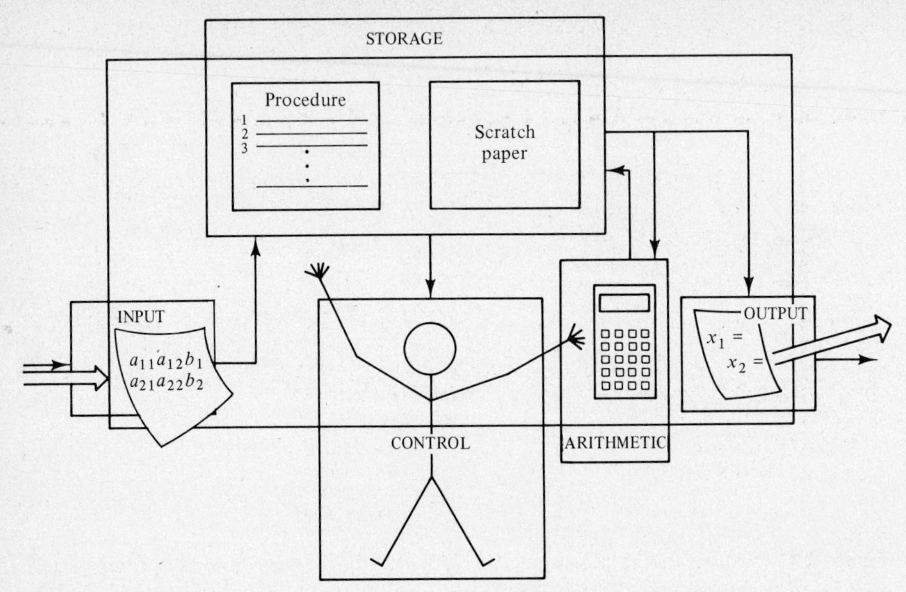

Figure 2.3 Relationship between the von Neumann computer model and a simple model of human computation.

and tend to obscure the basic ideas involved, which are indeed quite simple. Therefore, at this introductory level, it is preferable to have a system that exhibits most of the features of existing computers in a *simple,* uncluttered way, best suited for our *educational* purposes. This system we shall call Simplistic Educational Computer, or briefly, SEC. There are numerous ways to improve on the present design, and the alert reader will discover these ways after gaining some experience.

At this point, we shall be concerned only with a portion of SEC: the memory, the control unit, and the arithmetic unit. We deliberately avoid the input/output functions (to be considered much later in our study) and we assume that all of the necessary data already reside in memory.

We begin by describing the memory at the functional level. (We shall consider it at the hardware level after developing the pertinent notions.) The memory is to be viewed as a collection of *cells* [Figure 2.4(a)], each of which is capable of storing a string of bits of fixed length, called a computer *word.* The cells are consecutively numbered, from 0 to $K - 1$, where the integer K is the *memory size;* the number assigned to a cell uniquely identifies its position in memory, and is therefore called the *address,* or *location,* of the cell. The computer word is the basic unit of information exchanged between the various subsystems.

In memory we store both the data (initial data, intermediate results, and final results) and the procedures for solving specific problems. The latter are called *programs,* and the individual steps of a program are called *instructions.* Thus, a program is a sequence of instructions. Of course, instructions are not rather loosely phrased commands such as the steps of the procedure in Sec. 2.1;

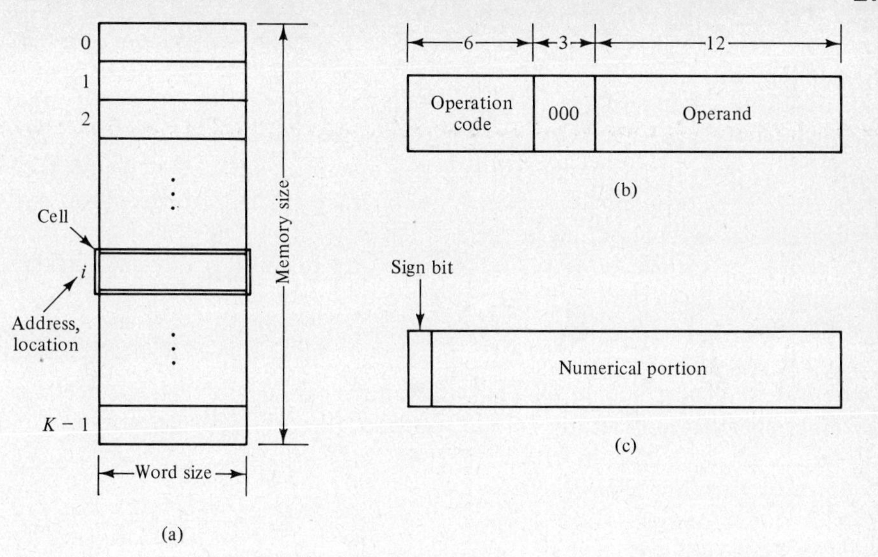

Figure 2.4 Functional organization of the computer memory and formats of instructions and numbers. (a) Memory format. (b) Instruction format. (c) Number format.

rather, they have to be precisely coded as strings of bits. In SEC, all instructions are strings of bits of length equal to the word size. The bits of an instruction are subdivided into substrings, called *fields*. A particular subdivision into fields is called *instruction format*. The typical SEC instruction format is shown in Figure 2.4(b): it consists of three fields of 6, 3, and 12 bits, respectively. Therefore, the word size of SEC is 21 bits and each such word will be succinctly expressed as a *string of 7 octal digits* (rather than 21 binary digits!). The leftmost 6-bit field is reserved for the *instruction code* (also called "operation code" or "machine code") and specifies the type of instruction. The central 3 bits, called b-field, are not used in the stripped-down version of SEC we are now considering and are conventionally set to 000_2 (or 0_8); the rightmost 12 bits, called Y-field, normally specify the *operand* of the instruction.

For many instructions, the "operand" is an address in memory. Therefore, the number of different addresses that we can specify is $2^{12} = 4096$, and this will be the size K of our memory. (Therefore, the latter consists of 4096 words of 21 bits each.)

Memory cells are used to store either instructions or data (typically numbers). Therefore, each number is represented with 21 bits. Since it is necessary to represent both positive and negative numbers, we choose the number format shown in Figure 2.4(c), where 1 bit is used to denote the sign (0 for nonnegative, 1 for negative) and 20 bits are used for the number size.

At this point we could undertake a detailed discussion of the representations of negative numbers which are most frequently used in digital computers. The topic, however, is not simple and is largely irrelevant to our present discussion. Therefore, in keeping with our general philosophy to separate the principles from details, we adopt *for the time being* a representation of signed numbers which is closest to our experience with decimal numbers: this is the so-called *modulus-and-sign representation.* Specifically, the rightmost 20 bits represent the absolute value of the number; to represent the sign, the *two* alternatives, "+" and "−," could be encoded by an additional *bit,* where we can make the entirely conventional choice of representing "+" with bit 0 and "−" with bit 1. Thus, assuming our numbers to be integers, the 21-bit modulus and sign enable us to represent all integers in the interval $[-(2^{20} - 1), (2^{20} - 1)]$. The reader is warned, however, of the necessity to unlearn this comfortable convention, when we shall be ready to delve into the structure of digital systems in Chapter 6.

The set of instructions of a computer is usually called the *instruction repertoire.* It is convenient to subdivide the repertoire into families of instructions of rather homogeneous nature. For our purposes, these families are as follows:

Basic Families of Instructions

1. Input/output.
2. Data transfer.
3. Arithmetic and logic.
4. Control.

2.3 THE INSTRUCTION REPERTOIRE OF SEC (STRIPPED-DOWN VERSION)

As mentioned earlier, we will not discuss at this stage the family of input/output instructions. We will consider, however, the other three families and we introduce the notion that, besides the main memory previously described, there are in SEC other storage devices called *registers,* each capable of storing (at most) one computer word. Data transfers occur between memory and registers or among registers. Very important among these additional registers is one, singled out by the name of *accumulator.*

At this point it is very important to stress the fundamental features of data transfers in digital machines, either as the result of specific data transfer instructions or as a necessary part of other instructions (as for example, addition):

> The transfer of the content of register A to register B leaves *unaltered* the content of register A and *replaces* the previous content of register B with the content of register A.

In other words, a data transfer is not to be viewed as a physical move, but rather as a *copying operation.* (This behavior is realized even when, for technological reasons, the access to the content of the source register—register A above—entails erasing it: the digital system is normally designed so that the original content is restored after this "destructive access.")

The basic data transfer instructions are shown in Table 2.1. They involve transfers between memory and the accumulator. The machine code for the Load

TABLE 2.1 TRANSFER INSTRUCTIONS

Instruction	Operation	Mnemonic
01,0,Y	Load the accumulator with the content of memory location Y	LDA
02,0,Y	Store in memory location Y the content of accumulator	STA
10,0,Y	Enter the nonnegative number 000_8Y into the accumulator	ENT

Accumulator (LDA) instruction is $000001_2 = 01_8$, while $000010_2 = 02_8$ is the machine code for Store Accumulator.

Notice that due to the arbitrariness of their choice, it is quite awkward to memorize the operation codes as octal numbers for all instructions; indeed, it is quite preferable to substitute for each 2-digit octal operation code a *mnemonic code,* normally consisting of two or three letters, which immediately reminds us of the operation involved (for example, LDA is the mnemonic for "LoaD the Accumulator" and we prefer to write LDA, 0, Y rather than 01, 0, Y). Keep in mind, however, that the mnemonic code is just a useful aid when writing programs on paper: the instructions are obviously stored in memory with machine codes. Quite different from these is the instruction ENT, for which the operand field contains a *number* rather than the *address of a number.* We say that ENT specifies its operand *immediately,* whereas LDA (as well as other instructions to be described next) specifies it by *memory reference* (or, in current jargon, *directly*). Obviously, the operand entered into the accumulator by ENT is the bit string Y extended with 9 binary zeros on the left (i.e., if Y = 4106, the number 0004106_8 is entered).

In Table 2.2 we display the arithmetic-logic instructions. The ADD and SUB(tract) instructions are self-explanatory. The AND and OR instructions are typical logic instructions and are quite useful for data manipulation. Specifically, for the AND instruction the ith bit of the result is 1 if the ith bits of *both* operands are 1, while for the OR instruction the ith bit of the result is 1 if the ith bit of *either* operand is 1 (or both are). For example,

```
OPERAND 1    001011101011001010110
OPERAND 2    010010111010101111010O
             ---------------------
   AND       000010101010001010010O
   OR        011011111011101111111O
```

TABLE 2.2 ARITHMETIC-LOGIC INSTRUCTIONS

Instruction	Operation	Mnemonic
03,0,Y	Add content of location Y to content of accumulator and leave sum in accumulator	ADD
04,0,Y	Subtract content of location Y from content of accumulator and leave difference in accumulator	SUB
05,0,Y	"And" content of location Y with content of accumulator and leave result in accumulator	AND
06,0,Y	"Or" content of location Y with content of accumulator and leave result in accumulator	OR

TABLE 2.3 CONTROL INSTRUCTIONS

Instructions	Operation	Mnemonic
00,0,–	Halt, no operation	HLT
20,0,Y	Jump to Y to fetch next instruction	JMP
21,0,Y	Jump to Y if content of accumulator is zero	JZA
22,0,Y	Jump to Y if content of accumulator is positive (>0)	JPA

Notice that all four instructions—ADD, SUB, AND, OR—take one operand from the accumulator and the other from memory; in other words, they are all memory reference instructions.

Finally, we consider the family of control instructions (Table 2.3). Some of these are extremely important, because they provide branching (i.e., decision-making) capabilities in the execution of programs. The instruction HLT is self-explanatory. The other three instructions are *transfers of control,* and are now described. Our computer, as most computers, carries out the computation specified by a program by executing a sequence of instructions. As soon as the current instruction has been executed, the system must know the address of the next instruction. One way to specify it is by *explicitly* having in the instruction format a "next instruction" field; a more economical approach consists of *implicitly* specifying the next instruction address, by incrementing by 1 the current instruction address (this eliminates the need for an extra field). This is the "normal" transfer of control. There are cases, however, in which we must deviate from this rule, because the next instruction is not to be found in the subsequent location, hence the need for the Jump instructions. As we see, instruction JMP is an unconditional transfer of control, whereas JZA and JPA effect a jump depending upon specific conditions on the content of the accumulator.

We shall now see how this basic instruction set can be used to write some simple programs.

2.4 SOME SIMPLE PROGRAMS IN MACHINE LANGUAGE

We begin with some elementary problems. In all of these cases we place the first instruction of our programs in address 0000.

Problem 1—Transfer the Content of 0020_8 to 0021_8

ADDRESS	INSTRUCTION	COMMENT
0000	LDA, 0, 0020	(the content of 0020 is in the accumulator)
0001	STA, 0, 0021	(the content of the accumulator is stored in 0021)
0002	HLT	

Problem 2—Subtract the Content of 0020_8 from That of 0021_8 and Place the Result in 0022_8

0000	LDA, 0, 0021	(content of 0021 in the accumulator)
0001	SUB, 0, 0020	(the difference is in the accumulator)
0002	STA, 0, 0022	(the content of the accumulator is stored)
0003	HLT	

Problem 3—Add the Numbers in Locations 0100_8 Through 0137_8 (32_{10} Distinct Numbers) and Place the Result in 0140_8

A brute force solution to this problem is shown below[1]:

```
0000    LDA, 0, 0100
0001    ADD, 0, 0101
0002    ADD, 0, 0102
  .       .
  .       .
  .       .
0037    ADD, 0, 0137
0040    STA, 0, 0140
0041    HLT
```

This program uses 34 instructions, that is, two more than the number of numbers being tallied. Of course, we would not think of using a program of this kind—a "straight-line program" with no transfer of control—to add, say, 1000 different numbers. Indeed, recalling our programming experience, in the latter case we would use a program incorporating a well-known structure, a DO LOOP. Consider for example the following simple FORTRAN program for solving Problem 3 [here A(1, 32) is an array of 32 memory cells containing the given numbers]:

```
      S = 0
      DO P I = 1, 32
      S = S + A(I)
    P CONTINUE
```

The remarkable feature of this extremely simple program is that it is written in a form that is insensitive to the number of items being added. We now analyze it carefully and identify the following important notions [Figure 2.5(a)]: there are two *variables* S and I that are initialized, respectively, to 0 and 1. The variable I is used to keep track of, or count, how many times the main operation, called *action,* is performed. Thus I is called a *counter.* The program will run until the counter, initialized to its first value 1, reaches the value 32, called the *bound.* The variable I is also used to point where the data, the number A(I), is to be found; the address of A(I) is therefore called a *pointer.* A flowchart of a typical loop program is shown in Figure 2.5(b), where the sequence of the basic actions is illustrated.

Before examining the implementation of the loop program in SEC machine language, an important comment is in order:

A memory location contains a 21-bit string. This bit string is interpreted as an instruction if control is transferred to it; it is interpreted as data if it is specified as the operand of an instruction (the same bit string could be used for both purposes in a program).

[1]Note that, once the instruction in location 0037 has been executed, the accumulator has indeed "accumulated" the final sum. This should motivate the choice of the name given to this register.

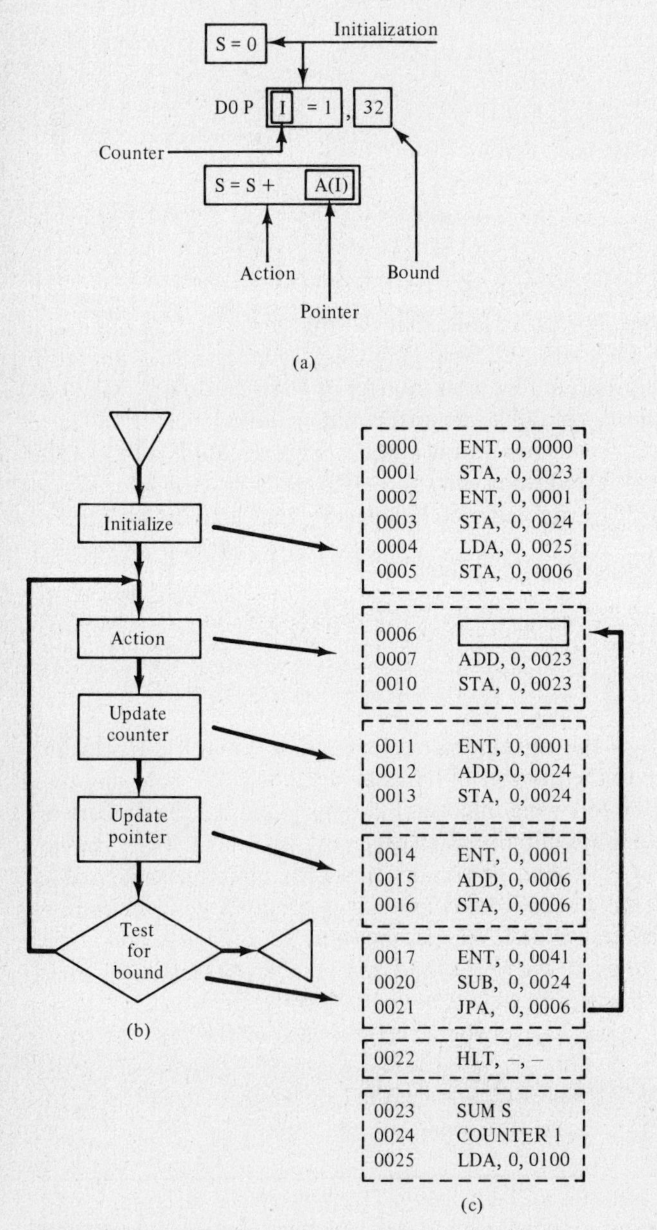
Initialization
S = 0
DO P
I
= 1
32
Counter
S = S +
A(I)
Action
Bound
Pointer
(a)
Initialize
Action
Update counter
Update pointer
Test for bound
(b)
0000 ENT, 0, 0000
0001 STA, 0, 0023
0002 ENT, 0, 0001
0003 STA, 0, 0024
0004 LDA, 0, 0025
0005 STA, 0, 0006
0006
0007 ADD, 0, 0023
0010 STA, 0, 0023
0011 ENT, 0, 0001
0012 ADD, 0, 0024
0013 STA, 0, 0024
0014 ENT, 0, 0001
0015 ADD, 0, 0006
0016 STA, 0, 0006
0017 ENT, 0, 0041
0020 SUB, 0, 0024
0021 JPA, 0, 0006
0022 HLT, –, –
0023 SUM S
0024 COUNTER 1
0025 LDA, 0, 0100
(c)

Figure 2.5 A simple FORTRAN program, the flowchart scheme of a loop program, and an actual loop program.

We now return to our loop program. We begin, starting in location 0000, by first initializing S and I and storing the initialized results at addresses 0023 and 0024, respectively (these two cells are "working memory" for our program):

```
0000    ENT, 0, 0000 }
0001    STA, 0, 0023 }   initialize S = 0
0002    ENT, 0, 0001 }
0003    STA, 0, 0024 }   initialize I = 1
```

The initialization is completed by setting up the pointer. The pointer is in the form of the address field of an instruction LDA, which loads the accumulator with the number A(I) (I is the counter that runs from 1 to 32). Since the address of A(1) is, by hypothesis, 0100, the initial form of the pointer instruction is LDA, 0, 0100. This starting configuration is kept in location 0025 and copied into 0006 for its execution. (Note that 0006 is another "working memory" location.) This copying is effected by

```
0004    LDA, 0, 0025
0005    STA, 0, 0006
```

The next address reached in the execution, 0006, is the beginning of the program segment realizing the operation S = S + A(I); indeed 0006 is the address of the pointer instruction

```
0006    [POINTER INSTR.]   (this cell contains LDA, 0, 0100 + I − 1)
0007    ADD, 0, 0023       (updated sum S)
0010    STA, 0, 0023       (updated sum is stored)
```

Next we must update the counter, stored in 0023, and the pointer, the address portion Y of instruction LDA, 0, Y in location 0006. The former is updated by the following instructions

```
0011    ENT, 0, 0001
0012    ADD, 0, 0024    (counter is updated)
0013    STA, 0, 0024    (updated counter is stored)
```

The pointer is now to be updated (notice that here the content of 0006 is used as a number, while before it was used as an instruction):

```
0014    ENT, 0, 0001
0015    ADD, 0, 0006    [LDA, 0, (updated pointer) is now in accumulator]
0016    STA, 0, 0006
```

Finally we must test for termination. This happens when I reaches the value 33_{10}; so if we subtract from 33_{10} the current value of the counter, the bound is reached when the difference attains for the first time the value 0. This is accomplished by the following instructions:

```
0017    ENT, 0, 0041    (33₁₀ = 41₈ is in accumulator)
0020    SUB, 0, 0024    (33 − I in accumulator)
0021    JPA, 0, 0006    (return to first instruction in loop)
0022    HLT, −, −
```

The entire program is displayed in Figure 2.5(c). Notice that various tricks can be used to slightly reduce the number of memory cells used (in our version, 22); however, our objective was to clearly identify the various elements in a loop program, rather than to optimize the program itself.

The loop program we just discussed, however, has an important shortcoming represented by the use we make of location 0006. The content of this location (which we have denoted as the "pointer instruction") is periodically modified during the execution of the program by the program itself. Although no incorrect behavior arises in our case (for the content of 0006 is correctly initialized), the situation where a program modifies itself—referred to as "self-modifying code"—is regarded as terrible programming practice, because it is eminently error prone. However, in a simple computer structure like the SEC, where instructions to be executed are fetched *only* from memory, we have practically no alternative for the realization of loop programs. In Chapter 7 we shall introduce an addition to the system that will enable us to systematically avoid this shortcoming.

2.5 APPLICATIONS OF DIGITAL COMPUTERS

Although numerical calculations are among the most important applications of digital computers, the preceding example and discussion should not leave the erroneous impression that this is all a computer can do. Indeed a computer has the capabilities to carry out any task that can be expressed as a sequence of well-defined steps; and without attempting a discussion of the meaning of "well-defined steps," we may just refer to a sequence of machine instructions, as introduced in Sec. 2.3. For these reasons, a computer is frequently and appropriately called a *processor,* thus freeing its name from the somewhat implicit association with numerical calculations.

For example, suppose that we wish to sort a set of numbers, that is, to arrange them in ascending order in consecutive locations of memory. The fact that we are ordering numbers is not essential. In fact we could be ordering characters in alphabetical order; all that is needed is the capability to enforce this order, for example, to arrange a C before an F. To give the simplest instance of such an application, suppose we have two numbers, N_1 and N_2, in locations 0101 and 0102, respectively, and we wish to sort them in ascending order, that is, to place $\min(N_1, N_2)$ in 0101 and $\max(N_1, N_2)$ in 0102; location 0100 can be used as "scratchpad." To compare two numbers we subtract one from the other and test the sign of the result. The following straightforward program, which consists of a comparison portion and an interchange portion, does the job:

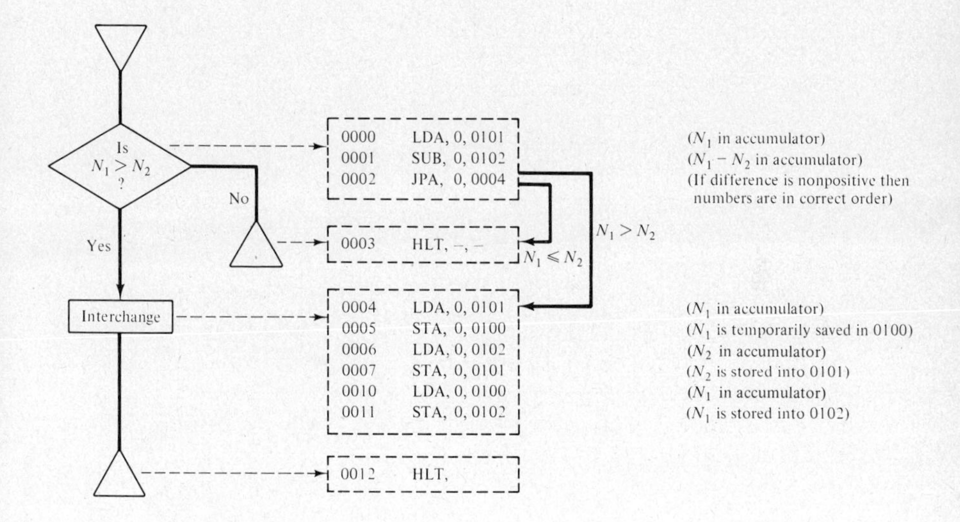

One could continue with several examples of a nonnumerical nature. However, the scope of many important applications can be fully appreciated only with a knowledge of the input/output operations of a digital system. We refer here to uses of a processor as a communication dispatcher (to receive, store, and forward messages), a process controller (to measure the state of a physical process, such as a foundry, to compare it with an ideal situation, and to generate the required actions), and so on. However, we must defer such references to a more advanced stage of our study.

2.6 LEVELS OF DIGITAL SYSTEM DESIGN

It should now be apparent that executing a program corresponds to executing a sequence of instructions, one after the other, until an instruction HLT is reached. The execution of HLT causes all activities to stop and therefore lasts until the computer is restarted by some external intervention.

To carry out an instruction, we realize that the computer must:

1. *Fetch the instruction* from memory and transfer it to an appropriate register in the control unit [see Figure 2.6(a)].
2. *Execute the instruction,* that is, once the instruction is stored in the control unit, its command can be interpreted and executed. This is done by generating control signals that effect the required transfers of data between registers and, if necessary, through the arithmetic-logic unit [see Figure 2.6(b), where we have sketched the flow of data occurring in the execution of the ADD instruction].

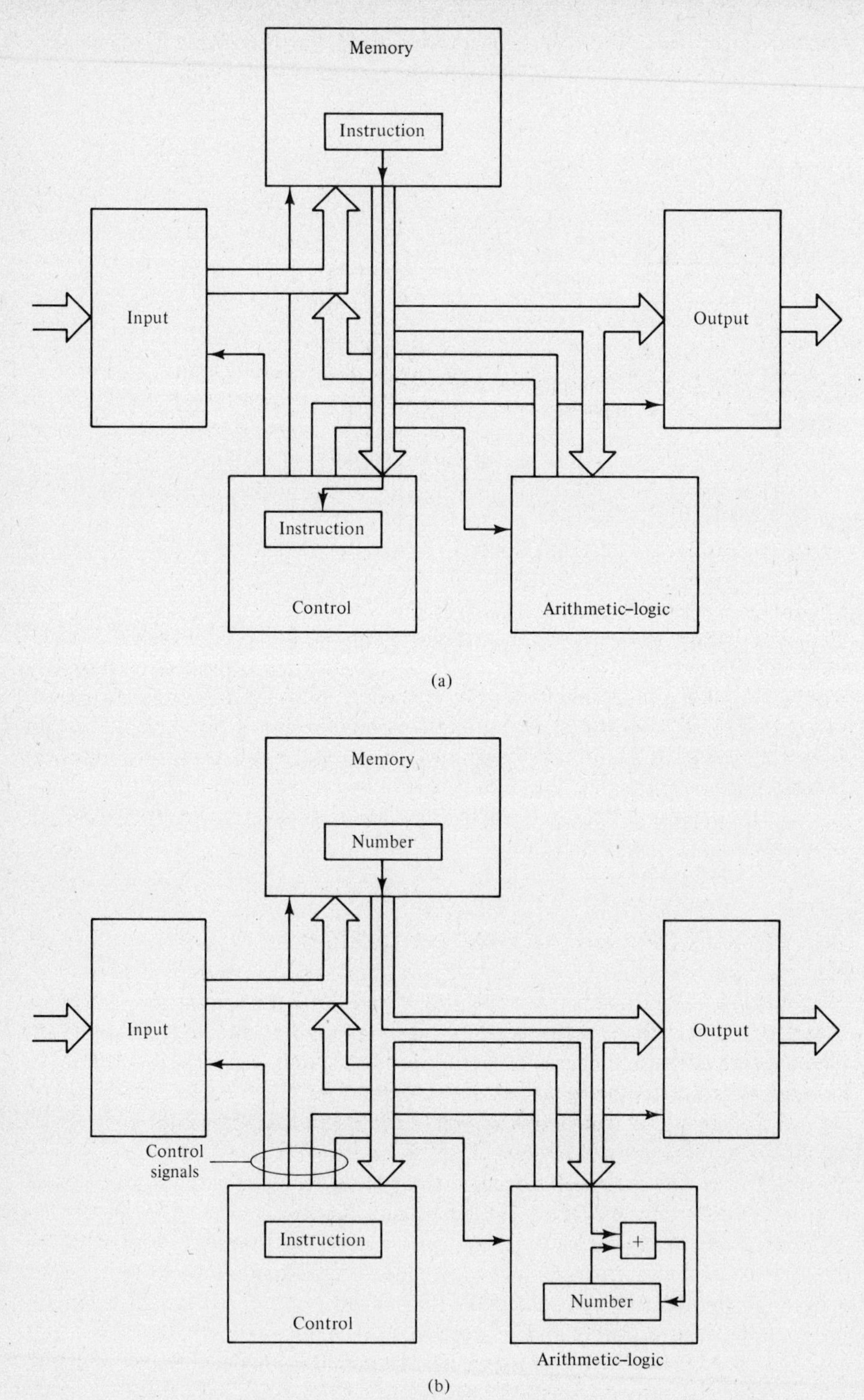

Figure 2.6 Data flows during the instruction fetch and instruction execute phases.

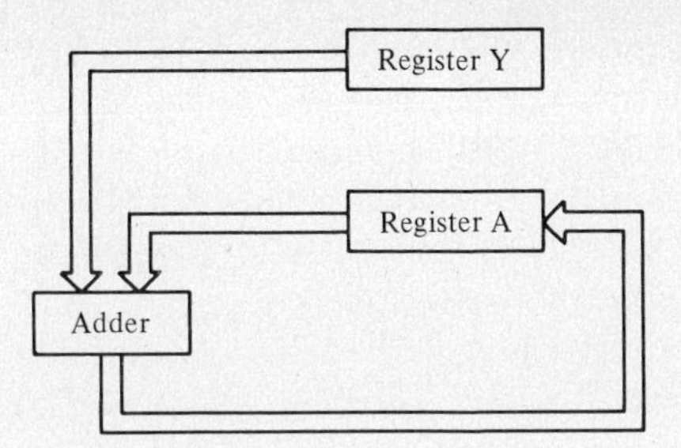

Figure 2.7 Interconnection to be established to execute the ADD instruction.

The preceding outline briefly describes the activity of a computer—or, for that matter, of any complex digital system—as the *sequencing of transfers of information from register to register through processing modules.* In other words, *registers, processing modules,* and *sequencing* are the essential items.

The design of a digital system is a typical *top-down* process, that is, one that proceeds from great generality to increasing specificity and detail. We shall now attempt to outline the various phases of this process.

The design begins by defining the specifications of the system (for concreteness, we shall refer to a computer such as our SEC). This involves the memory size, the registers, the processing modules (such as the arithmetic-logic unit), the input/output units, and—very important—the instruction repertoire.[2] This is the *functional,* or *architectural, design level,* and here is where we presently stand in the design of our very simple machine.

The next natural step is the realization of the instruction repertoire, that is, a definition of the interconnection of the various components (registers and processing units) and the design of the temporal sequences of the actions required, not only to fetch instructions from memory, but also to execute each individual instruction. For example, suppose we want to execute the ADD instruction of SEC, that is, add the content of memory cell Y to the content of the accumulator register and place the sum in the accumulator. For ease of reference, we shall say that the two addends are respectively stored in register Y (the memory cell of address Y) and in register A (the accumulator) (refer to Figure 2.7). Then, one must establish connections from the outputs of registers Y and A to the adder input terminals (so that the addends may be transmitted to the adder) and a connection from the adder output terminals to the input of register A (so that the sum is correctly stored). The realization of the instruction repertoire as sketched in Figure 2.7 is the *register-transfer level* of digital design.

Note, however, that the three functional components illustrated in Figure 2.7 are still at the crudest "black box" level. Indeed we have no feeling for their internal structure, except the knowledge that registers are designed to store information (and hence have *memory*), while the adder simply "transforms"

[2]A note of caution. The instruction repertoire of an actual processor is considerably richer and more complex than the one we have presently proposed for SEC. Its selection depends on a host of engineering design considerations, which we shall "discover" later in our study.

information (without storing it): such memoryless digital components are referred to as *combinational*. Although we could proceed at this point with the discussion of the techniques used in register transfer design, our lack of knowledge of the detailed structure of the digital modules is a serious drawback, since this knowledge would apprise us of relative complexities, costs, limitations, and so on, which are such fundamental aspects of any engineering design. Thus we prefer to defer for the time being the register-transfer design phase, and proceed in the top-down analysis of our units.

We see that these units—registers and combinational modules—can be designed as assemblies of simpler modules, such as storage devices with 1-bit capacity (called flip-flops) and "logic" gates. Examples of the logic diagrams of two typical modules are shown in Figure 2.8. Although the reader is not expected

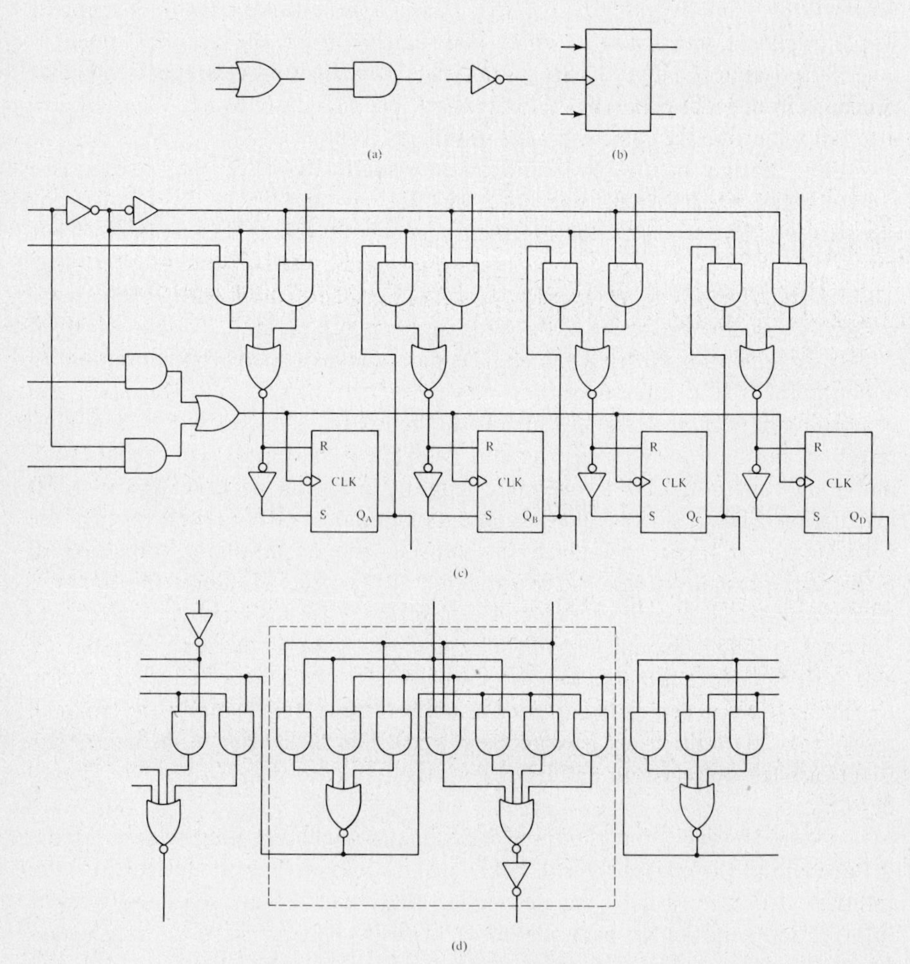

Figure 2.8 Logical diagram of (a) gates, (b) flip-flop, (c) a register, and (d) a binary adder.

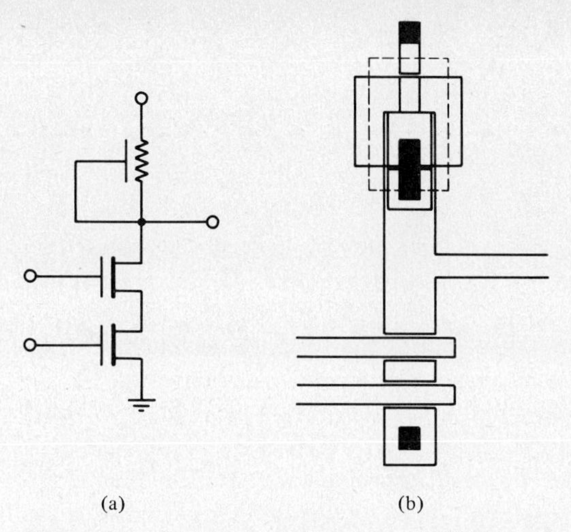

Figure 2.9 Electronic design level. A NAND gate (for positive logic) as an electrical diagram (a) and as an integrated circuit layout (b) (NMOS technology).

to appreciate any of their details at this point, he or she should at least notice that logic gates are used both to control access to flip-flops [Figure 2.8(c)] and to realize the "information processing" function of an adder [Figure 2.8(d)].

The deepest (and most detailed) level of design is that in which flip-flops and gates are realized as appropriate interconnections of basic devices and circuit elements, such as transistors, diodes, resistors, and capacitors. This is the *electronic design level,* or *device level.* In Figure 2.9(a) we have shown the electrical scheme of a NAND gate, which is a fundamental logic element, and—just as an example—what its realization looks like in current integrated NMOS circuit technology [Figure 2.9(b)]. This illustration is presented simply to give concreteness to each level of digital system design, and not to outline a topic with which this text will be concerned. Indeed, our analytical study of the structure of digital systems will stop short of the device level, except to the extent that is indispensable to understand the constraints placed by this level on the higher levels of design. For example, each device has normally the ability to drive a very limited number of similar devices, which severely constrains the flexibility to interconnect digital components. In the same vein, although basically no knowledge of electronics is expected at this point, we assume that the reader is aware at least of the behavior of such elementary devices as a diode or a transistor.[3]

We can now summarize our preceding discussion. From a conceptual standpoint, the computer design activity proceeds from general to specific,

[3]The concerned reader should be reassured. Knowledge of electronics is not expected to reach beyond the notion that a diode is a two-terminal device that can be traversed by current only in one direction, while a transistor is a three-terminal device in which the continuity between two main terminals (closed or open) is controlled by means of the signal on the third terminal.

through the following sequence of design levels:

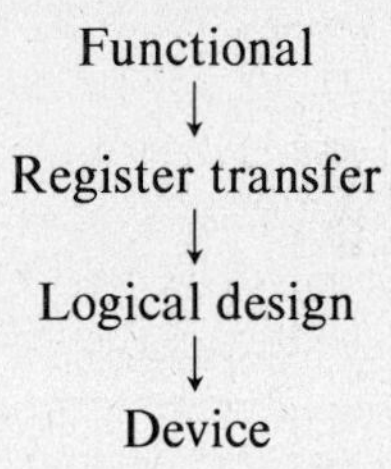

The *analysis* we have outlined is very useful in motivating and placing in the proper perspective the various facets of the design process. However, we are convinced that a more mature understanding of each level of design is attained if the reader has a detailed knowledge of the inner structure of the modules used at that level. Therefore, at this point, we shall further defer the discussion of the techniques used in register-transfer design, until an adequate knowledge of the techniques for logical design is obtained. Now that we have a clear perception of our destination, we can comfortably reverse the direction of our study, and *synthesize* larger modules from smaller components.

In the next chapters, we shall thus begin from the logical design level to return back, ultimately, to the complete specification of the internal structure of our SEC machine. At the end we shall be able to justify the computer down to its most elementary details (wires and gates).

NOTES AND REFERENCES

Recognizable ancestors of modern computers are, in some sense, the mechanical calculators of the seventeenth century. Two renowned scientists and philosophers, Blaise Pascal in France and Gottfried Leibnitz in Germany, are usually credited with building the first mechanical calculators. Those machines had very limited arithmetic capabilities, and we must wait about two centuries for a giant leap in the history of calculating machines. This occurred when the English scientist Charles Babbage (1792–1871) proposed his "analytical engines." These were conceived as mechanical machines with complex arithmetic capabilities, operating on a sequence of data under the control of a sequence of instructions; data and instructions were to be represented on punched cards, of the type used earlier to control mechanical looms. Therefore, Babbage's "engines" had the major traits of a modern computer—input/output, control unit, arithmetic and storage facilities—including the crucial feature of conditional branching. (It must be noted that the "program" was not capable to modify itself, as in modern computers.) Babbage's idea, however, was too ambitious for the technology of the times, did not lead to any realization, and went regrettably into oblivion.

It was only the availability of reliable electromechanical devices (relays) and the advent of electron tubes that permitted the modern development. Apparently independently, and with no knowledge of Babbage's work, simple versions of the "computer" were reinvented in Germany (Zuse), England (Manchester group), and the United States (Atanasoff at Iowa State University, and groups at Harvard University and the University of Pennsylvania) in the late thirties and early forties. It was only in 1945 that the first recognized electronic computer, the ENIAC, was built at the Moore School. The ENIAC, however, did not have the organization of a processor as we conceive it today. Such organization was exhibited by its successor, the EDVAC, the first machine where program and data were stored in the same medium (stored-program computer). This type of machine is today universally known as the *von Neumann computer,* since the renouned American mathematician John von Neumann authored a report describing the system. There is, however, considerable controversy on the paternity of the stored-program concept. Computers have developed immensely since 1945, and we shall make reference to these developments as we go along.

The history of computing machines is still the subject of research and controversy. The reader interested in this topic is referred to the interesting account of B. Randell (1973).

Introductions to the concepts of computer and programs in machine language can be found in many textbooks, such as Taub (1982), Mano (1976), Sloan (1976), and Booth (1971). The reader should be cautioned, however, that there is no universal machine language, and that each such language pertains to a specific machine. Although the mnemonic codes are today reasonably standardized, the instruction formats vary widely. To cope with this problem a few machine-independent languages (the *high-level languages,* such as FORTRAN, Pascal, and BASIC) have been devised, so that the users' community need not always be cognizant of the machine language. Indeed a special program, the compiler, automatically translates a user's program expressed in a (universal) high-level language into the (specific) machine language of the computer chosen by the user.

PROBLEMS

2.1. A memory of computer XYZ has 16384 words of 21 bits each.
- **(a)** How many bits are required for specifying the address?
- **(b)** Which is the maximum number of instruction codes that XYZ can have, if addresses are directly specified?
- **(c)** Which is the largest positive integer representable in an XYZ word?

2.2. Write a *short* sequence of SEC instructions which produces a jump to location 0142_8 on negative accumulator.

2.3. Write a *short* sequence of SEC instructions which produces a jump to location 1435_8 if the contents of location 0126_8 and the accumulator are identical.

2.4. Assume that numbers A and B are stored respectively in locations 0100_8 and 0101_8. For each of the following simple programs express the content of the accumulator when HLT has been executed as a simple arithmetic function of A and B.

```
0000   LDA, 0, 0100     0000   ENT, 0, 0003     0000   LDA, 0, 0100
0001   ADD, 0, 0100     0001   SUB, 0, 0101     0001   SUB, 0, 0101
0002   STA, 0, 0102     0002   HLT              0002   JPA, 0, 0005
0003   SUB, 0, 0101                             0003   LDA, 0, 0101
0004   ADD, 0, 0102                             0004   SUB, 0, 0100
0005   HLT                                      0005   HLT
```

2.5. Write a SEC program to count the number of 1s in the bit string stored in the accumulator, assuming that the leftmost bit is 0 (constants may be used in implementing the program).

2.6. Analyze this simple SEC program and explain what it does.

```
0010   JZA, 0, 0013
0011   JPA, 0, 0013
0012   JMP, 0, XXXX
0013   HLT
```

2.7. Find what the following SEC programs do. That is, express the contents of the accumulator as a function of A and B, after the programs have been executed. Numbers A and B are stored in 0010_8 and 0011_8, respectively.

```
(a) 0101   LDA, 0, 0010      (b) 0103   LDA, 0, 0010
    0102   SUB, 0, 0011          0104   ADD, 0, 0010
    0103   STA, 0, 0100          0105   STA, 0, 0101
    0104   ADD, 0, 0100          0106   ADD, 0, 0101
    0105   HLT, 0, 0000          0107   ADD, 0, 0011
                                 0110   HLT, 0, 0000
(c) 0000   LDA, 0, 0010
    0001   SUB, 0, 0011
    0002   JPA, 0, 0004
    0003   ENT, 0, 0000
    0004   ADD, 0, 0011
    0005   HLT, 0, 0000
```

2.8. Consider the following program in SEC machine language.

```
0020   1000000
0021   0400030
0022   0300031
0023   0200024
0024   0000002
0025   2100027
0026   0400027
0027   0000001
0030   0200131
0031   0600157
```

All other locations contain 0000000. Answer the following questions:

(a) Rewrite the program using mnemonic codes for the instructions *alone* (not for data, of course).

(b) Give the memory address where the program halts.

(c) Give the number in the accumulator when it halts.

2.9. Write a short SEC program to determine whether the number in the accumulator is even or odd.

2.10. Let two numbers A and B be respectively stored in locations 0010_8 and 0011_8. Write two sequences of SEC instructions to implement the following tasks:
(a) Store the value A − 2B in location 0100_8.
(b) If A − B > 0 then jump to location 0100_8.

2.11. Write a SEC program, starting in location 0100_8, to examine the contents of locations 0075_8 and 0076_8 and do the following: "If either location (or both) contains a negative number then location 0070_8 will contain −1, else 0070_8 will contain 0." Initially 0070_8 contains +1.

2.12. Write a SEC program, starting in location 0200_8, which exchanges the contents of memory locations 0100_8 and 0101_8 as shown below:

	After	Before
0100	A	B
0101	B	A

You are allowed to use location 0102_8 only as "scratch memory."

2.13. Write a SEC program, starting in location 0100, which searches locations 0200_8 through 0207_8 to find the largest number stored in them: this number is to be stored in location 0210_8.
(a) Write a straight-line (no loops) program.
(b) Write a loop program.

2.14. Write a SEC program to enter zero in all the memory locations from 0100_8 to 0200_8.

2.15. Ten different positive numbers are stored in the consecutive locations 0010_8 to 0021_8. We must check if they are stored in increasing order. If this is the case, we must enter the value 0 in location 0100_8; otherwise we must enter the value 1 in such a location. Write a SEC program to perform the above task.

Part Two

GENERAL TECHNIQUES

Chapter 3

Combinational Networks and Switching Algebra

3.1 INTRODUCTION TO BINARY FUNCTIONS OF BINARY VARIABLES

As we have seen in the preceding chapter, a digital system can be analyzed as an interconnection of functional components of basically two types:

1. *Storage* components, called memory or registers, as appropriate, which store information.
2. *Combinational* components, which do not have any memory capability and whose function is the implementation of the required information processing activities (computing).

In this chapter we shall begin with the study of combinational components, while the more complex components with memory will be considered in Chapter 5. Referring back to Figure 2.7, which illustrates an addition subsystem, we shall now be exclusively concerned with the ADDER module, which is repeated for convenience in Figure 3.1. The adder has two sets of input lines, one for each of the two operands A and B, and one set of output lines for the sum. Each set of lines (input and output) consists of as many wires as there are bits in the number it is designed to carry.

As we saw in connection with the addition of two numbers represented in binary (Sec. 1.4), such an adder could be realized by means of a collection of simpler blocks, called *adder cells,* each of which is assigned to a fixed position of the operands (that is, the ith cell receives the ith bits of the addends). The adder is completed by connecting the carry-out output of the ith cell to the carry-in input of the $(i + 1)$st cell and setting c_0 permanently to 0 [see Figure 3.2(a)].

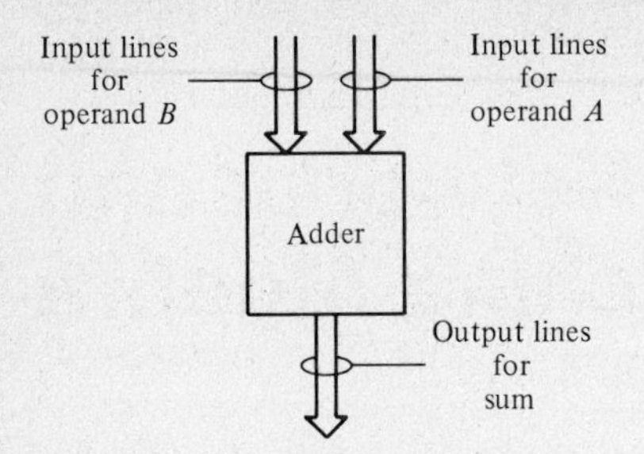

Figure 3.1 A binary adder.

Thus all we need do to design the adder is to design the adder cell, whose behavior is specified in Figure 1.3 [repeated here for convenience as Figure 3.2(b)]. The adder cell receives three binary inputs A, B, and C, of which A and B are the operand bits and C is the carry-in bit, and produces two binary outputs C' and S, of which S is the sum bit and C' is the carry-out bit. A, B, and C (the input variables) are independent and therefore can appear in any one of the eight configurations shown in Figure 3.2(b); C' and S, instead, depend entirely upon the binary values of A, B, and C. Specifically, we saw that the ordered pair (C', S) corresponding to a given triple (A, B, C) will represent in binary the

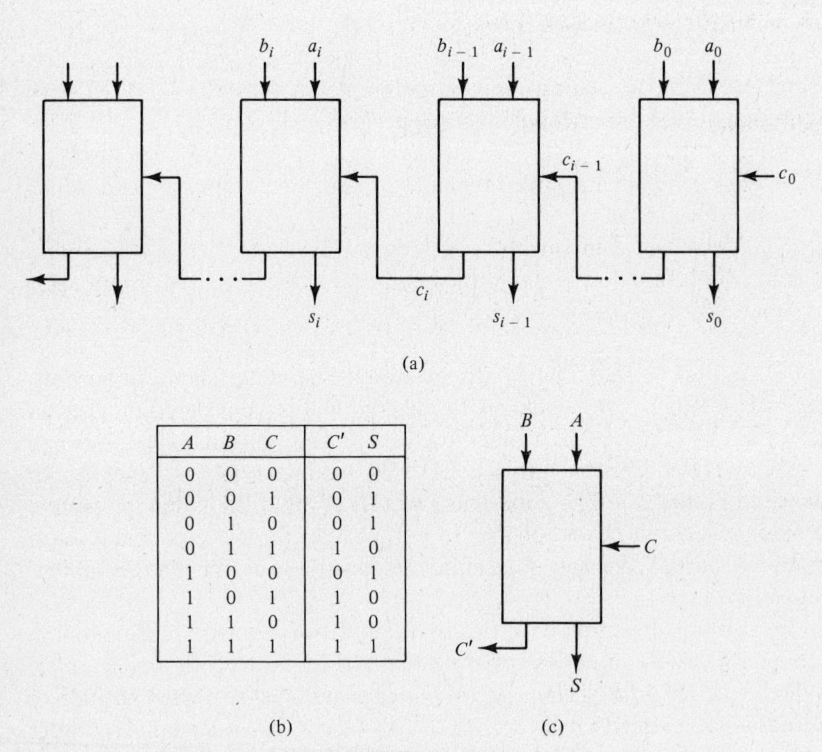

A	B	C	C'	S
0	0	0	0	0
0	0	1	0	1
0	1	0	0	1
0	1	1	1	0
1	0	0	0	1
1	0	1	1	0
1	1	0	1	0
1	1	1	1	1

(b) (c)

Figure 3.2 An adder (a), an adder cell (b), and its behavior (c).

number of 1s appearing in the binary string ABC; for example, in $ABC = 110$ there are two 1s, whence (C', S) will represent in binary the number 2, that is, $C'S = 10$ [see Figure 3.2(b)].

We recognize a familiar notion; both C' and S are *functions* of (A, B, C), that is, for each of them we have a *domain,* consisting of the eight possible triples of binary values for (A, B, C) [Figure 3.2(b), three left columns], and a *range* consisting of the two values $\{0, 1\}$. Thus C' and S are each binary functions of binary variables; now, we have the following:

A binary function of binary variables is called a *switching function.* A *combinational circuit* or *network* is a digital subsystem which realizes a switching function.

The conventional way to display a switching function f is that shown in Figure 3.2(b). Specifically, the combinations of 0s and 1s are ordered so that, when viewed as binary numbers, they are in natural order. Next to each combination the value of the function is given: this table of function values is called the *truth table* of f. The reason for this name is that a binary variable is said to be *true* when equal to 1, and *false* otherwise. So, the function table gives the "truth" values of the function.

Our objective is to develop a methodology for the design of a combinational network which realizes a given switching function.

3.2 SWITCHING FUNCTIONS OF ONE AND TWO VARIABLES

To gain insight into the nature of switching functions, we begin by considering binary functions of one binary variable x. This variable can assume only two values, 0 and 1, which form the domain (see Figure 3.3). All the functions of one variable are obtained by filling a two-place truth table in all possible ways, that is, in four ways, shown in Figure 3.3.

We say that a function is *degenerate* if it does not depend upon all of its arguments, and *nondegenerate* otherwise. So, we see that function f_0 is degener-

Functions / Domain	f_0	f_1	f_2	f_3
0	0	0	1	1
1	0	1	0	1

Figure 3.3 Truth tables of all functions of one variable.

x_2	x_1	g_0	g_1	g_2	g_3	g_4	g_5	g_6	g_7	g_8	g_9	g_{10}	g_{11}	g_{12}	g_{13}	g_{14}	g_{15}
0	0	0	0	0	0	0	0	0	0	1	1	1	1	1	1	1	1
0	1	0	0	0	0	1	1	1	1	0	0	0	0	1	1	1	1
1	0	0	0	1	1	0	0	1	1	0	0	1	1	0	0	1	1
1	1	0	1	0	1	0	1	0	1	0	1	0	1	0	1	0	1

Figure 3.4 Truth tables of all functions of two variables.

ate: in fact, it is constant and equal to 0, so we will call it the *constant* 0; similarly, f_3 will be called the *constant 1*. Instead, f_1 and f_2 are nondegenerate functions: notice that $f_1(0) = 0$ and $f_1(1) = 1$, whence $f_1(x) = x$, so f_1 will be called the *identity;* $f_2(0) = 1$ and $f_2(1) = 0$ (f_2 maps 0 to 1 and 1 to 0) and will be called the *complement function* and denoted by $f_2(x) = \overline{x}$.

We are now ready to consider binary functions of two binary variables x_2 and x_1. Here the pair (x_2, x_1) can assume four possible values (00, 01, 10, 11) which form the domain (see Figure 3.4). All functions are now obtained by filling a truth table with four entries in all possible ways (Figure 3.4). Obviously, this can be done in 16 ways, that is, we have 16 binary functions of 2 binary variables. Let us examine these functions $g_0, g_1, \ldots, g_{15}$. We realize that g_0 and g_{15} are, respectively, the constants 0 and 1; moreover we notice that $g_3 = x_2$, $g_5 = x_1$, $g_{10} = \overline{x}_1$, and $g_{12} = \overline{x}_2$, that is, the latter are actually nondegenerate functions of *only one* variable. The remaining 10 functions $\{g_1, g_2, g_4, g_6, g_7, g_8, g_9, g_{11}, g_{13}, g_{14}\}$ are nondegenerate functions of two variables (i.e., *each* of them depends upon *both* variables). We could analyze all of them, but temporarily we content ourselves with the study of g_1 and g_7.

For convenience, we redraw the truth tables of functions g_1 and g_7 in Figure 3.5. Function g_1 is equal to 1 only when both x_2 *and* x_1 are equal to 1; for this reason it is called the AND function. Function g_7 is equal to 1 when either x_2 *or* x_1, or both, are equal to 1; for this reason it is called the OR function.

We can now imagine that special devices are available for the realization of some of the functions we have considered. Specifically we have a one-input–one-output device, called an *inverter* [in Figure 3.6(a) we give the conventional symbol for this device], which realizes the function COMPLEMENT; a two-input–one-output device [Figure 3.6(b)], called *AND gate,* which realizes the function AND, and an analogous device [Figure 3.6(c)], called *OR gate,* which realizes the function OR.

The output of an AND gate with inputs x_1 and x_2 will be denoted by $x_1 \cdot x_2$ or simply x_1x_2; analogously, the output of an OR gate with inputs x_1 and x_2 will be denoted by $x_1 + x_2$. (The context will avoid confusion with the symbols "$\cdot$" and "+" when used in ordinary arithmetic.)

x_2	x_1	g_1	g_7
0	0	0	0
0	1	0	1
1	0	0	1
1	1	1	1

Figure 3.5 Study of the AND and OR functions.

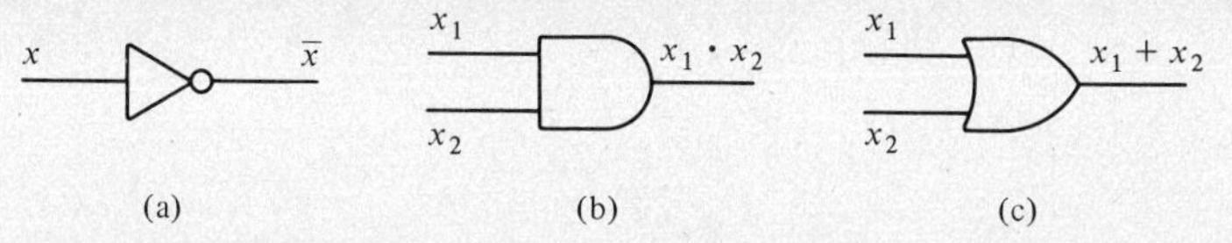

Figure 3.6 Symbols for inverter (a), AND gate (b), and OR gate (c).

3.3 NETWORKS AND EXPRESSIONS

Consider an interconnection of the basic building blocks, AND gates, OR gates, and inverters, such as that shown in Figure 3.7(a). Such an interconnection we call a *network*. Notice that each gate output, except the single network output, feeds exactly one gate input, and that there are no loops, that is, when tracing a path in the obvious way, in no case will this path close itself. The input terminals are all the unconnected gate inputs. We may think of constructing this network by connecting to the two input terminals of gate G_1 [see Figure 3.7(b)] the output terminals of two smaller networks. In turn the latter networks could be decomposed into even smaller networks, and so on until we reach the simplest networks of all: terminals. This analysis actually enables us to give an (inductive) definition of a network:

Definition of Combinational Networks

1. Input terminals are networks (elementary networks);
2. If N_1 and N_2 are networks (represented as black boxes), so are the following:

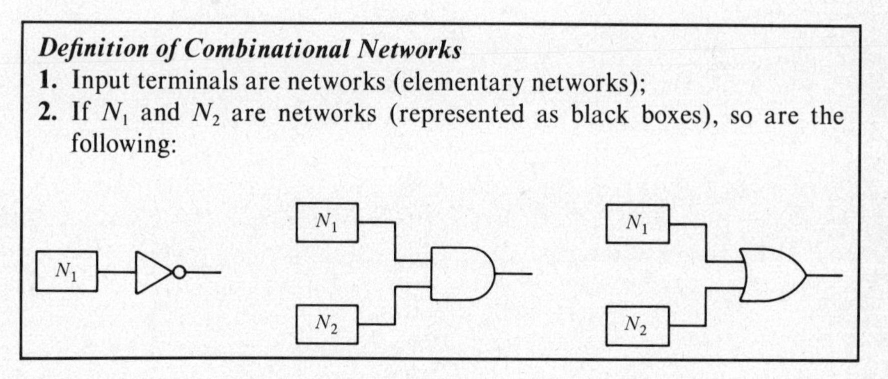

In analogy with the preceding definition, consider now the following definition of "boolean expressions":

Definition of Boolean Expressions

1. Variables (both complemented and uncomplemented) and constants are expressions (elementary expressions);
2. If E_1 and E_2 are expressions, so are $\overline{E}_1$, $E_1 \cdot E_2$, and $E_1 + E_2$.[1]

EXAMPLE 3.1

x_1, 0, x_2 are elementary expressions; $[x_1 + (\overline{x}_2 x_3)] \cdot x_4$ is an expression, specifically, it is the AND of expressions $E_1 = x_1 + (\overline{x}_2 x_3)$ and $E_2 = x_4$.

[1]To avoid any ambiguity, the new expressions $E_1 . E_2$ and $E_1 + E_2$ should both be parenthesized as $(E_1 \cdot E_2)$ and $(E_1 + E_2)$. However, we shall conform here to the familiar rules for parentheses adopted in ordinary algebra for "+" and "·."

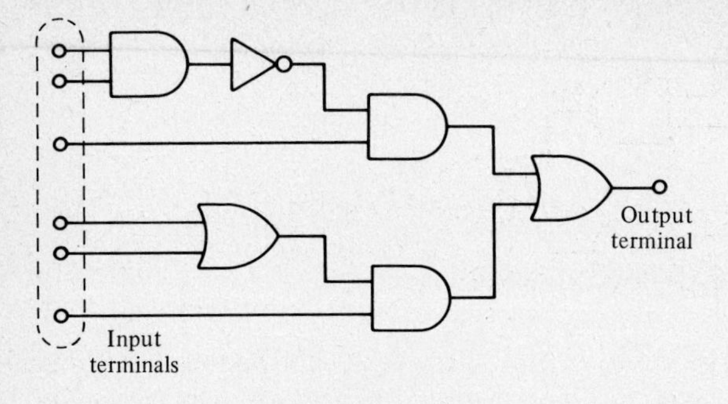

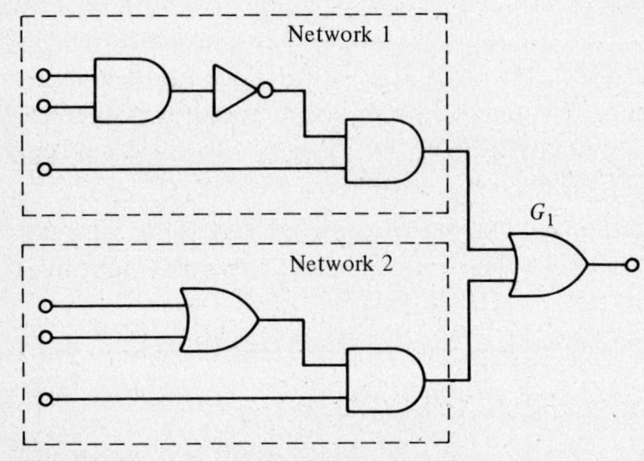

Figure 3.7 A combinational network.

Suppose that we now assign either a variable (complemented or otherwise) or a constant to each input terminal of a nonelementary network. Then, with the output of each gate whose inputs are connected to the input terminals, we can associate an expression, and so on downstream until we associate an expression with the output of the network. [As an example see Figure 3.8(a).] Therefore, we see that there is a one-to-one correspondence between expressions and networks whose inputs have been assigned (input-assigned network). Specifically we say that

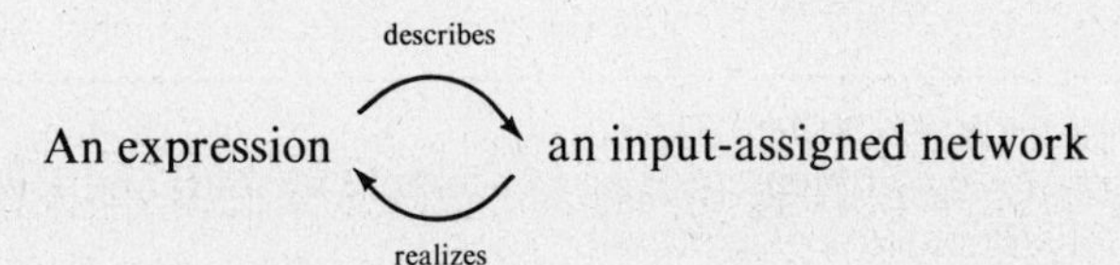

Normally we will drop the qualifier "input-assigned" whenever the context makes it obvious.

Consider the network of Figure 3.8(a). We may now assign to each binary input variable, that is, to x_1, x_2, and x_3, one of the two possible values 0 or 1. Once this assignment has been made we can easily trace the network downstream and calculate a binary value on each internal wire of the network until we obtain a binary value at the output terminal [Figure 3.8(b)]. Consider what we have done:

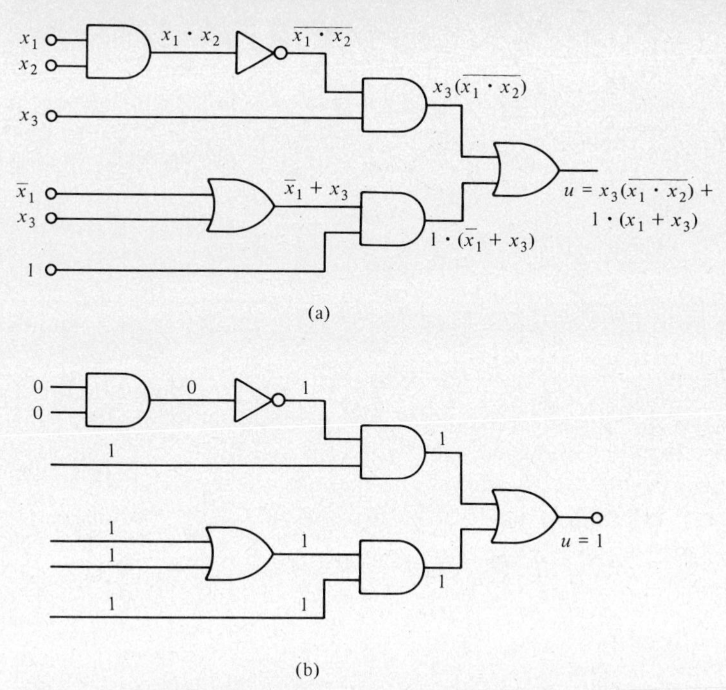

Figure 3.8 Determination of the output expression of a network.

we have chosen a set of binary values for (x_1, x_2, x_3) (in our example 001) and have obtained a binary value of u; and for each different choice of input values the network embodies a well-defined rule for obtaining a value of u. This means that u is a function of (x_1, x_2, x_3), from which we conclude the following:

Any given switching network computes a switching function of its inputs.

A most remarkable fact—to be shown later—is that the converse of the above statement is also true, that is, for any given switching function we can design a network that realizes it. The corresponding design techniques we shall present in the next sections.

The notion of combinational networks can be slightly generalized to encompass the class of networks in which a gate output may feed *more than one* gate.[2] A gate output feeding *two or more* gate inputs is said to have *multiple fan-out*. Notice however that a network with multiple-fan-out gates does not correspond to a single boolean expression: indeed, if we want to be able to reconstruct the network from its description, we must have a distinct expression for each gate having a multiple fan-out.

[2]The number of inputs driven by the output of a gate is called the *fan-out* of that gate. Also, the number of inputs of a gate is called the *fan-in* of the gate.

3.4 SWITCHING ALGEBRA

The techniques for designing combinational networks rest on the properties of a fundamental formal system called *switching algebra* (which is a special case of more general systems called boolean algebras, although frequently switching algebra is referred to as "boolean algebra").

The objects that switching algebra deals with are the (boolean) expressions defined in the preceding section. The basic axiom is as follows:

Axiom. Each expression assumes either the value 0 or the value 1 for all assignments of values (0 or 1) to its variables.

We begin by regarding the function of one variable, COMPLEMENT, as a *unary* operation, that is, as an operation with *one* operand x, the function's argument, which is itself to be regarded as an expression and can only assume the two values 0 and 1. The table of the operation is repeated below.

Operand	COMPLEMENT
0	1
1	0

Notice that the complement of 0 is 1, that is, $\overline{0} = 1$; similarly $\overline{1} = 0$. It follows that

$$\overline{(\overline{0})} = \overline{1} = 0, \qquad \overline{(\overline{1})} = \overline{0} = 1$$

This is summarized by the identity

$$\overline{\overline{x}} = x \qquad \text{Involution} \qquad (3.1)$$

which describes a fundamental property of COMPLEMENT. Notice also that the constants 0 and 1 are mutually complementary.

We now regard the functions AND and OR of two variables x_1 and x_2 as *binary* operations, that is, as operations with *two* operands. Here again x_1 and x_2 are to be regarded as expressions, and, by the axiom, each can only assume either the value 0 or the value 1. The transformation of each of the function tables to the corresponding operation table, shown in Figures 3.9(a) and 3.9(b), should be self-explanatory.

(a)

x_2	x_1	AND
0	0	0
0	1	0
1	0	0
1	1	1

		x_2	
AND		0	1
x_1	0	0	0
	1	0	1

(b)

x_2	x_1	OR
0	0	0
0	1	1
1	0	1
1	1	1

		x_2	
OR		0	1
x_1	0	0	1
	1	1	1

Figure 3.9 Operation tables for AND and OR.

From the inspection of these operation tables, we can now deduce their characteristic properties. First of all, both tables are *symmetric* with respect to the main diagonal, that is, we can exchange the role of x_1 and x_2; we shall summarize this as follows

$$x_1 \cdot x_2 = x_2 \cdot x_1 \qquad x_1 + x_2 = x_2 + x_1 \qquad \text{Commutativity} \qquad (3.2)$$

Next, we notice that $0 \cdot 0 = 0 + 0 = 0$ and $1 \cdot 1 = 1 + 1 = 1$, which leads to the property, for any expression x,

$$xx = x \qquad x + x = x \qquad \text{Idempotency} \qquad (3.3)$$

Since $0 \cdot 1 = 0$ and $1 \cdot 1 = 1$ we extract the rule $x \cdot 1 = x$; similarly $0 + 0 = 0$ and $1 + 0 = 1$ gives $x + 0 = x$, and we have the properties

$$x \cdot 1 = x \qquad x + 0 = x \qquad (3.4)$$

Also, $0 \cdot 0 = 0$ and $1 \cdot 0 = 0$ yields $x \cdot 0 = 0$; similarly $0 + 1 = 1$ and $1 + 1 = 1$ yields $x + 1 = 1$, whence the properties

$$x \cdot 0 = 0 \qquad x + 1 = 1 \qquad (3.5)$$

Finally, considering the off-diagonal elements in both operation tables, we see that $0 \cdot 1 = 1 \cdot 0 = 0$ and $0 + 1 = 1 + 0 = 1$, whence

$$x \cdot \overline{x} = 0 \qquad x + \overline{x} = 1 \qquad \text{Complementarity} \qquad (3.6)$$

There is now a collection of additional properties that can be very easily derived by means of a very useful proof mechanism called *perfect induction,* which is stated as follows:

Perfect Induction. Two expressions E_1 and E_2 on the same set of variables are equivalent (denoted by $E_1 = E_2$) if, for all possible assignments of values to the variables, the values of E_1 and E_2 coincide.

Perfect induction will now be used to prove the following identities:

$$x \cdot (y + z) = xy + xz \qquad x + yz = (x + y)(x + z) \qquad \text{Distributivity} \qquad (3.7)$$

Indeed, the claim is proved by the following tables:

x	y	z	$y + z$	$x(y + z)$	xy	xz	$xy + xz$	yz	$x + yz$	$x + y$	$x + z$	$(x + y)(x + z)$
0	0	0	0	0	0	0	0	0	0	0	0	0
0	0	1	1	0	0	0	0	0	0	0	1	0
0	1	0	1	0	0	0	0	0	0	1	0	0
0	1	1	1	0	0	0	0	1	1	1	1	1
1	0	0	0	0	0	0	0	0	1	1	1	1
1	0	1	1	1	0	1	1	0	1	1	1	1
1	1	0	1	1	1	0	1	0	1	1	1	1
1	1	1	1	1	1	1	1	1	1	1	1	1

By perfect induction we can also prove the identities

$x(x + y) = x$	$x + xy = x$	Absorption	(3.8)
$(xy)z = x(yz)$	$(x + y) + z = x + (y + z)$	Associativity	(3.9)
$\overline{xy} = \overline{x} + \overline{y}$	$\overline{x + y} = \overline{x} \cdot \overline{y}$	De Morgan's Law	(3.10)

(Notice that because associativity holds, we will omit parentheses when writing the AND or the OR of more than two variables.)

Consider now the identities (3.2)–(3.10) which we have established. They are offered in pairs such that one pair is obtained from the other by interchanging AND and OR and by interchanging the constants 0 and 1. This fact is summarized as follows:

Principle of Duality. Given a valid identity, we obtain another valid identity by:

1. Interchanging the operators AND and OR.
2. Interchanging the constants 0 to 1.

We can now concisely summarize the properties of switching algebra which we have just established.

Switching Algebra is a set B of elements (boolean expressions) containing the constants 0 and 1, with the following operations:

1. Two binary operations, AND and OR, which are commutative (3.2), associative (3.9), idempotent (3.3), absorptive (3.8), and mutually distributive (3.7),
2. A unary operation, COMPLEMENT (or NEGATION), with the properties of involution (3.1), complementarity (3.6), De Morgan's law (3.10).

The constants 0 and 1 have the following properties (3.4), (3.5):

$$\overline{1} = 0$$

$$x \cdot 1 = x \qquad x + 0 = x$$

$$x \cdot 0 = 0 \qquad x + 1 = 1$$

Identities (3.1)–(3.10) given above represent a set of rules—given in dual pairs—which can be applied to transform an expression into an equivalent expression. It can be shown that rules (3.1)–(3.10) are not independent and that we can select five of them and derive the others from these: this, however, is outside our present scope.

EXAMPLE 3.2

Prove the following identities, without using perfect induction and by transforming the left side to the right side.

$$a + \overline{a}b = a + b; \qquad (3.11)$$

$$
\begin{aligned}
a + \bar{a}b &= a \cdot 1 + \bar{a}b && \text{[by (3.4)]} \\
&= a(b + \bar{b}) + \bar{a}b && \text{[by (3.6)]} \\
&= ab + a\bar{b} + \bar{a}b && \text{[by (3.7)]} \\
&= a\bar{b} + ab + \bar{a}b && \text{[by (3.2)]} \\
&= a\bar{b} + ab + ab + \bar{a}b && \text{[by (3.3)]} \\
&= a(\bar{b} + b) + (a + \bar{a})b && \text{[by (3.7)]} \\
&= a \cdot 1 + 1 \cdot b && \text{[by (3.6)]} \\
&= a + b && \text{[by (3.4)]}
\end{aligned}
$$

An alternative and simpler proof of (3.11) runs as follows:

$$
\begin{aligned}
a + \bar{a}b &= (a + \bar{a})(a + b) && \text{[by (3.7)]} \\
&= 1 \cdot (a + b) && \text{[by (3.6)]} \\
&= a + b && \text{[by (3.4)]}
\end{aligned}
$$

$$ab + bc + \bar{a}c = ab + \bar{a}c; \tag{3.12}$$

$$
\begin{aligned}
ab + bc + \bar{a}c &= ab + 1 \cdot bc + \bar{a}c && \text{[by (3.4)]} \\
&= ab + (a + \bar{a})bc + \bar{a}c && \text{[by (3.6)]} \\
&= ab + abc + \bar{a}bc + \bar{a}c && \text{[by (3.7)]} \\
&= ab \cdot 1 + abc + \bar{a}bc + \bar{a}c \cdot 1 && \text{[by (3.4)]} \\
&= ab(1 + c) + \bar{a}c(b + 1) && \text{[by (3.7)]} \\
&= ab \cdot 1 + \bar{a}c \cdot 1 && \text{[by (3.5)]} \\
&= ab + \bar{a}c && \text{[by (3.4)]}
\end{aligned}
$$

Identities (3.11) and (3.12) are actually theorems which have been proved by using the valid identities (3.1)–(3.10); (3.11) is sometimes, but improperly, called "absorption" because of its similarity with (3.8), and (3.12) is known as the "consensus" identity. These two identities are quite convenient because they are relatively easy to memorize and can themselves be applied to accomplish transformations of boolean expressions. We could continue deriving identities of this kind to be included in our bag of valid rules: however, the burden of memorization will rapidly reach the point of diminishing return.

We conclude with a summary of the manipulative rules of switching algebra.

TABLE 3.1 SWITCHING ALGEBRA SUMMARY

(P1) $XY = YX$		(S1) $X + Y = Y + X$	Commutativity
(P2) $X(YZ) = (XY)Z$		(S2) $X + (Y + Z) = (X + Y) + Z$	Associativity
(P3) $XX = X$		(S3) $X + X = X$	Idempotency
(P4) $X(X + Y) = X$		(S4) $X + XY = X$	Absorption
(P5) $X(Y + Z) = XY + XZ$		(S5) $X + YZ = (X + Y)(X + Z)$	Distributivity
(P6) $X\bar{X} = 0$		(S6) $X + \bar{X} = 1$	Complementarity
	(C1) $\bar{\bar{X}} = X$		Involution
(P7) $\overline{XY} = \bar{X} + \bar{Y}$		(S7) $\overline{X + Y} = \bar{X}\bar{Y}$	De Morgan's
(P8) $X(\bar{X} + Y) = XY$		(S8) $X + \bar{X}Y = X + Y$	
	(B1) $\bar{1} = 0$		
(P10) $X \cdot 0 = 0$		(S10) $X + 1 = 1$	
(P11) $X \cdot 1 = X$		(S11) $X + 0 = X$	
(P13) $(X + Y)(Y + Z)(\bar{X} + Z) = (X + Y)(\bar{X} + Z)$		(S13) $XY + YZ + \bar{X}Z = XY + \bar{X}Z$	Consensus

3.5 BOOLEAN EXPRESSIONS: NORMAL AND CANONICAL FORMS

We saw earlier that every boolean expression involving n distinct variables describes a switching function of those n variables. We shall now show a very important fact, namely, the converse of the above statement: *given any binary function of n binary variables we can construct a boolean expression describing that function.* Since every boolean expression corresponds to a combinational circuit consisting of single-fan-out gates, we obtain the far-reaching result that every switching function is realizable by means of a combinational network.

An expression involves variables in *uncomplemented* or *complemented* forms: we call *literal* any occurrence of a variable in either form. For example, the expression $\{[x_1 + x_2\overline{(x_3 + x_4\overline{x}_1)}]\ \overline{x}_3 + x_2x_4\}$ has four variables and eight literals.

An expression is in *normal SOP* (sum-of-products) *form* when it is the OR (sum) of ANDs (products) of literals. We shall now describe how an arbitrary boolean expression E can be transformed into an equivalent expression in normal SOP form.

3.5.1 Reduction to Normal SOP Form

Let the expression $E = [x_1 + x_2\overline{(x_3 + x_4\overline{x}_1)}]\ x_3 + \overline{\overline{x}_2x_4}$ be given.

1. Place all complements directly on variables (by using De Morgan's laws).

In our example

$$\begin{aligned} E &= (x_1 + x_2 \cdot \overline{x}_3 \cdot \overline{x_4\overline{x}_1})x_3 + (\overline{\overline{x}}_2 + \overline{x}_4) \\ &= (x_1 + x_2 \cdot \overline{x}_3(\overline{x}_4 + \overline{\overline{x}}_1))x_3 + x_2 + \overline{x}_4 \\ &= [x_1 + x_2\overline{x}_3(\overline{x}_4 + x_1)]x_3 + x_2 + \overline{x}_4 \end{aligned}$$

2. Apply the distributive law.

In our example

$$\begin{aligned} E &= (x_1 + x_2\overline{x}_3\overline{x}_4 + x_1x_2\overline{x}_3)x_3 + x_2 + \overline{x}_4 \\ &= x_1x_3 + x_2\overline{x}_3x_3\overline{x}_4 + x_1x_2\overline{x}_3x_3 + x_2 + \overline{x}_4 \end{aligned}$$

3. Eliminate redundant terms (using idempotency and complementarity).

In our example notice that, by (3.6), $x_3\overline{x}_3 = 0$ and that, by (3.5), all product terms containing a factor 0 are 0 themselves, whence

$$\begin{aligned} E &= x_1x_3 + 0 + 0 + x_2 + \overline{x}_4 \\ &= x_1x_3 + x_2 + \overline{x}_4 \end{aligned}$$

The latter expression is in normal SOP form and is equivalent to the given expression.

With reference to expressions on n variables, a special type of product term (AND term) is one that contains as a factor each variable, either uncomplemented or complemented: these terms are called *fundamental products* or *minterms*. For example, for $n = 4$, $\bar{x}_1\bar{x}_2\bar{x}_3x_4$ and $\bar{x}_1x_2\bar{x}_3x_4$ are minterms but $\bar{x}_1\bar{x}_3x_4$ is not. A normal SOP expression is said to be in *canonical* (SOP) *form* if its product terms are all minterms. We shall now describe how to transform a normal form expression into a canonical form expression.

3.5.2 Transformation from Normal Form to Canonical Form

Let the normal form expression $x_1x_3 + x_2 + \bar{x}_3$ be given.

1. If a product term contains neither x_i nor $\bar{x}_i$, "multiply" it by $(x_i + \bar{x}_i)$. [Notice that this transforms the product term into an equivalent expression since $x_i + \bar{x}_i = 1$ by (3.6).]

In our example x_1x_3 does not contain a literal with index 2; x_2 does not contain literals with indices 1 and 3; $\bar{x}_3$ does not contain literals with indices 1 and 2. Thus

$$x_1x_3 + x_2 + \bar{x}_3 = x_1x_3(x_2 + \bar{x}_2) + x_2(x_1 + \bar{x}_1)(x_3 + \bar{x}_3) + \bar{x}_3(x_1 + \bar{x}_1)(x_2 + \bar{x}_2)$$

2. Apply the distributive law.

In our example

$$\begin{aligned} E = {} & x_1x_3\bar{x}_2 + x_1x_3x_2 \\ & + x_2\bar{x}_1\bar{x}_3 + x_2\bar{x}_1x_3 + x_2x_1\bar{x}_3 + x_2x_1x_3 \\ & + \bar{x}_3\bar{x}_1\bar{x}_2 + \bar{x}_3\bar{x}_1x_2 + \bar{x}_3x_1\bar{x}_2 + \bar{x}_3x_1x_2 \end{aligned}$$

3. Eliminate repeated product terms using idempotency.

In our example, the following sets of terms are sets of identical terms: (2nd, 6th) (3rd, 8th) (5th, 10th). Thus, after eliminating the repeated terms and rearranging the order of the indices as (3, 2, 1), we obtain the canonical expression for $x_1x_3 + x_2 + \bar{x}_3$:

$$E = x_3\bar{x}_2x_1 + x_3x_2x_1 + \bar{x}_3x_2\bar{x}_1 + x_3x_2\bar{x}_1 + \bar{x}_3x_2x_1 + \bar{x}_3\bar{x}_2\bar{x}_1 + \bar{x}_3\bar{x}_2x_1$$

We begin by introducing a useful notation for minterms. Consider, for example, the minterm $\bar{x}_4x_3x_2\bar{x}_1$; we associate with this minterm an ordered binary 4-tuple $(b_4b_3b_2b_1)$, where b_i corresponds either to x_i or $\bar{x}_i$, and $b_i = 1$ if x_i is uncomplemented and is 0 otherwise. In our example, with $\bar{x}_4x_3x_2\bar{x}_1$ we associate

0110: this string is the binary equivalent of the integer 6, so that we shall denote $\overline{x}_4x_3x_2\overline{x}_1$ by m_6. Referring to the previous example, the expression E in this new notation becomes $m_5 + m_7 + m_2 + m_6 + m_3 + m_0 + m_1$, or, equivalently, OR $(m_0, m_1, m_2, m_3, m_5, m_6, m_7)$.

Suppose now that we have combinational networks F and G which respectively compute switching functions $f(x_1, \ldots, x_n)$ and $g(x_1, \ldots, x_n)$, and that we connect the outputs of these networks to the inputs of gates or inverters. Clearly if f and g are fed, say, to an AND gate, then the function u at the output of this gate will be 1 only when both f and g are 1, and similarly for the other cases. Obviously u is a function of the same set of variables $\{x_1, \ldots, x_n\}$ as f and g, so we obtain the following simple rules for its truth table:

> The truth table of $\overline{f}$ is the entry-by-entry (componentwise) complement of the truth table of f; the truth table of $f \cdot g$ [or $(f + g)$] is the componentwise AND (or OR) of the truth tables of f and g.

Given a minterm m of $x_1, x_2, \ldots, x_n$ (the minterm itself obviously describes a function of these variables), we recognize that $m = 1$ exactly when the following conditions hold: if x_i appears in m in uncomplemented form, then $x_i = 1$; if x_i appears in m in complemented form, then $\overline{x}_i = 1$, that is, $x_i = 0$. Thus $m = 1$ only when each variable attains a specific value, that is, $m = 1$ for a unique combination of the variables or, equivalently, *the truth table of a minterm has exactly one "1."* (Incidentally, this explains the denomination "minterm": a nondegenerate function with the minimum number of 1s in its truth table.)

EXAMPLE 3.3

For $n = 4$, $\overline{x}_4\overline{x}_3\overline{x}_2x_1 = m_1$ is 1 when and only when $\overline{x}_4 = 1$, $\overline{x}_3 = 1, \overline{x}_2 = 1, x_1 = 1$, that is, when $(x_4x_3x_2x_1) = (0001)$.

Now, let f be a switching function of arguments $x_1, \ldots, x_n$, given by means of its truth table (see Figure 3.10 where $n = 3$). We may view the truth table of f as the OR of as many distinct truth tables as it has 1s, each of the latter truth tables having exactly a single 1 [Figure 3.10(a)]. But each such table is the

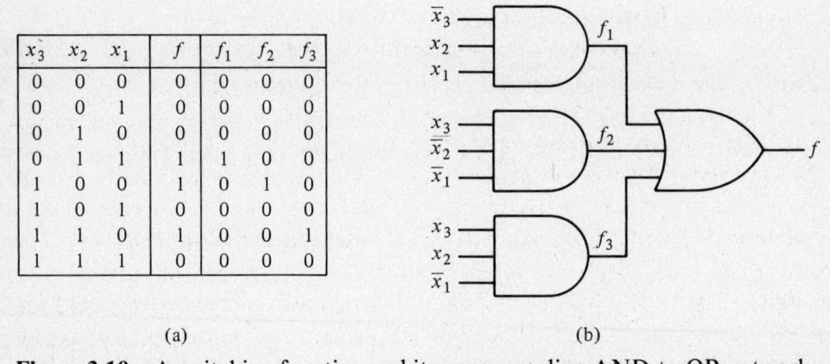

x_3	x_2	x_1	f	f_1	f_2	f_3
0	0	0	0	0	0	0
0	0	1	0	0	0	0
0	1	0	0	0	0	0
0	1	1	1	1	0	0
1	0	0	1	0	1	0
1	0	1	0	0	0	0
1	1	0	1	0	0	1
1	1	1	0	0	0	0

Figure 3.10 A switching function and its corresponding AND-to-OR network.

table of a minterm! Moreover, each minterm is a product of literals, which for $i = 1, 2, \ldots, n$ contains either x_i or $\overline{x}_i$, depending upon whether in the combination corresponding to the single 1 in the table the x_i entry is 1 or 0. In Figure 3.10, $f_1 = 1$ in correspondence to $(x_3x_2x_1) = (011)$, whence $f_1 = \overline{x}_3x_2x_1$. Similarly we obtain $f_2 = x_3\overline{x}_2\overline{x}_1$ and $f_3 = x_3x_2\overline{x}_1$. In conclusion, since $f = f_1 + f_2 + f_3$, we have

$$f = \overline{x}_3x_2x_1 + x_3\overline{x}_2\overline{x}_1 + x_3x_2\overline{x}_1$$

Notice that this is a most remarkable finding: given a function f by means of its truth table (i.e., as a binary function of binary variables), we have obtained an expression (actually, a canonical expression) describing that function!

Once we have an expression for the given function, we shall design the corresponding combinational network. Before proceeding, however, we recall that in Sec. 3.2 we have introduced AND gates and OR gates as two-input–one-output devices; since we have proved [identity (3.9)] that the AND and OR operations are associative, we may think of using in our networks devices which realize the AND (or OR) of *more than two* inputs: this is indeed technically possible, although, for physical reasons, the number of inputs may not be too large. Therefore, we see that by using these newly introduced gates, we can construct the network of Figure 3.10(b) which computes the given function. (This network is a collection of AND gates feeding a single OR gate, and is therefore called an AND-to-OR network.) Notice that we have achieved the objective set forth at end of Sec. 3.3 and summarized below:

Given a switching function f by means of its truth table (i.e., as a binary function of binary variables), we can construct a switching network that computes it.

In Table 3.2, left side, we summarize the important notions concerning SOP-canonical expressions. A discussion analogous to the one just completed can be carried out with reference to *normal POS* (product-of-sums) expressions, that is, an AND of ORs of literals. Indeed, all we need in the preceding discussion is a set of substitutions as dictated by the principle of duality:

AND	↔	OR
Product	↔	Sum
Minterm	↔	Maxterm (see below)
0	↔	1

The conclusions are summarized in Table 3.2, right side. Notice the perfect duality of corresponding statements in this table. The only novel term in this table is "maxterm," the dual of minterm: the reason for the denomination is that a maxterm describes a nondegenerate function with the maximum number of 1s in its truth table. A maxterm is usually denoted by the symbol M_j: specifically $M_j = \overline{m}_j$, that is, $M_j = 0$ if and only if $m_j = 1$, and *vice versa*. For example, for variables x_3, x_2, x_1, $M_5 = \overline{m}_5 = \overline{x_3\overline{x}_2x_1} = (\overline{x}_3 + x_2 + \overline{x}_1)$, that is, M_5 is the maxterm which is 0 exactly for $(x_3x_2x_1) = (101)$ and is 1 otherwise.

TABLE 3.2

Canonical SOP expressions	Canonical POS expressions
Minterm = a *product* of literals which has as a "factor" each of the n variables either true or complemented	*Maxterm* = a *sum* of literals which has as an "addend" each of the n variables either true or complemented
Canonical *SOP* expressions = *OR* of *minterms*	Canonical *POS* expression = *AND* of *maxterms*
A *minterm* is a canonical *SOP* expression which is 1 for exactly one combination of the variables	A *maxterm* is a canonical *POS* expression which is 0 for exactly one combination of the variables
A *minterm* corresponds to one switching function whose truth table has exactly one "1"	A *maxterm* corresponds to one switching function whose truth table has exactly one "0"
There exists a one-to-one correspondence between switching functions and canonical *SOP* expressions	There exists a one-to-one correspondence between switching functions and canonical *POS* expressions

In conclusion a boolean function can be specified either as an *OR of minterms* (corresponding to the 1s in the truth table) or as an *AND of maxterms* (corresponding to the 0s in the truth table).

EXAMPLE 3.4

Decimal equivalent	x_3	x_2	x_1	f
0	0	0	0	0
1	0	0	1	0
2	0	1	0	0
3	0	1	1	1
4	1	0	0	1
5	1	0	1	0
6	1	1	0	1
7	1	1	1	0

$f = \text{OR}\ (m_3, m_4, m_6)$, SOP
$f = \text{AND}\ (M_0, M_1, M_2, M_5, M_7)$, POS

Consider now a boolean function $f(x_1, x_2, \ldots, x_n)$ expressed in canonical SOP form. Each minterm of f contains either $\overline{x}_n$ or x_n; therefore, we associate into two separate expressions, F_0 and F_1, the minterms of f depending upon whether they contain $\overline{x}_n$ or x_n, that is,

$$f = F_0 + F_1$$

Now from all terms of F_0 we can factor out $\overline{x}_n$; notice that the factored expression, which we shall call f_0, consists exactly of minterms over the variables $x_1, x_2, \ldots, x_{n-1}$, that is, $F_0 = \overline{x}_n f_0(x_1, x_2, \ldots, x_{n-1})$. Similarly we can express F_1 as $F_1 = x_n f_1(x_1, x_2, \ldots, x_{n-1})$. It follows that f can be expressed as

$$f(x_1, \ldots, x_{n-1}, x_n) = \overline{x}_n f_0(x_1, \ldots, x_{n-1}) + x_n f_1(x_1, \ldots, x_{n-1}) \quad (3.13)$$

In the above relation, we now set $x_n = 0$ and obtain

$$f(x_1, \ldots, x_{n-1}, 0) = 1 \cdot f_0(x_1, \ldots, x_{n-1}) + 0 \cdot f_1(x_1, \ldots, x_{n-1})$$

that is, $f_0(x_1, \ldots, x_{n-1}) = f(x_1, \ldots, x_{n-1}, 0)$. Similarly, if we set $x_n = 1$ in (3.13) we have

$$f(x_1, \ldots, x_{n-1}) = 0 \cdot f_0(x_1, \ldots, x_{n-1}) + 1 \cdot f_1(x_1, \ldots, x_{n-1})$$

that is, $f_1(x_1, \ldots, x_{n-1}) = f(x_1, \ldots, x_{n-1}, 1)$. This result is called the fundamental theorem of boolean algebra and can be stated as follows.

Fundamental Theorem of Boolean Algebra. Every function $f(x_1, \ldots, x_n)$ of $x_1, \ldots, x_n$, for any x_i can be expressed as

$$f = \overline{x}_i f_0 + x_i f_1$$

where $f_0 = f(x_1, \ldots, x_{i-1}, 0, x_{i+1}, \ldots, x_n)$ and $f_1 = f(x_1, \ldots, x_{i-1}, 1, x_{i+1}, \ldots, x_n)$ are both functions of the $(n-1)$ variables $x_1, \ldots, x_{i-1}, x_{i+1}, \ldots, x_n$.

3.6 OTHER IMPORTANT BOOLEAN CONNECTIVES

Although the operators AND, OR, and NOT are perfectly adequate for the realization of any combinational network, there are other connectives which are quite important and are now introduced.

3.6.1 The NAND and NOR Connectives

The first of these connectives, called NAND (AND followed by NOT), realizes the function $\overline{x \cdot y}$ of two variables x and y; its circuit symbol is given, for two inputs, in Figure 3.11(a). (Note that, as already has been done for the AND and OR connectives, the function NAND can be generalized to any number of variables, as $\overline{x \cdot y \ldots \cdot w}$.) The connective NAND is interesting because alone it can be used to realize any combinational network. (We refer to this property by saying that NAND is *logically complete*.) Indeed, since we know that AND, OR, and NOT are adequate for realizing combinational networks, all we need to show is that each of these three connectives can, in turn, be realized by an expression involving only NAND. This is readily shown below:

$$a \cdot b = \overline{\overline{a \cdot b}} = \overline{\overline{a \cdot b} \cdot \overline{a \cdot b}}$$

$$= \text{NAND}\,[\text{NAND}(a, b), \text{NAND}(a, b)] \qquad \text{(rules C1 and P3, Table 3.1)}$$

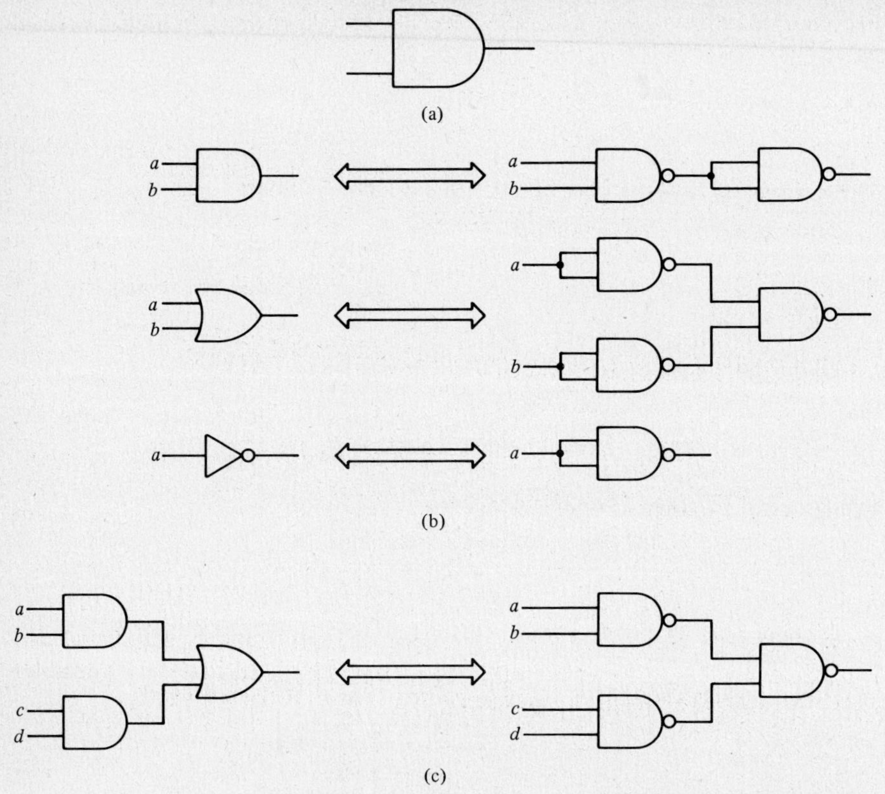

Figure 3.11 (a) Circuit symbol of NAND gate. (b) and (c) NAND gate realizations of AND, OR, and NOT gates and SOP expression.

$$a + b = \overline{\overline{a} \cdot \overline{b}} = \overline{\overline{aa} \cdot \overline{bb}} = \text{NAND}\,[\text{NAND}(a, a), \text{NAND}(b, b)]$$

(rules P7 and P3, Table 3.1)

$$\overline{a} = \overline{aa} = \text{NAND}(a, a) \qquad \text{(rule P3, Table 3.1)}$$

These transformations are illustrated in Figure 3.11(b). Thus, given a network consisting of AND, OR, and NOT gates, by using the above rules, one can transform it into an equivalent one consisting of NAND gates alone. But besides this rather cumbersome transformation, there is a more direct and useful correspondence between NAND expressions and SOP expressions, as shown below:

$$ab + cd = \overline{\overline{ab + cd}} = \overline{\overline{ab} \cdot \overline{cd}} = \text{NAND}[\text{NAND}(a, b), \text{NAND}(c, d)]$$

So we see that any SOP expression can be realized by a network consisting of NAND gates alone [see Figure 3.11(c)], by simply replacing with NAND gates both the AND gates and the OR gate in the standard AND-to-OR realization of the given expression.

As we may expect from duality, there is another connective, NOR, which enjoys analogous properties. This connective NOR (OR followed by NOT)

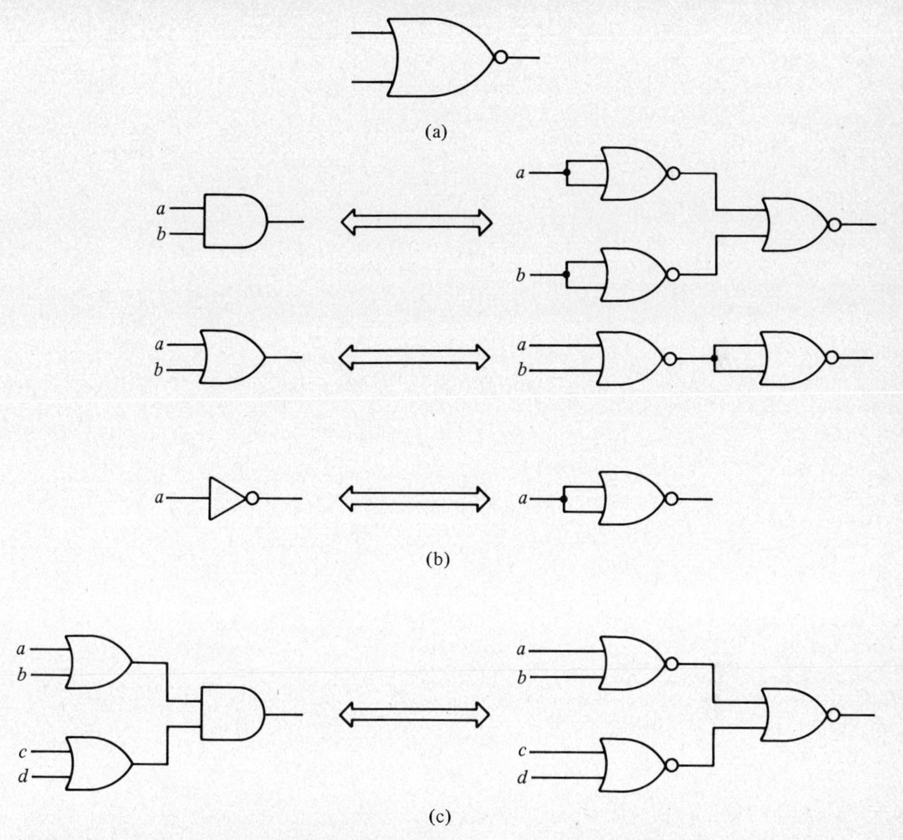

Figure 3.12 (a) Circuit symbol of NOR gate. (b) and (c) NOR gate realizations of AND, OR, and NOT gates and POS expression.

realizes the function $\overline{x + y}$ of x and y, and its symbol is given, for two inputs, in Figure 3.12(a). Transformations analogous to those obtained above can be easily derived and the results are shown in Figures 3.12(b) and 3.12(c). Notice that a two-level NAND network corresponds to a SOP expression, and a two-level NOR network corresponds to a POS expression. These properties make NOR and NAND gates very attractive and popular in digital design, since entire systems can be realized by using just one type of component. This is a very desirable feature from a technological viewpoint, and we shall return to this point in Chapter 4.

3.6.2 The XOR Connective

Finally, we introduce the connective EXCLUSIVE OR (frequently abbreviated as XOR), which realizes the function $(x\overline{y} + \overline{x}y)$ of two variables x and y. The symbol used for this connective is $\oplus$, while the circuit symbol is given in Figure 3.13. The reason for the name EXCLUSIVE OR is that $(x \oplus y)$ is equal to 1 if and only if either x or y, *but not both,* is equal to 1. (By contrast, the ordinary OR

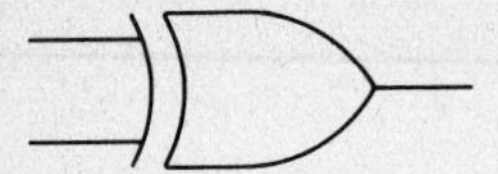

Figure 3.13 Symbol for the EXCLUSIVE OR gate.

is correctly called "inclusive" OR, although the adjective "inclusive" is normally omitted.)

The EXCLUSIVE OR has several interesting properties, whose proof is left as an exercise. First, EXCLUSIVE OR is associative

$$(x \oplus y) \oplus z = x \oplus (y \oplus z)$$

so that we may omit parentheses and write $x \oplus y \oplus z$. Therefore, the EXCLUSIVE OR is generalized to an arbitrary number of variables, and we shall have EXCLUSIVE OR gates with correspondingly many input lines. Other important properties, also left as an exercise, are

$$x(y \oplus z) = xy \oplus xz$$
$$x \oplus 1 = \overline{x}$$

TABLE 3.3

Name	Symbol set 1	Symbol set 2
AND		
OR		
NOT		
NAND		
NOR		
EXCLUSIVE OR		

It is appropriate to introduce at this point an alternative set of standard symbols for the logic gates discussed in this chapter. They are displayed in Table 3.3, vis-à-vis their by now familiar counterparts (Symbol Set 1), and deserve no additional comments. Although in this text we shall only use the gate symbols of set 1, the reader should also be acquainted with the alternative representations.

3.7 REVIEW EXAMPLE

Find canonical SOP and POS expressions for the following expression:

$$f(x, y, z) = \overline{(y \oplus \bar{x}) + \overline{(\bar{x} + z)}} + \overline{(\bar{y} + \bar{z})}$$

As a preliminary step we express the connective "$\oplus$" in terms of $\{+, \cdot, \bar{\ }\}$, that is,

$$\begin{aligned} f &= \overline{(xy + \bar{x}\bar{y}) + \overline{(\bar{x} + z)}} + y \cdot z \\ &= \overline{xy} \cdot \overline{\bar{x}\bar{y}} \cdot (\bar{x} + z) + yz \\ &= (\bar{x} + \bar{y})(x + y) \cdot (\bar{x} + z) + yz \end{aligned}$$

At this point, we perform a sequence of steps, or their duals, depending upon whether we seek a SOP or a POS expression.

Let us consider first the SOP transformation. Then we apply the distribution of AND on OR [relation (P5)],[3] and obtain

$$\begin{aligned} f &= (\bar{x} + \bar{y})(x + y)(\bar{x} + z) + yz = (\bar{x}x + \bar{x}y + x\bar{y} + \bar{y}y)(\bar{x} + z) + yz \\ &= \bar{x}x\bar{x} + \bar{x}xz + \bar{x}y\bar{x} + \bar{x}yz + x\bar{y}\bar{x} + x\bar{y}z + \bar{x}\bar{y}y + \bar{y}yz + yz \end{aligned}$$

Applying idempotency and complementarity [relations (P3) and (P6)], we obtain the normal form expression

$$f = \bar{x}y + \bar{x}yz + x\bar{y}z + yz$$

This can now be transformed to canonical SOP form by multiplying each term by the appropriate $(a + \bar{a})$-type factors; that is,

$$f = \bar{x}y(z + \bar{z}) + \bar{x}yz + x\bar{y}z + (x + \bar{x})yz$$

and, after distributing AND on OR [relation (P5)]

$$f = \bar{x}yz + \bar{x}y\bar{z} + \bar{x}yz + x\bar{y}z + xyz + \bar{x}yz$$

and after eliminating repeated terms using idempotency (S3), we obtain

$$f = \bar{x}yz + \bar{x}y\bar{z} + x\bar{y}z + xyz$$

Assuming that the variables be taken in the order (x, y, z), then in compact minterm notation we obtain

$$f = m_3 + m_2 + m_5 + m_7$$

Let us now consider the POS transformation. Here we apply the distribu-

[3]See Table 3.1.

tion of OR on ANDs [relation (S5)] and obtain

$$
\begin{aligned}
f &= (\bar{x} + \bar{y})(x + y)(\bar{x} + z) + yz \\
&= [(\bar{x} + \bar{y})(x + y)(\bar{x} + z) + y][(\bar{x} + \bar{y})(x + y)(\bar{x} + z) + z] \\
&= [(\bar{x} + y + \bar{y})(x + y + y)(\bar{x} + y + z)] \\
&\cdot [(\bar{x} + \bar{y} + z)(x + y + z)(\bar{x} + z + z)]
\end{aligned}
$$

Again, applying idempotency (S3) and complementarity (S6), we obtain the normal POS expression

$$f = (x + y)(\bar{x} + y + z)(\bar{x} + \bar{y} + z)(x + y + z)(\bar{x} + z)$$

This expression can be transformed to a canonical one by "adding" to each sum the appropriate $a\bar{a}$-type terms, that is,

$$f = (x + y + z\bar{z})(\bar{x} + y + z)(\bar{z} + \bar{y} + z)(x + y + z)(\bar{x} + y\bar{y} + z)$$

and, after distributing OR on AND, we obtain

$$f = (x + y + z)(x + y + \bar{z})(\bar{x} + y + z)(\bar{x} + \bar{y} + z) \cdot (x + y + z)(\bar{x} + y + z)(\bar{x} + \bar{y} + z)$$

Elimination of repeated factors [idempotency (P3)] yields the final expression

$$f = (x + y + z)(x + y + \bar{z})(\bar{x} + y + z)(\bar{x} + \bar{y} + z)$$

or

$$f = M_0 \cdot M_1 \cdot M_4 \cdot M_6$$

NOTES AND REFERENCES

Boolean algebra, which—as we saw—provides the formalism for the description of (binary) digital networks, was developed in the last century, originating with the English mathematician George Boole, who in 1854 published his fundamental work, *An Investigation of the Laws of Thought*. Boole's goal was essentially the development of a formalism to compute the truth or falsehood (i.e., the "truth value") of complex compound statements from the truth values of their component statements. The discipline developed later into a more complex body of knowledge, known as symbolic logic.

Apparently, early in this century, more than one scientist perceived the applicability of boolean algebra to the design of telephone circuits (Ehrenfest, 1910). It was only in the thirties that the potential was fully realized, when C. E. Shannon (1938) published his paper "A Symbolic Analysis of Relay and Switching Circuits," which became the foundation of switching theory and logical design. Due to the context in which it was originally used (telephone networks, also called switching networks), the name "switching algebra" has become standard for the algebra of functions of two-valued variables. Although initially the interests of researchers focused on relay networks (also called contact networks), as mechanical devices were gradually replaced by electronic devices the techniques were tailored to gate networks of the type described in this and the next chapter. The term "gate" was already in use in the forties to denote the logical elements discussed earlier.

There are very many good references on boolean algebra and we may quote only a selected few of them. Suffice it to mention the texts by Hill and Peterson (1968), Kohavi (1970), and Hohn (1966). These books give a sufficiently rigorous formulation of the subject, tailored to the analysis and the design of combinational networks. In addition, like most of the earlier books, Hohn's and Kohavi's texts contain also a discussion of the boolean techniques used in connection with relay circuits. (Some of the more recent works completely omit this topic, which has been but totally overshadowed by the impressive development of electronic networks.) The reader interested in studying the relation of switching algebra to boolean algebras in general, is referred to Preparata-Yeh (1973) for an elementary introduction.

PROBLEMS

3.1. Demonstrate, without using perfect induction, whether or not each of the following equations is valid.

(a) $(x + y)(\overline{x} + y)(x + \overline{y})(\overline{x} + \overline{y}) = 0$

(b) $xy + \overline{x}\overline{y} + \overline{x}yz = xy\overline{z} + \overline{x}\overline{y} + yz$

(c) $xy + \overline{x}\overline{y} + x\overline{y}z = xz + \overline{x}\overline{y} + \overline{x}yz$

In each of the following three problems, mention at each step the rule (Table 3.1) you have used.

3.2. Simplify the following expressions using the axioms and theorems of switching algebra. (Try to obtain expressions with as few literals as possible.)

(a) $deh + \overline{e}g\overline{h} + \overline{h}e + hf\overline{e} + \overline{h}ej$

(b) $ab + \overline{a}c + \overline{b}cd$

(c) $ab + \overline{a}c + \overline{b}d + cd$

3.3. Using the properties of switching algebra, prove whether or not each of the following equations holds:

(a) $y + \overline{x}\overline{y}\overline{z} = \overline{x}\overline{z} + y\overline{z} + yz$

(b) $xy + xz + yz = (x + y) \cdot (x + z) \cdot (z + y)$

3.4. Using the properties of switching algebra, simplify the following expressions:

(a) $\overline{x}\overline{y} + \overline{x}y + y\overline{z} + x\overline{z}$

(b) $yw + \overline{x}y\overline{z} + x\overline{z}w + xyz + \overline{x}zw$

(c) $(x + z)(x + \overline{y})(y + z)$

3.5. In the algebra of numbers (here + means PLUS)

$$a + b = a + c \text{ implies } b = c \text{ (cancellation law)}$$

Prove, by a counterexample, that in switching algebra the statement

$$a + b = a + c \text{ implies } b = c$$

is false (here + means OR).

3.6. Is the NAND operator
(**a**) Commutative?
(**b**) Associative?
Prove each of these results.

3.7. (**a**) Prove by perfect induction that the following expressions are equivalent.

$$\bar{a}\bar{b}\bar{c} + b\bar{c} + [a \cdot (b + \overline{bc})] = \bar{c} + a$$

(**b**) Prove this again without using perfect induction citing the axioms and theorems of switching algebra listed in Sec. 3.4.

3.8. For each of the following combinational networks:

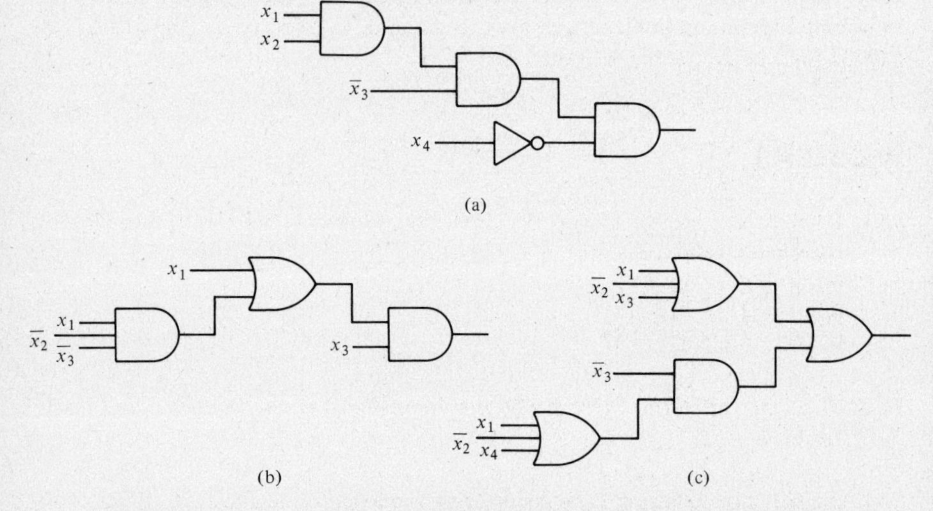

(**a**) Find the boolean expression corresponding to the output.
(**b**) Find an equivalent network constructed exclusively from NAND gates (i.e., a network whose output is described by an expression equivalent to the one describing the output of the given network).
(**c**) Find an equivalent network constructed exclusively from NOR gates.

3.9. For the following network:

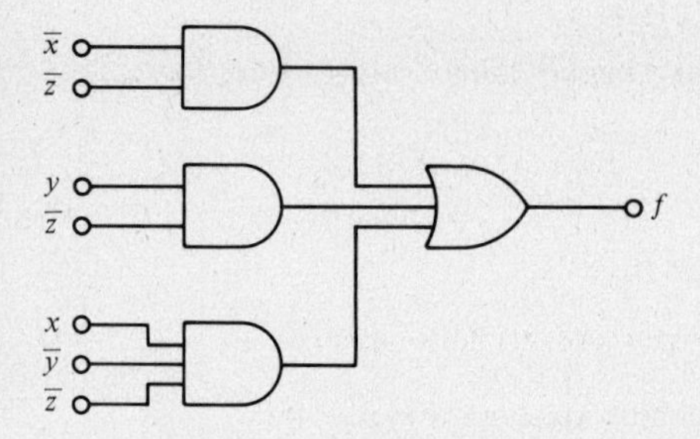

(**a**) Find the boolean expression corresponding to the output f.
(**b**) Find a simpler network that realizes the same output function.

3.10. A safe has four locks v, w, x, and y, all of which must be unlocked for the safe to open. The keys to the locks are distributed among three persons A, B, and C as follows:

A has keys for v and y.
B has keys for v and x.
C has keys for w and y.

Let variables A, B, C be 1 if the corresponding person is present, and 0 otherwise.

(a) Construct the truth table of the function $f(A, B, C)$ which is equal to 1 if and only if the safe can be opened.
(b) Express f as a canonical SOP expression.

3.11. Convert the following expressions to SOP normal form and simplify them using algebraic methods:

(a) $AB + ABCD + ABC\overline{D}$
(b) $(A + B + C) \cdot (\overline{A} + \overline{B} + C)$
(c) $(AB + \overline{BC + \overline{A}C})$

3.12. Let $m_i(x, y)$ denote the m_i minterm with reference to two variables x and y (obviously $0 \le i \le 3$). Give a POS expression for $m_2(x, y)$.

3.13. Find the canonical sum-of-product and product-of-sum expressions for the functions $f(x_1, x_2)$ and $g(x_1, x_2)$ defined as

$$f(x_1, x_2) = 0 \quad \text{if and only if} \quad x_1 = 1 \quad \text{and} \quad x_2 = 0$$

$$g(x_1, x_2) = \begin{cases} x_1 & \text{if} \quad x_2 = 0 \\ \overline{x}_1 & \text{if} \quad x_2 = 1 \end{cases}$$

(a) Determine whether $g(x_1, x_2)$ is equivalent to

$$\overline{[f(x_1, x_2) \cdot f(x_2, x_1)]}$$

(b) Show that

$$g[f(x_1, x_2), f(x_2, x_1)] = g(x_1, x_2)$$

3.14. A *majority function* $f(X, Y, Z)$ has value 1 whenever *two* or more of the three variables have value 1. Write the canonical product-of-sum (POS) form for f.

3.15. For each of the following expressions:

$$a\overline{b}c + \overline{ac}d + a\overline{c}$$
$$(a + b + cd)(\overline{a} + b)(a + b + c)$$
$$ab + \overline{a}c + \overline{b}cd$$
$$ab + \overline{a}c + \overline{b}d + cd$$
$$(a + b)(\overline{a} + c)(b + c + d)$$
$$\overline{a}cd + bd$$
$$b\overline{c} + \overline{a}b + bc\overline{d} + \overline{a}\overline{b}d + a\overline{b}\overline{c}d$$

(a) Find its canonical SOP expression.
(b) Find its canonical POS expression.

Chapter 4

Elements of Logical Design: Combinational Networks and Modules

4.1 INTRODUCTION

In the preceding chapter we have developed the foundations of switching algebra, which is the basic formalism for the analysis and the design of combinational networks. While carrying out this task, we have only temporarily forgotten the original motivation for our study, that is, the ability to design a computer system. In this chapter, which is entirely devoted to combinational networks, we shall see in some examples how the techniques can be applied to the design of specific portions of a digital computer. On the other hand, the reader will realize that logical design techniques reach beyond a digital computer into the wider domain of digital systems in general.

As we know, combinational networks are capable of realizing any binary function of binary variables (indeed, we have seen in Chapter 3 that any such function is described by a boolean expression, which, in turn, is realized by a combinational network). So far, "time" has never been mentioned. If we examine a physical gate (of the types introduced in Chapter 3), we will always measure a *nonzero propagation delay,* that is, a nonzero difference between the time the inputs are applied and the time the output is available. While this property is crucial for the realization of memory devices (as we shall see in Chapter 5), its only consequence in the context of combinational network is "to delay" the operation of the network, without affecting its switching function (i.e., the network functionality). This observation has two significant consequences.

1. If we are concerned with the *speed* of the network (especially, when the network is inserted into a larger system), we should keep in mind that

the smaller the number of gates through which signals have to propagate before the output is produced, the faster is the network's response.

2. If we are concerned exclusively with the *functionality* of the network (i.e., with the switching function realized by the network), then we are perfectly justified to make the convenient assumption that each network gate, and thus the entire network, responds *instantaneously*.

We also know that binary variables are generally[1] represented by voltages on wires (lines). Typically there are *two nominal voltages,* V_0 and V_1, which are chosen to represent the logical values 0 and 1, respectively. These two voltages are chosen sufficiently apart (a few volts in transistor technology) to avoid those random disturbances that may cause one nominal voltage to be mistaken for the other. Within this guideline, which ensures reliable operation, we are still free to choose which of V_0 and V_1 is the larger. This leads to two distinct conventions which are summarized below:

Positive logic means that $V_1 > V_0$.
Negative logic means that $V_1 < V_0$.

In this text, we shall work with positive logic, that is, a high voltage means logical 1, and a low voltage means a logical 0.

4.2 ANALYSIS AND DESIGN OF COMBINATIONAL NETWORKS

4.2.1 Analysis

The analysis of a single output combinational network consists of obtaining the boolean function realized by that network. This task is organized as a sequence of steps, as shown in Figure 4.1.

1. Given a combinational network with assigned inputs, for any gate directly connected to the input terminals we determine the expression corresponding to its output; next we repeat this operation for all gates whose input expressions have been determined, and so on, until we obtain the expression of the network output. (Note that this procedure is not limited to tree networks but applies also to networks with gates having multiple fan-out—refer to Sec. 3.3.)

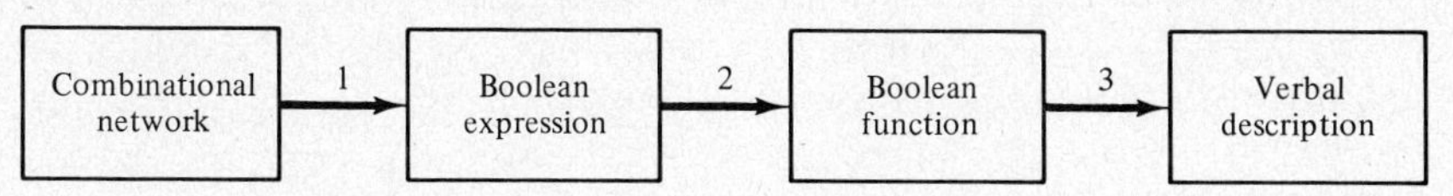

Figure 4.1 Steps in the analysis of a combinational network.

[1]The exceptions to this statement occur in different media, such as magnetic memories, punched paper, and so on.

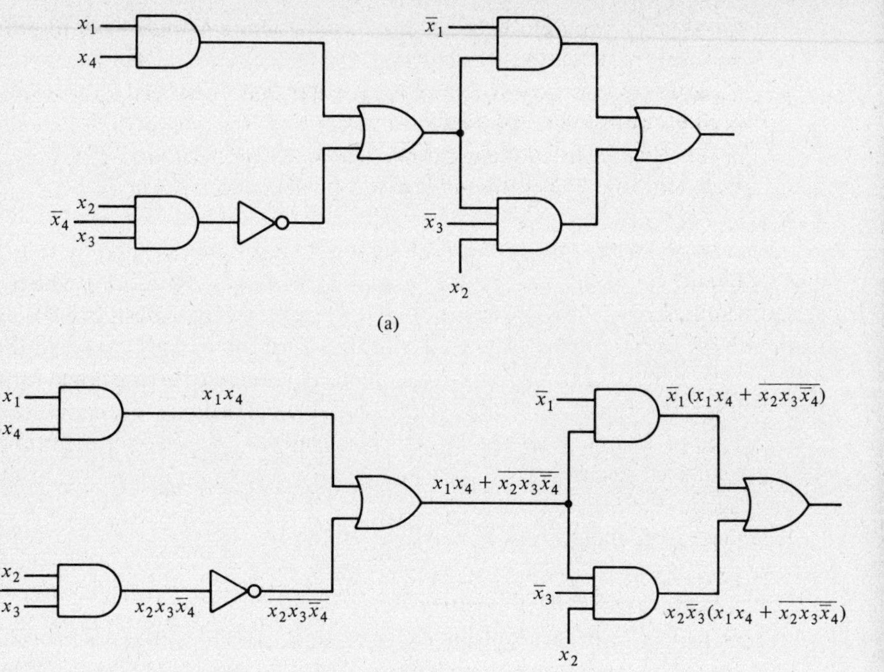

Figure 4.2 Obtaining the expression of a network.

EXAMPLE 4.1

Given the network shown in Figure 4.2(a), the expressions corresponding to each gate output are shown in Figure 4.2(b). The output expression is $\overline{x}_1(x_1x_4 + \overline{x_2x_3\overline{x}_4}) + x_2\overline{x}_3(x_1x_4 + \overline{x_2x_3\overline{x}_4})$.

2. Once the output expression has been obtained, we successively transform it to normal form and to canonical form (equivalent to the truth table), using the procedures described in Sec. 3.5 (frequently, as in our running example, to SOP forms).

EXAMPLE 4.2

$$
\begin{aligned}
& x_1(x_1x_4 + \overline{x_2x_3\overline{x}_4}) + x_2\overline{x}_3(x_1x_4 + \overline{x_2x_3\overline{x}_4}) \\
&\quad = \overline{x}_1(x_1x_4 + \overline{x}_2 + \overline{x}_3 + x_4) \\
&\quad + x_2\overline{x}_3(x_1x_4 + \overline{x}_2 + \overline{x}_3 + x_4) && \text{De Morgan} \\
&\quad = \overline{x}_1(\overline{x}_2 + \overline{x}_3 + x_4) + x_2\overline{x}_3(\overline{x}_2 + \overline{x}_3 + x_4) && \text{Absorption} \\
&\quad = \overline{x}_1\overline{x}_2 + \overline{x}_1\overline{x}_3 + \overline{x}_1x_4 + x_2\overline{x}_3 \\
&\quad + x_2\overline{x}_3x_4 && \text{Distribution} \\
&\quad = \overline{x}_1x_4 + \overline{x}_1\overline{x}_2 + \overline{x}_1\overline{x}_3 + x_2\overline{x}_3 \qquad \text{(normal form)} && \text{Absorption} \\
&\quad = x_4(x_3 + \overline{x}_3)(x_2 + \overline{x}_2)\overline{x}_1 + (x_4 + \overline{x}_4)(x_3 + \overline{x}_3)\overline{x}_2\overline{x}_1 \\
&\quad + (x_4 + \overline{x}_4)\overline{x}_3(x_2 + \overline{x}_2)\overline{x}_1 + (x_4 + \overline{x}_4)\overline{x}_3x_2(x_1 + \overline{x}_1) \\
&\quad = x_4\overline{x}_3\overline{x}_2\overline{x}_1 + \overline{x}_4\overline{x}_3x_2\overline{x}_1 + x_4x_3\overline{x}_2\overline{x}_1 + x_4x_3x_2\overline{x}_1
\end{aligned}
$$

$$+ \bar{x}_4\bar{x}_3\bar{x}_2\bar{x}_1 + \bar{x}_4x_3\bar{x}_2\bar{x}_1 + x_4\bar{x}_3\bar{x}_2\bar{x}_1 + x_4x_3\bar{x}_2\bar{x}_1$$
$$+ \bar{x}_4\bar{x}_3\bar{x}_2\bar{x}_1 + \bar{x}_4\bar{x}_3x_2\bar{x}_1 + x_4\bar{x}_3\bar{x}_2\bar{x}_1 + x_4\bar{x}_3x_2\bar{x}_1$$
$$+ \bar{x}_4\bar{x}_3x_2\bar{x}_1 + \bar{x}_4\bar{x}_3x_2x_1 + x_4\bar{x}_3x_2\bar{x}_1 + x_4\bar{x}_3x_2x_1$$

and, eliminating repeated terms,

$$= x_4\bar{x}_3\bar{x}_2\bar{x}_1 + x_4\bar{x}_3x_2\bar{x}_1 + x_4x_3\bar{x}_2\bar{x}_1 + x_4x_3x_2\bar{x}_1 + \bar{x}_4\bar{x}_3\bar{x}_2\bar{x}_1$$
$$+ \bar{x}_4x_3\bar{x}_2\bar{x}_1 + \bar{x}_4\bar{x}_3x_2\bar{x}_1 + x_4\bar{x}_3x_2x_1$$
$$+ \bar{x}_4\bar{x}_3x_2x_1 \qquad \text{(canonical form)}$$
$$= m_8 + m_{10} + m_{12} + m_{14} + m_0 + m_4 + m_2 + m_{11} + m_3$$

or also

$$= \text{OR}\ (m_0, m_2, m_3, m_4, m_8, m_{10}, m_{11}, m_{12}, m_{14})$$

3. Once we have obtained the truth table of the function, we may want to interpret verbally what the function accomplishes. This task may be difficult and, sometimes, vacuous. However, as we shall see below, the synthesis of a combinational network always starts from a verbal statement of what is to be accomplished.

4.2.2 Synthesis

The synthesis of a combinational network is the reverse process of the one just described, and is correspondingly organized as follows (Figure 4.3):

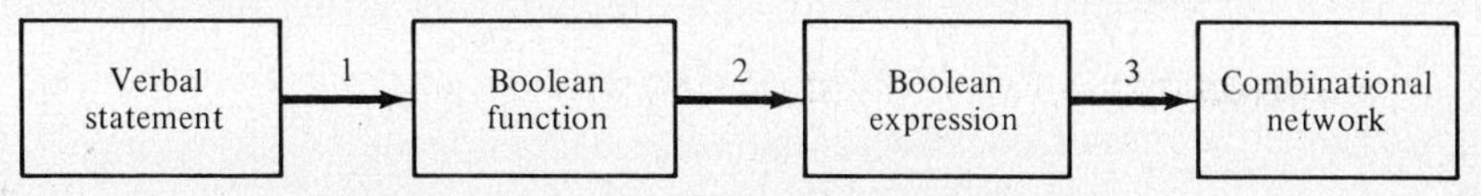

Figure 4.3 Steps in the synthesis of a combinational network.

1. Normally we start from a verbal statement expressing how a certain event (the function) depends upon the occurrences of other independent events (the variables). For example, consider the following case:

> In a chemical process the water must be turned off if and only if: (1) the tank is full or (2) the output is shut off, the concentration is below 1%, and the water level is not below the minimum level.

In the preceding sentence, there are several *simple statements,* such as "water must be turned off," "the tank is full," "the output is shut off," and so on, which are combined to make what we call a *compound statement.* The combination is done by using three connectives: AND, OR, NOT. In other words we can slightly formalize the sentence above as follows:

> NOT (the water must be on) if and only if (the tank is full) OR ((the output is shut off) AND (the concentration is below 1%) AND NOT (the water level is below the minimum level)).

Here we identify the function "the water is on," and the variables "the tank is full," "the output is shut off," "the concentration is below 1%," and "the water level is below the minimum level." For brevity, we shall designate them with the letters W, T, O, C, L, respectively. Also if we make the correspondences

If and only if	$=$
AND	$\cdot$
OR	$+$
NOT	$\overline{}$

the previous compound statement is totally formalized into the following boolean expressions:

$$\overline{W} = T + OC\overline{L}$$

which is equivalent to

$$W = \overline{T + OC\overline{L}}$$

From this expression, we can obtain the truth table of the corresponding boolean function by the methods described in Sec. 3.5 [Figure 4.4(a)].

In other cases, the truth table of the function is directly constructed, since the form of the statement directly lends itself to this purpose. Consider the following example: "Design a combinational network with four inputs x_1, x_2, x_3, x_4 and one output u, such that the latter is 1 if and only if exactly two of the inputs are 1." In this case we list the 16 configurations for the input variables and write "1" in the function table next to each configuration having the precribed number of 1s, and 0 otherwise [Figure 4.4(b)].

2. Once the output boolean function has been obtained, we may want to transform the corresponding canonical form expression by means of the valid

TOCL	*W*
0000	1
0001	1
0010	1
0011	1
0100	1
0101	1
0110	0
0111	1
1000	0
1001	0
1010	0
1011	0
1100	0
1101	0
1110	0
1111	0

(a)

$x_4x_3x_2x_1$	u
0000	0
0001	0
0010	0
0011	1
0100	0
0101	1
0110	1
0111	0
1000	0
1001	1
1010	1
1011	0
1100	1
1101	0
1110	0
1111	0

(b)

Figure 4.4 Truth tables for the functions of the previous examples.

identities of switching algebra, into an equivalent expression that best suits our design purposes. Once this equivalent expression has been obtained, we shall draw the desired combinational network in a straightforward manner (notice that since we normally have a *single* expression, the gates of the corresponding network will have fan-out 1.)

We must now temporarily digress on "design purposes," before tackling the question of how to transform the canonical expression of the given function.

Any design activity aims at achieving a given objective (*performance*) with the least expenditure of resources (*cost*). Logical design is no exception. Fabrication cost, of course, depends both upon the amount of resources used (materials and processes) and upon the amount of labor used in this activity. The interplay of these two items is very much dependent upon the adopted technology. In the 1950s, when transistors or vacuum tubes were individually deployed, the design objective was essentially the minimization of the number of such components used in the realization of a given circuit. (Since typically each gate input requires one active component, this design objective translates into the minimization of the number of gate inputs.) Subsequently, technology evolved toward multitransistor packages (integrated circuits), that is, toward the simultaneous fabrication within the same semiconductor chip (typically a silicon chip) of complete circuits containing from a few tens to several hundreds (medium-scale integration, MSI), to several thousands (large-scale integration, LSI), to tens of thousands of transistors and above (very-large-scale integration, VLSI). Correspondingly, two major design philosophies emerged:

1. (MSI). The identification of complex standard modules, sufficiently versatile so that most functions foreseeable in a digital system could be realized by interconnecting such modules; these modules are to be deployed and interconnected on a printed-circuit board by the user. A related methodology consists of fabricating on a single chip a large array of basic devices whose interconnection can be specified by the user and physically realized by the manufacturer.
2. (LSI and VLSI). The global fabrication of a complete digital system on a single chip. Here the objective is the minimization of silicon chip area used by the design. Note that both electrical components (transistors, resistors, etc.) and wires use chip area, so that component count is no longer an absolute criterion. However, for given types of design there is a statistical correlation between component count and chip area, so that component count can still be used as a useful design guideline.

Due to the technological progress in integrated circuits, networks whose identifiable constituents are small components—such as individual devices or even individual logic gates—are all but obsolete. Surprisingly, however, the corresponding design techniques, which are based on the component count criterion, are relevant to the more advanced technologies. This is due not only to the observed relation between component count and chip area in LSI and VLSI, but also for the direct applicability to the design of an extremely important MSI module, the programmable logic array (PLA), to be discussed in Sec. 4.8.3. Therefore, we shall begin with the discussion of the minimization techniques of discrete-component networks (discrete logic).

4.3 MINIMIZATION TECHNIQUES

In the context of logical design, performance is normally expressed by fabrication cost and by operation cost, which is reflected by the computation speed of the complete system. Thus our design objectives will be *primarily* the *speed* and the *cost* (number of components) of the network.

Speed of a combinational network is a somewhat new notion in our study. Indeed, although we have conveniently assumed so far that a combinational network operates instantaneously, this is not so. Each change of logical value propagates through a gate in nonzero time, as we shall see more extensively in Chapter 5. Thus the delay of a combinational network is roughly proportional to the length of the chain of gates an input change must traverse to manifest itself at the output.

General systematic design techniques under different choices of network parameters are not available. However, if we restrict ourselves to networks corresponding to normal expressions [that is, two-level AND-to-OR (SOP) and OR-to-AND (POS) networks], then elegant techniques for the solution of the problem are known. Note that two-level networks constitute the class of fastest networks, since any switching function is realizable by a network of at most two levels (e.g., the network corresponding to canonical SOP expression of f) and, on the other hand, very few switching functions are realizable with a single gate (one-level network). For this class, we now state the optimality criterion (stated, for convenience, for AND-to-OR networks):

> A two-level AND-to-OR network is *minimal* if it has (1) the minimum number of AND gates and, (2) among all networks having the minimum number of AND gates, it has the minimum number of AND gate inputs.

An analogous definition can be given for OR-to-AND networks, by just interchanging the words AND and OR. Since an AND gate corresponds to a product of literals, the previous definition translates into the following for the minimal expression.

> A normal SOP expression is *minimal* if there exists (1) no other equivalent expression involving fewer products, and (2) no other equivalent expression involving the same number of products but a smaller number of literals in these products.

An analogous definition can be given for minimal POS expressions, by interchanging the words "sum" and "product." The following discussion, however, will refer to SOP expressions for the purpose of concreteness.

Minimization starts from the SOP canonical expression of a given function. The basic identities involved are distributivity and $x + \overline{x} = 1$. In fact, if we have an expression consisting of two product terms which differ in *exactly one* variable, say $\overline{x}_3x_2x_1 + \overline{x}_3\overline{x}_2x_1$ (they differ in x_2), then we can factor the common part using distributivity [Eq. (3.7)], and obtain

$$\overline{x}_3x_2x_1 + \overline{x}_3\overline{x}_2x_1 = \overline{x}_3x_1(x_2 + \overline{x}_2)$$

But $(x_2 + \overline{x}_2) = 1$, whence

$$\overline{x}_3x_2x_1 + \overline{x}_3\overline{x}_2x_1 = \overline{x}_3x_1$$

We see therefore that we have succeeded in replacing two three-literal terms with one two-literal term—a considerable simplification. We want to be able to carry out this kind of simplification in a systematic way.

Two minterms that differ in exactly one variable are called *adjacent:* adjacent minterms can be combined as explained above. Therefore, it is convenient to organize processing so that this adjacency is easily exploited. One method is based on the so-called "Karnaugh map," a graphical presentation of switching functions which gives immediate pictorial evidence of adjacent minterms. The map is an extremely effective tool for a small number of variables (typically, up to four). For larger numbers of variables one resorts to an equivalent tabular method, the Quine–McCluskey procedure. These two methods will be discussed in the next three sections. Only the Karnaugh map method, however, is a prerequisite for subsequent material.

4.4 MINIMIZATION TECHNIQUE BASED ON THE KARNAUGH MAP

4.4.1 Karnaugh Map Representation of Functions

The *Karnaugh map* (K-map) is simply a way of rearranging the truth table of a function f in a way that will greatly facilitate the minimization task. Thus, for a function f of n variables, it consists of 2^n squares. Each square of the Karnaugh map contains the value f_j of the function f corresponding to a given minterm m_j (i.e., $f_j = 1$, if m_j *is* a minterm of f, and $f_j = 0$ otherwise). The standard correspondence between map squares and function values is shown in Figure 4.5.

A K-map for functions of one variable x_1 is shown in Figure 4.5(a). It consists of two squares, labeled f_0 and f_1. By adjoining two one-variable K-maps, as shown in Figure 4.5(b), we obtain a K-map for functions $f(x_2, x_1)$ of two variables x_1 and x_2 (two-variable K-map); each square is labeled for convenience with the corresponding f_j. Next, by adjoining two two-variable K-maps we obtain a three-variable K-map, shown in Figure 4.5(c) for functions $f(x_3, x_2, x_1)$ of three variables x_1, x_2, x_3 and labeled with the same conventions. Finally by adjoining two three-variable K-maps we obtain a four-variable K-map for functions $f(x_4, x_3, x_2, x_1)$ [Figure 4.5(d)]. We could continue building K-maps

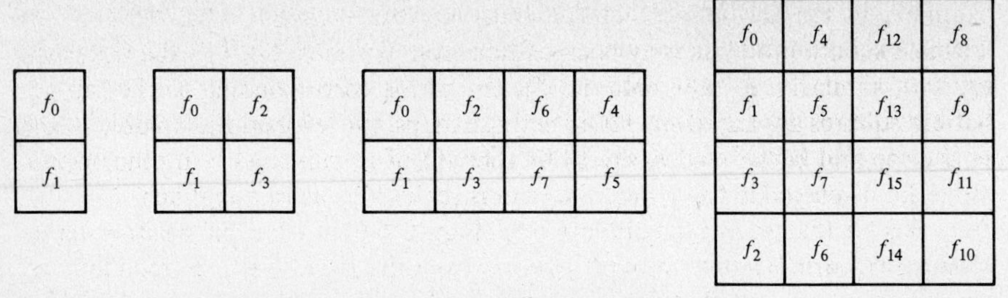

Figure 4.5 K-maps for one, two, three, and four variables.

for five, six, and so on, variables; however, the K-map, which is a remarkable tool for up to four variables, becomes increasingly clumsier for a larger number and loses most of its effectiveness.

In Figure 4.6(a) we have copied the K-maps of Figure 4.5, replacing each symbol f_j with the binary representation of the integer j. By so doing, each square is now labeled with the binary combination of the corresponding row in the truth table, for functions $f(x_1)$, $f(x_2, x_1)$, $f(x_3, x_2, x_1)$, and $f(x_4, x_3, x_2, x_1)$, respectively. Note that any two squares sharing an edge in a map correspond to two binary combinations that differ *in exactly one* position, that is, they correspond to adjacent minterms (as defined on p. 79). Moreover, in the three-variable and four-variable maps we note the following: in each row, the minterms corresponding to the leftmost and rightmost squares are adjacent (for

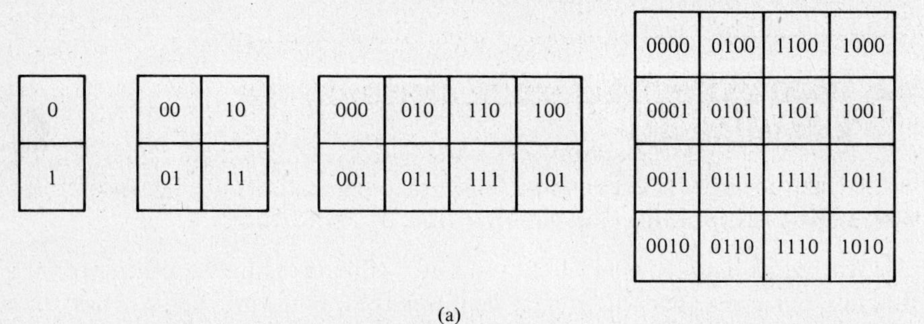

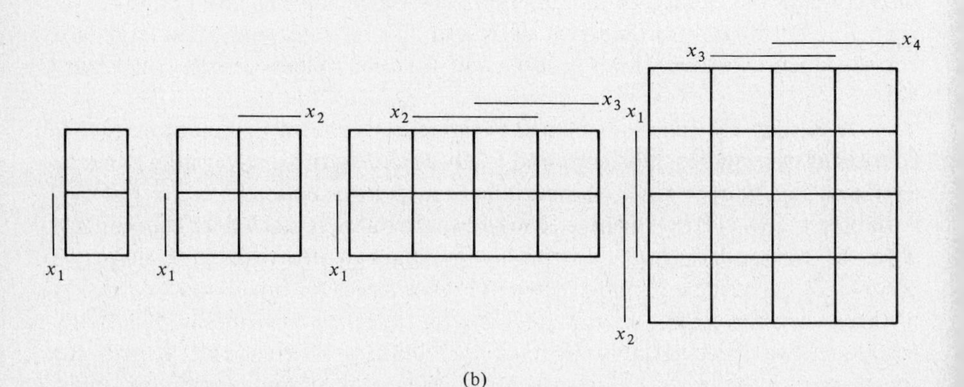

Figure 4.6 Region labeling of the K-maps.

example, in the second row of the four-variable map the label 0001 of the leftmost square differs in only one position from the label 1001 of the rightmost square); similarly, in each column, the minterms corresponding to the top and bottom squares are adjacent. Thus, in these maps, the left and right sides, as well as the top and bottom sides, are to be thought of as coincident; in other words, although displayed in the plane, the map is really a torus (a doughnut).

Recall that in any combination in Figure 4.6(a) each bit position corresponds to a variable (i.e., the first position from the right to x_1, the second to x_2, and so on); so we see that for any selected bit position (or equivalently, for its corresponding variable x_j), there is a "band" of squares whose binary combinations all have value $x_j = 1$. For example, in the three-variable map, squares labeled f_2, f_6, f_3, f_7 form a band in which $x_2 = 1$:

$$(0\underset{\uparrow}{1}0,\ 1\underset{\uparrow}{1}0,\ 0\underset{\uparrow}{1}1,\ 1\underset{\uparrow}{1}1)$$

As another example, in the four-variable map squares labeled $f_1, f_5, f_{13}, f_9, f_3, f_7, f_{15}, f_{11}$ form a band in which $x_1 = 1$:

$$(000\underset{\uparrow}{1},\ 010\underset{\uparrow}{1},\ 110\underset{\uparrow}{1},\ 100\underset{\uparrow}{1},\ 001\underset{\uparrow}{1},\ 011\underset{\uparrow}{1},\ 111\underset{\uparrow}{1},\ 101\underset{\uparrow}{1})$$

Precisely, we note that each variable x_j divides the map into two *regions* with the same number of squares, the x_j *region* and the $\overline{x}_j$ *region,* which are those in which the x_j variable is 1 and 0, respectively. As a consequence, an alternative—or additional—way to label a K-map is by indicating the regions: specifically a segment drawn close to one of the map sides and labeled x_j illustrates the width of the x_j region. Such labeling of the K-maps for one, two, three, and four variables is shown in Figure 4.6(b).

So far the K-map is simply a new way of displaying a truth table, with the notion that adjacent squares correspond to combinations differing in only one variable.

EXAMPLE 4.3

Given the function $f = \text{OR}(m_0, m_2, m_3, m_4, m_5)$ of variables x_3, x_2, x_1, its conventional truth table and K-map are shown below:

minterms	$x_3x_2x_1$	f	
m_0	0 0 0	1	f_0
m_1	0 0 1	0	f_1
m_2	0 1 0	1	f_2
m_3	0 1 1	1	f_3
m_4	1 0 0	1	f_4
m_5	1 0 1	1	f_5
m_6	1 1 0	0	f_6
m_7	1 1 1	0	f_7

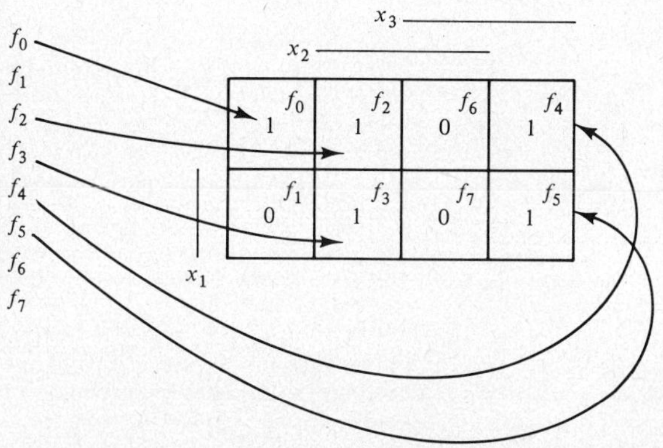

x_3 (segment over the right two columns), x_2 (segment over the middle two columns), x_1 (segment beside the bottom row)

f_0 1	f_2 1	f_6 0	f_4 1
f_1 0	f_3 1	f_7 0	f_5 1

In Sec. 3.5.2 we have established that the truth table of $\overline{f}$ is the entry-by-entry complement of the truth table of f, and that the truth table of $f \cdot g$ (or $f + g$) is the entry-by-entry AND (or OR) of the truth tables of f and g. Since the K-map is just a rearrangement of the truth table, these properties hold intact if we replace "K-map" for "truth table." This is illustrated by the following important examples.

EXAMPLE 4.4

Entry-by-entry AND and OR of K-maps:

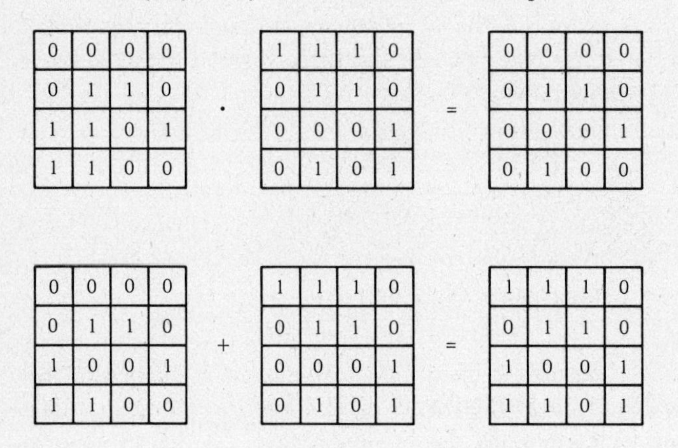

0	0	0	0
0	1	1	0
1	1	0	1
1	1	0	0

$\cdot$

1	1	1	0
0	1	1	0
0	0	0	1
0	1	0	1

=

0	0	0	0
0	1	1	0
0	0	0	1
0	1	0	0

0	0	0	0
0	1	1	0
1	0	0	1
1	1	0	0

+

1	1	1	0
0	1	1	0
0	0	0	1
0	1	0	1

=

1	1	1	0
0	1	1	0
1	0	0	1
1	1	0	1

4.4.2 Product Terms on a Karnaugh Map

We now consider very special switching functions, that is, those whose expression is a single product of literals (possibly, a single literal), and ask what they look like on the K-map. For example, consider the product $f = \overline{x}_2 x_3$ on the four-variable map (see Figure 4.7); f is equal to 1 if both $\overline{x}_2$ and x_3 are equal to 1, that is, on the map the 1s of f appear exactly in the squares or *cells* that belong

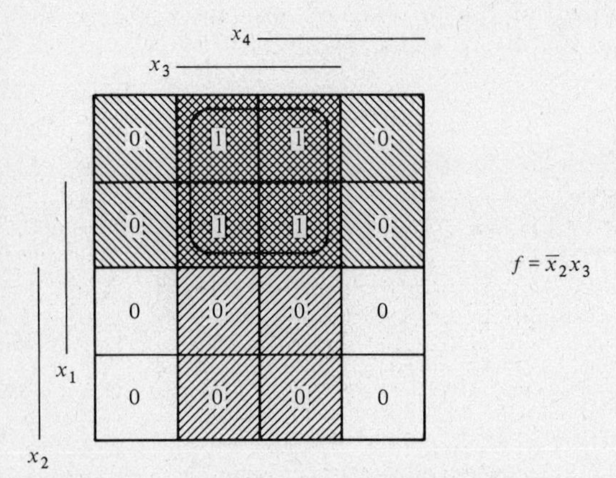

Figure 4.7 A 2-cube region corresponding to the product of two variables.

simultaneously to the $\overline{x}_2$ region (shaded \\\\) and to the x_3 region (shaded ////), or, in other words, in the *intersection* of these two regions. In general:

> A product of literals appears on the map as the intersection of the regions corresponding to each of the literals. (All the squares in this intersection bear a 1 and all other squares bear a 0.)

Thus, in the four-variable map, a single literal corresponds to an eight-cell rectangle [Figure 4.8(a)], a two-literal product to four-cell squares or rectangles [Figure 4.8(b)], a three-literal product to a two-cell rectangle [Figure 4.8(c)],

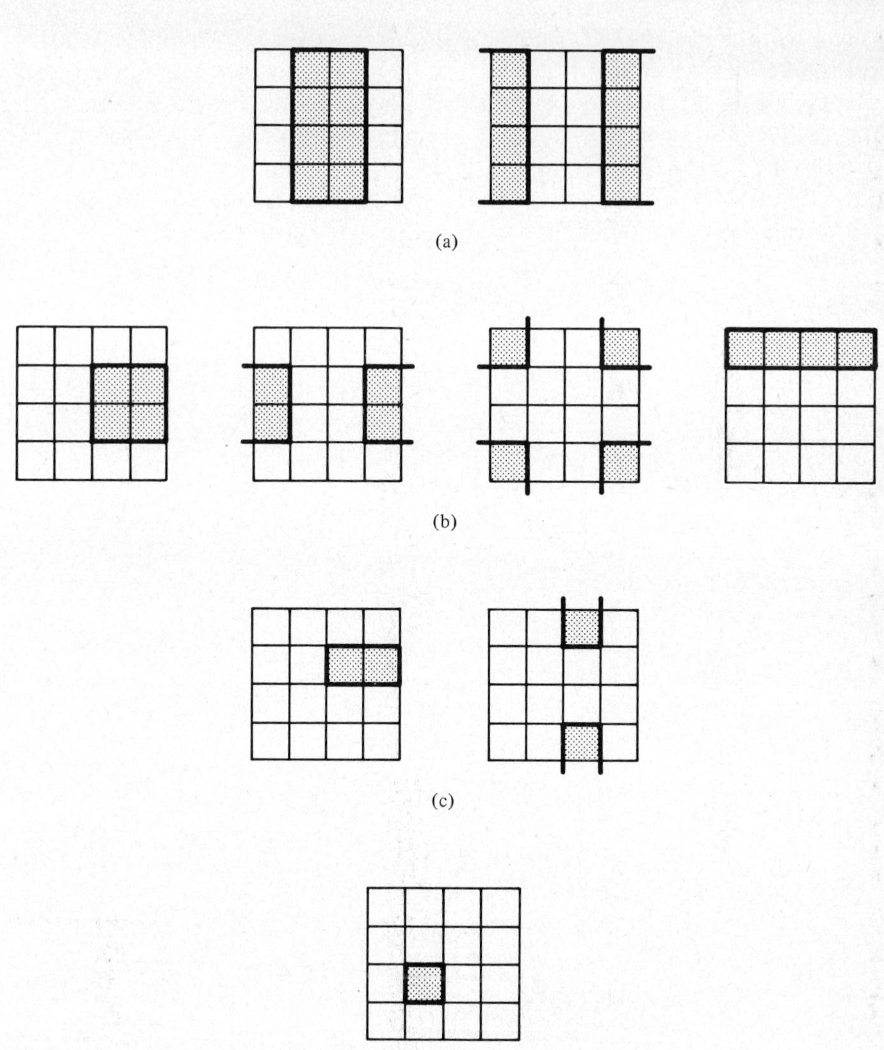

Figure 4.8 Illustration of the map representation of product terms. (a) Single literals. (b) Two-literal terms. (c) Three-literal terms. (d) Four-literal terms.

and a four-literal product, a minterm, to a single cell [Figure 4.8(d)]. Conventionally, the map configurations corresponding to one-, two-, three-, and four-literal terms are called 3-, 2-, 1-, and 0-cubes, respectively; clearly, a k-cube contains 2^k squares. All possible shapes (not all possible positions!) of such cubes in the map are displayed in Figure 4.8; keep in mind that opposite sides of the map are to be thought of as coincident.

4.4.3 A Procedure to Obtain a Minimal Normal Expression

As for truth tables, the OR of K-maps of two functions is a K-map obtained as the square-by-square OR of their entries. If we find a collection of cubes, such that the OR of their K-maps is the K-map of a given function f, then the OR of the product terms corresponding to each of those cubes is a SOP normal expression of f, and conversely.

Given the map of f, we can find several such collections; one that satisfies our requirements of minimality must exhibit the following properties:

1. There is no smaller collection of cubes so that the OR of their maps gives the map of f.
2. No cube is contained in a larger cube, such that in each square of the latter $f = 1$.

Indeed, some of the cubes correspond to single literal terms and the others to product terms (AND gates). We shall see later (Sec. 4.6) that *all* minimal expressions of f must contain the same set of single literals, whence property 1 guarantees the minimality of the number of AND gates. Property 2 guarantees that no AND gate can be replaced by one having fewer inputs. Thus, to find a minimal SOP expression for a given function f, represented by means of a K-map, we proceed as follows (below a 1 of f is said to be *covered* by a cube if the corresponding map square is contained in that cube):

Sop Minimization Procedure

Step 1. For each 1 in the map of f we determine the *largest* cubes containing it and such that in each of their squares we have $f = 1$.[2]

Step 2. If a 1 is covered by just one largest cube, this cube must be included in the minimal collection (*essential*).

Step 3. We find a minimal set of largest cubes needed to cover the 1s in the map not covered by the essential cubes.

[2]These cubes are called "prime implicants" of f. The rationale of such a phrase will be explained in Sec. 4.6.

While steps 1 and 2 are quite straightforward to implement, step 3 is not given in clear procedural terms. Of course, the map is a great aid in carrying out the task. We shall illustrate it by some examples.

EXAMPLE 4.5

Given the four-variable function $f = \mathrm{OR}(m_0, m_1, m_3, m_6, m_7, m_8, m_{11}, m_{12}, m_{14}, m_{15})$, find a minimal SOP expression for it.

Using the map labeling given in Figure 4.5(d), we obtain the following K-map [Figure 4.9(a)]. In this map we quickly find the *largest* cubes, as shown in Figure 4.9(b). Of these cubes, only two are essential: x_1x_2 and x_2x_3. The remaining 1s of the function can be covered by any of the four collections of terms $\{\bar{x}_2\bar{x}_3\bar{x}_4, \bar{x}_1\bar{x}_2x_4\}$, $\{x_1\bar{x}_3\bar{x}_4, \bar{x}_1\bar{x}_2\bar{x}_3, \bar{x}_1x_3x_4\}$, $\{x_1\bar{x}_3\bar{x}_4, \bar{x}_1\bar{x}_2\bar{x}_3, \bar{x}_1\bar{x}_2x_4\}$, $\{\bar{x}_2\bar{x}_3\bar{x}_4, \bar{x}_1\bar{x}_2\bar{x}_3, \bar{x}_1x_3x_4\}$; of course we shall choose the first one, because it contains only two terms, yielding the selection illustrated in Figure 4.9(c). In this case there is a *unique* minimal expression.

EXAMPLE 4.6

Solve the same problem as in Example 4.5 for the function

$$f = \mathrm{OR}(m_0, m_1, m_3, m_6, m_7, m_{10}, m_{11}, m_{14})$$

The procedure is illustrated in Figure 4.10. Note that the function has two minimal SOP expressions.

A minimal POS expression of a function f can also be found by means of a K-map. The procedure is the perfect "dual" of the one just described and it is best justified as follows. We know, by involution (Sec. 3.4) that $f = \bar{\bar{f}}$. Therefore suppose we seek a minimal *SOP* expression for the function $\bar{f}$, whose K-map is obtained from the one of f by interchanging 0s and 1s. Once a minimal SOP expression of $\bar{f}$ is available, by complementing it we obtain an expression of f, which is the desired one, since by applying De Morgan laws twice (first to the OR of terms, next to the terms themselves), we obtain exactly a POS expression.

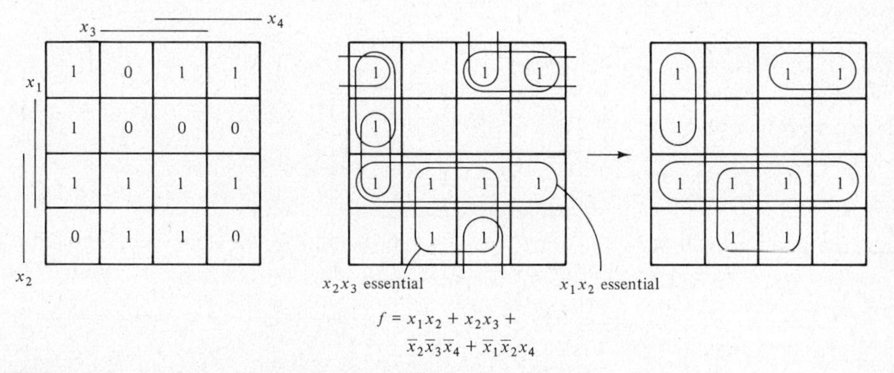

Figure 4.9 A map and its minimal set of covering cubes.

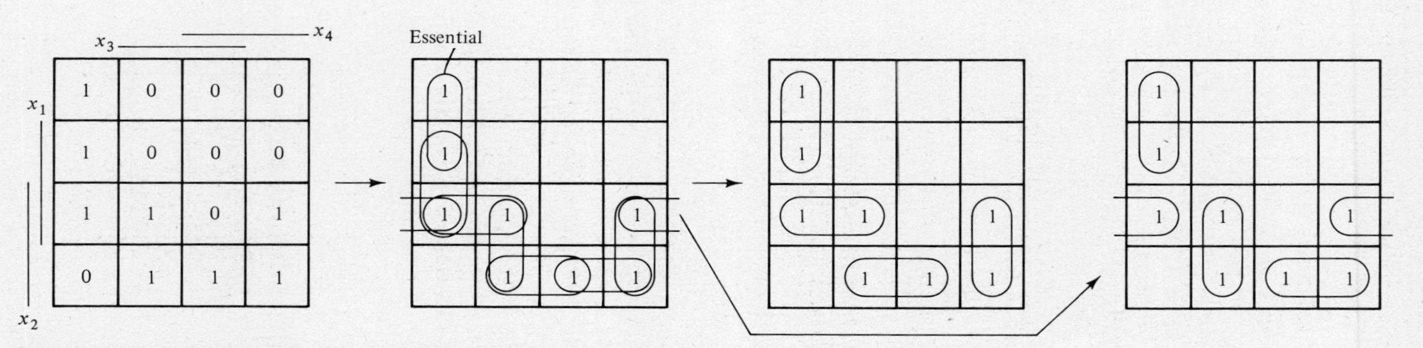

Figure 4.10 A map and its two minimal sets of covering cubes.

EXAMPLE 4.7

Suppose we have already found, by the known techniques, a minimal SOP expression of $\bar{f}$, that is,

$$\bar{f} = \bar{x}_2\bar{x}_4 + \bar{x}_1 x_3 \qquad \text{(minimal SOP expression of } \bar{f}\text{)}$$

$$f = \overline{(\bar{f})} = \overline{\bar{x}_2\bar{x}_4 + \bar{x}_1 x_3} = \overline{\bar{x}_2\bar{x}_4} \cdot \overline{\bar{x}_1 x_3} = (x_2 + x_4)(x_1 + \bar{x}_3)$$

In other words, each product term of $\bar{f}$ transforms into a sum term of f, where each literal is replaced by its complement. Clearly, the resulting POS expression is minimal, since there is no expression with fewer sum terms and none of them can be replaced by a simpler one (otherwise the SOP expression of $\bar{f}$ would not be minimal).

In summary, to obtain a minimal POS expression of f, we have the procedure:

Pos Minimization Procedure

1. Obtain the K-map of $\bar{f}$ from the K-map of f.
2. Obtain a minimal SOP expression of $\bar{f}$.
3. Complement the expression obtained in step 2 and apply De Morgan laws twice.

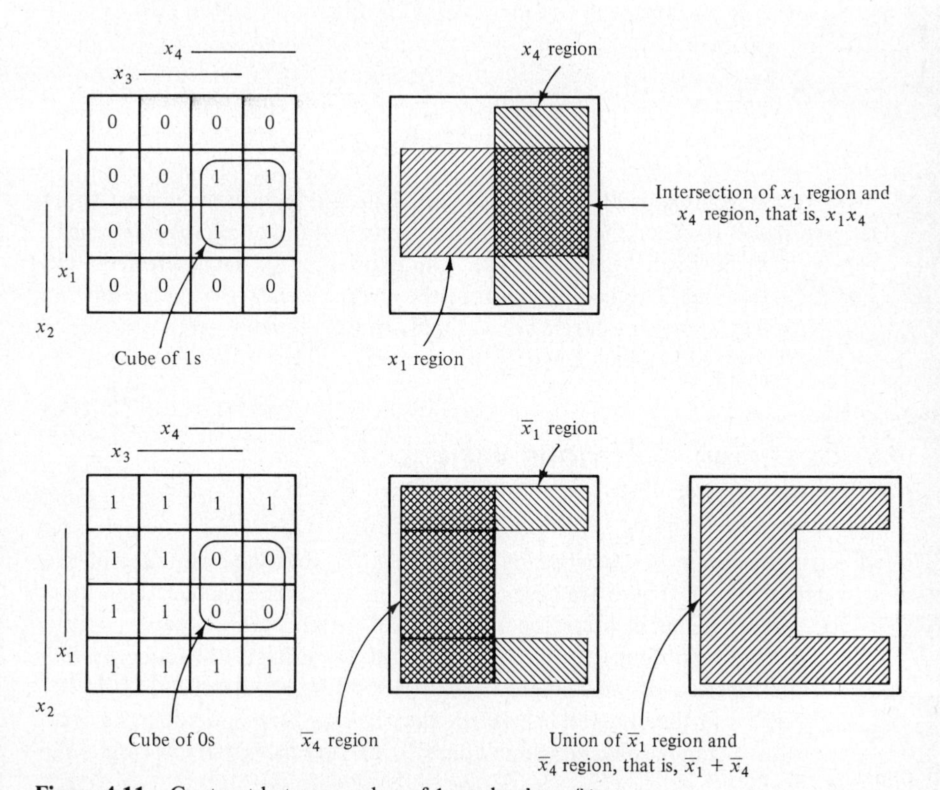

Figure 4.11 Contrast between cubes of 1s and cubes of 0s.

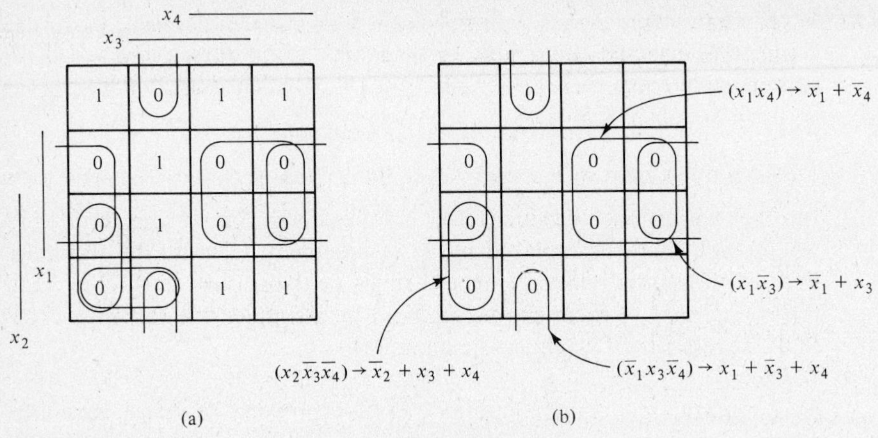

Figure 4.12 A K-map and a minimal POS expression of it.

Alternatively, an equivalent procedure to obtain a minimal POS expression of f—and a direct dual of the one that obtains a minimal SOP expression—begins by determining largest cubes covering the 0s of f (rather than its 1s). Once a minimal collection of such cubes has been found, we must read these cubes off the map. The duality is reflected in the fact that while a cube of 1s is the *intersection* of x_i regions, the 1s surrounding a cube of 0s are the *union* of x_i regions (see Figure 4.11).

EXAMPLE 4.8

Given the function $f = (m_0, m_5, m_7, m_8, m_{10}, m_{12}, m_{14})$, find a minimal POS expression for it.

The K-map of f is shown in Figure 4.12, along with the largest cubes covering the 0s (i.e., the 1s of $\overline{f}$). There are two minimal expressions, one of which is shown in Figure 4.12(b). The expression of each selected POS cube is obtained by complementing the expression of the corresponding SOP cube (shown in parentheses). Thus the expression corresponding to Figure 4.12(b) is $(\overline{x}_1 + \overline{x}_4)(\overline{x}_1 + x_3)(x_1 + \overline{x}_3 + x_4)(\overline{x}_2 + x_3 + x_4)$.

4.5 SWITCHING FUNCTIONS WITH "DON'T CARE" CONDITIONS

A switching function is described by its truth table, that is, each entry of this table is either 0 or 1. There are cases, however, in which the problem statement does not provide a specification for all entries of a truth table. Typical in this respect is a function of four variables giving the BCD coding of the numerals 0–9 (Sec. 1.6): in these cases the configurations 1010, 1011, 1100, 1101, 1110, 1111 will never appear at the input and therefore the corresponding entries in the truth table can be arbitrarily selected ("don't care"). The symbol for don't care is the letter δ.

In general, a *don't care* is an unspecified entry in a truth table. It originates either because the corresponding input configuration is known never to occur, or, if it may occur, the output is known to be irrelevant on the basis of other design considerations.

Given a truth table with s don't cares, since each of them can be arbitrarily chosen, we are free to realize any of 2^s switching functions and still satisfy our design specification. Among these 2^s switching functions we shall obviously choose the simplest. Normally our problem statement consists of a set of input combinations for which the function f is 1 and a set for which f is don't care, as shown in the following example.

$f = \text{OR}(m_0, m_1, m_2, m_{14}), \delta(m_3, m_5, m_6)$ Incompletely Specified Function

To take advantage of don't cares, we think of them as 1s in carrying out step 1 of the minimization procedure, that is, we can use them if they help in forming larger cubes; however, we will not be required to cover any don't care in carrying out steps 2 and 3.

EXAMPLE 4.9

Find a minimal SOP expression for the following incompletely specified function of four variables.

$$f = \text{OR}(m_1, m_5, m_6, m_7, m_{15}), \delta(m_9, m_{10}, m_{11}, m_{13}, m_{14})$$

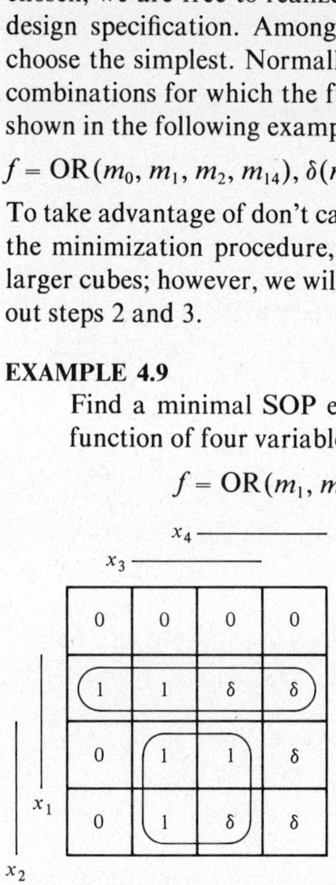

x_4, x_3 (columns); x_1, x_2 (rows)

0	0	0	0
1	1	δ	δ
0	1	1	δ
0	1	δ	δ

$f = x_1\bar{x}_2 + x_3x_2$

Thus we see that, in the actually realized function, we have chosen $m_9 = m_{13} = m_{14} = 1$, while $m_{10} = m_{11} = 0$. Notice the beneficial effect of don't cares as regards the simplicity of the expression: If we change the δs to 0s, then $f = x_1\bar{x}_2\bar{x}_4 + x_3x_2\bar{x}_4 + x_3x_2x_1$.

We conclude this section with a review example, which ties directly with our SEC system. It concerns the design of a comparator of two 20-bit numbers, the numerical portions of the SEC operands (see Sec. 2.2). Although we are not committed, at this point, to the deployment of one such unit in SEC, our objective is to show the direct applicability of the machinery developed so far. A more substantial example will be presented in Chapter 6 in the actual design of the arithmetic-logic unit of SEC.

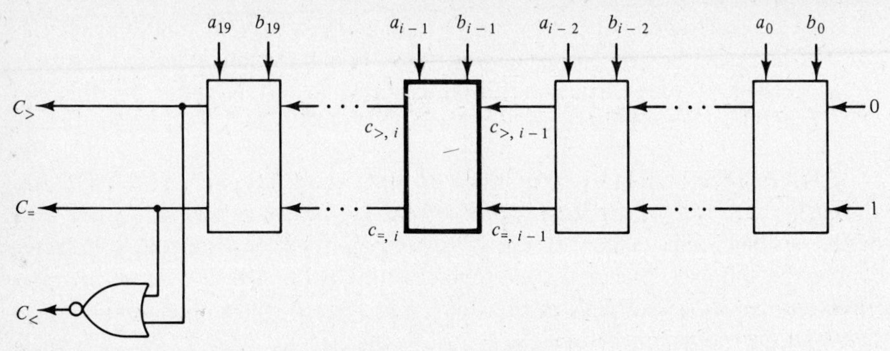

Figure 4.13 The structure of a 20-bit binary comparator.

EXAMPLE 4.10

We wish to design a comparator of two 20-bit numbers A and B. The unit will have 40 input lines (20 lines for each of A and B) and 3 output lines, corresponding respectively to the following situations: $(A > B)$, $(A = B)$, and $(A < B)$. We wish to realize a modular structure, that is, one that can be easily adapted to any operand length. Specifically, our structure should resemble the one proposed for the binary adder in Figure 3.2. This leads to a structure consisting of a bank of 20 *comparator cells,* each of which (except the leftmost and the rightmost) is connected to its two immediate neighbors (see Figure 4.13).

We begin by noting that, since in each comparison *exactly one* of the three situations $(A > B)$, $(A = B)$, $(A < B)$ must occur, one of them—say $(A < B)$—occurs when neither of the other two does. In other words, letting the binary variables $C_>$, $C_=$, and $C_<$ correspond respectively to $(A > B)$, $(A = B)$, and $(A < B)$, we have the relation $C_< = \overline{C_> + C_=}$. Thus only $C_>$ and $C_=$ need be computed.

The ith comparator cell (shown with heavier lines in Figure 4.13) receives the ith bits of both operands and two additional inputs $c_{>,i-1}$ and $c_{=,i-1}$ from the cell immediately to its right; these 2 bits are the result of the comparison restricted to the rightmost $(i - 1)$ bits of both operands.[3] The ith cell generates output $c_{>,i}$ and $c_{=,i}$, with analogous meanings. We can now obtain the truth tables for the generic comparator cell with inputs a, b, $c_>$, and $c_=$ and outputs $c'_>$ and $c'_=$. Notice that the input condition $c_>c_= = 11$ cannot occur, and thus gives rise to don't cares in the output. The construction of the truth tables, shown in Figure 4.14(a) is straightforward and is left as an exercise for the reader. In Figure 4.14(b) we show the Karnaugh maps of the two output functions $c'_>$ and $c'_=$ and their minimal SOP expressions. These expressions can be further manipulated by factoring out $c_>$ in the expression of $c'_>$ and $c_=$ in the expression of $c'_=$, thereby yielding

$$c'_> = c_> \cdot (a + \overline{b}) + a\overline{b}$$
$$c'_= = c_= \cdot (ab + \overline{a}\overline{b}) = c_= \cdot (a \oplus \overline{b}) \qquad \text{(see Sec. 3.6)}$$

[3]The reader will realize that the choice $c_{>,0} = 0$ and $c_{=,0} = 1$, shown in Figure 4.13, is the correct one.

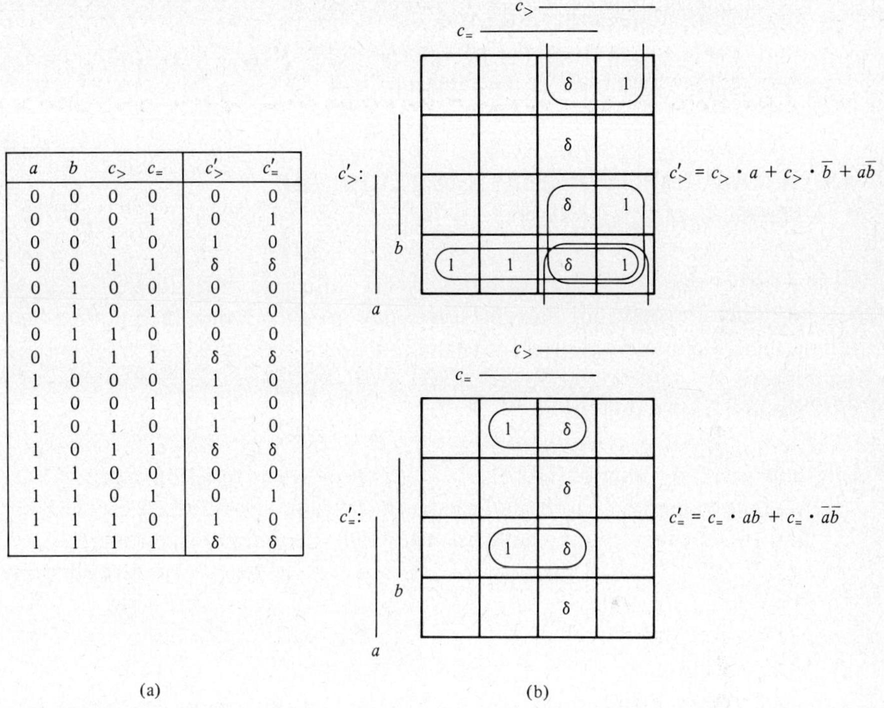

a	b	$c_>$	$c_=$	$c'_>$	$c'_=$
0	0	0	0	0	0
0	0	0	1	0	1
0	0	1	0	1	0
0	0	1	1	δ	δ
0	1	0	0	0	0
0	1	0	1	0	0
0	1	1	0	0	0
0	1	1	1	δ	δ
1	0	0	0	1	0
1	0	0	1	1	0
1	0	1	0	1	0
1	0	1	1	δ	δ
1	1	0	0	0	0
1	1	0	1	0	1
1	1	1	0	1	0
1	1	1	1	δ	δ

Figure 4.14 (a) Truth tables of the comparator cell. (b) Karnaugh-maps and minimum SOP expressions.

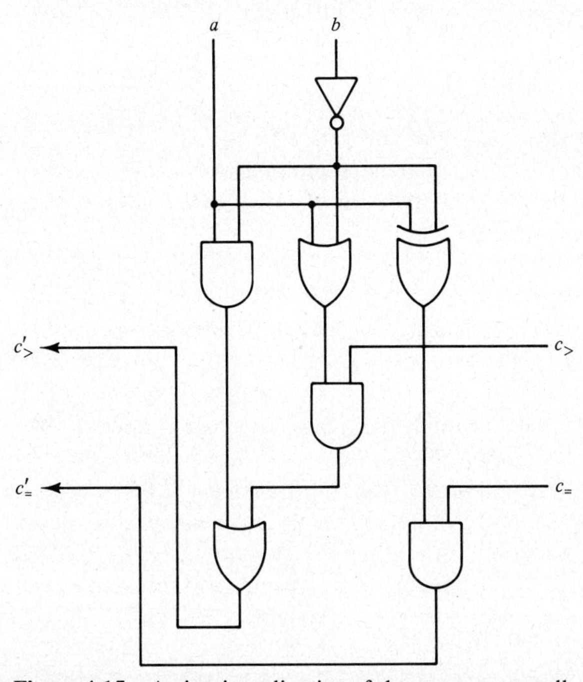

Figure 4.15 A circuit realization of the comparator cell.

This leads to the circuit realization shown in Figure 4.15. The design of the comparator unit is completed as shown in Figure 4.13 by the addition of a NOR gate, which realizes the function $C_<$.

*4.6 A TABULAR MINIMIZATION TECHNIQUE (QUINE–MCCLUSKEY)[4]

We saw earlier that the K-map technique is very effective with functions of up to four variables. Beyond four variables, we may resort to other techniques. One such method, technically equivalent to the K-map procedure, is due to Quine and McCluskey.

We first need a definition.

Definition 4.1. A function $g(x_1, \ldots, x_n)$ *implies* a function $f(x_1, \ldots, x_n)$ (equivalently, f *covers* g) if the truth table of f has a 1 wherever that of g has a 1 (of course, f may have a 1 where g has 0, but not vice versa).

EXAMPLE 4.11

x_3	x_2	x_1	g	f
0	0	0	0	0
0	0	1	1	1
0	1	0	1	1
0	1	1	0	1
1	0	0	1	1
1	0	1	0	0
1	1	0	0	1
1	1	1	0	0

For convenience, let $P_1, P_2, \ldots,$ denote product of literals (some of them may consist of a *single* literal). Obviously, a minimal SOP expression of f has the form

$$f = P_1 + P_2 + \cdots + P_t$$

Each of the Ps is itself a switching function and, obviously, implies f. In fact we have the following.

Definition 4.2. An *implicant* of a function $f(x_1, \ldots, x_n)$ is a product term which implies f.

In addition, from any product term above, say P_j, we cannot remove any literal without altering the function (otherwise the expression would not be minimal). Thus, the Ps form a special category of implicants, as expressed by the following.

[4] *An asterisk "*" denotes a topic of rather specialized character, to which no reference is made in what follows.*

Definition 4.3. A *prime implicant* of a function $f(x_1, \ldots, x_n)$ is an implicant of f which does not imply any other implicant of f (a prime implicant then coincides with a "largest subcube" as defined in Sec. 4.4.3). A prime implicant is *essential* if it appears in each minimal SOP expression of f.[5]

With these notions, we can now describe the minimization procedure, which consists of two phases:

Phase 1. Generation of the prime implicants of f.

Phase 2. Selection of a set of prime implicants yielding a minimal expression.

Phase 1 is carried out through the repeated application of the rule $x + \overline{x} = 1$. Since, however, dealing with literals is rather clumsy, this method adopts a simplified notation. In what follows, we assume that f is a function of the n variables $x_1, x_2, \ldots, x_n$. Each product term of n or fewer literals is made to correspond to a string of n symbols drawn from the set {0, 1, –}. In such strings each position corresponds to a unique variable (x_n the leftmost, x_1 the rightmost) and the symbols are chosen as follows:

0 → Complemented variable
1 → Uncomplemented variable
– → Absent variable

For example, for $n = 5$, –01–0 corresponds to $\overline{x}_4 x_3 \overline{x}_1$. We say that two strings are a mergeable *pair* if they coincide in all positions except *one*, where one string has 1 and the other has 0. For example, –01–0 and –11–0 are a mergeable pair. Note that a mergeable pair of strings can be replaced by a single string (merge string) where a "–" enters the position where the two original strings differed. For example,

$$\overline{x}_4 x_3 \overline{x}_1 + x_4 x_3 \overline{x}_1 = (\overline{x}_4 + x_4) x_3 \overline{x}_1 = x_3 \overline{x}_1$$

–01–0   –11–0 (Mergeable pair)   ––1–0 (Merge string)

Initially, the strings are those which correspond to the *minterms* of f, that is, they are *binary* strings (no variable is absent from any minterm). Note that a mergeable pair consists of two strings whose numbers of 1s differ by exactly 1; therefore the strings are partitioned into groups according to their number of 1s (the number of 1s of a string is called the *weight* of the string), and these groups are ordered by increasing weight. Such a rearranged list is called the *initial list* [see Example 4.12 and Figure 4.16(a)]. To this list we apply the following "Procedure Quine–McCluskey" which looks for all mergeable pairs and replaces each such pair by the resulting merge string. The result is a set of merge strings

[5] We can easily show that if a single literal, say x_i, is a prime implicant of f, then it is essential.

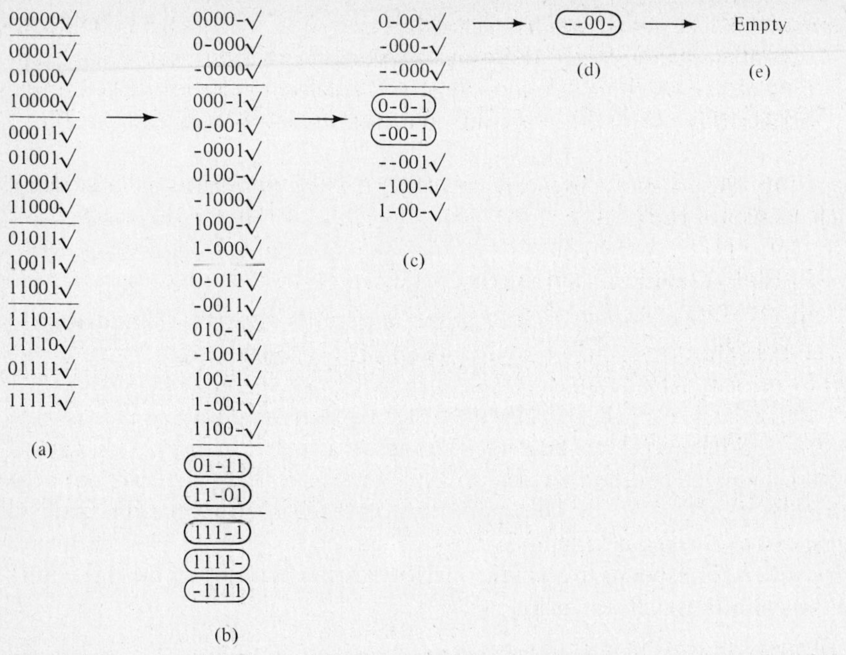

Figure 4.16 (a) Initial list. (b) First output list. (c) Second output list. (d) Third output list. (e) Fourth output list.

that form the output list. The procedure is further applied, recursively, to the output list [Figures 4.16(b), 4.16(c), 4.16(d), and 4.16(e)] until an empty list is produced [Figure 4.16(e)]. The strings that do not belong to any mergeable pair are the prime implicants of the function.

Procedure Quine–McCluskey
If the list is empty then terminate: else
scan the strings of the list and for each string u in the list:
scan the strings in the *next* group and for each string v in this group:
if $\{u, v\}$ are a mergeable pair then

1. Check (✓) both u and v.
2. Enter the merge string into the output list if not already present.

If the output list is not empty, apply the procedure to the output list; collect unchecked vectors in all processed lists (these are the prime implicants).

EXAMPLE 4.12
Find a minimal SOP expression for the function

$$f = \text{OR}(m_0, m_1, m_3, m_8, m_9, m_{11}, m_{15}, m_{16}, m_{17}, m_{19}, m_{24}, m_{25}, m_{29}, m_{30}, m_{31})$$

We begin by expressing each minterm as a 5-bit binary string and arrange these strings in the specified order (by weight) to obtain the initial list (consecutive groups are separated by a horizontal line). At the fourth iteration, the procedure produces an empty list and terminates. The prime implicants are the product terms corresponding to the unchecked (encircled) strings: 01–11, 11–01, 111–1, 1111–, –1111, 0–0–1, –00–1, —00–.

Once the prime implicants have been obtained we must select a subset of them to obtain a minimal expression. The basic ideas are that each prime implicant "covers" a collection of minterms (for example, 0–0–1 covers 00001, 00011, 01001, and 01011, i.e., m_1, m_3, m_9, and m_{11}) and that the selected prime implicants must collectively cover all the minterms of the function.

Suppose that minterm m_j is covered by prime implicants P_l, P_m, . . . , P_r. Then we have "minterm m_j is covered if and only if P_l is selected or P_m is selected . . . or P_r is selected." More concisely, if we let the shorthand C_j denote the statement "m_j is covered," and the shorthand P_l denote the statement "P_l is selected," and replace "if and only if" with "=" and "or" with "+" (logical OR), then the above verbal statement is replaced by the following statement

$$C_j = P_l + P_m + \cdots + P_r \tag{4.1}$$

This means (as we already saw in a similar instance in Sec. 4.2.2) that statement C_j is true if and only if at least one of the statements P_l, P_m, . . . , P_r is true. In addition, suppose that the function f has minterms m_i, m_j, . . . , m_k. Then, all minterms are covered if and only if

$$\Phi = C_i C_j \ldots C_k = \text{true} \tag{4.2}$$

We may then replace each C_i by the corresponding OR expression, and obtain therefore an expression that is in POS form. If we apply distributivity, we can transform this POS expression into a SOP expression, each term being a product of P_js, that is, a selection of a subset of prime implicants. Clearly each such selection corresponds to a valid expression for f, because it covers all and only the minterms of f.

EXAMPLE 4.13

Suppose that $f = \text{OR}(m_0, m_1, m_2, m_5, m_6, m_7)$. Applying the known methods, we obtain the prime implicants $P_1 = 00–$, $P_2 = 0–0$, $P_3 = –01$, $P_4 = –10$, $P_5 = 1–1$, $P_6 = 11–$. All prime implicants have the same cost. The minterm/prime-implicant coverings are expressed by

$$C_0 = (P_1 + P_2), \quad C_1 = (P_1 + P_3), \quad C_2 = (P_2 + P_4),$$
$$C_5 = (P_3 + P_5), \quad C_6 = (P_4 + P_6), \quad C_7 = (P_5 + P_6)$$

and

$$\text{true} = C_0C_1C_2C_5C_6C_7 = (P_1 + P_2)(P_1 + P_3)$$
$$(P_2 + P_4)(P_3 + P_5)(P_4 + P_6)(P_5 + P_6)$$

If we transform this expression into SOP form we obtain

$$\begin{aligned} \text{true} &= (P_1 + P_2P_3)(P_4 + P_2P_6)(P_5 + P_3P_6) \\ &= (P_1P_4 + P_1P_2P_6 + P_2P_3P_4 + P_2P_3P_6)(P_5 + P_3P_6) \\ &= P_1P_4P_5 + P_1P_4P_3P_6 + P_1P_2P_5P_6 + P_1P_2P_3P_6 + P_2P_3P_4P_5 \\ &\quad + P_2P_3P_4P_6 + P_2P_3P_5P_6 + P_2P_3P_6 \end{aligned}$$

Therefore there are two expressions with three prime implicants ($P_1P_4P_5$ and $P_2P_3P_6$) and six expressions with four prime implicants. Since in this case all prime implicants have the samc cost, the cost evaluation is simple and, obviously, we shall prefer an expression with three prime implicants.

The ideas just presented are the logical formulation for the selection of a set of prime implicants. However, in practical cases, even for Example 4.12, the actual implementation would be horrendously time consuming. However, since we are not seeking *all* minimal expressions, but just *one* minimal expression, we can take a few shortcuts. We begin by forming a table, the minterm/prime implicant table, where minterms are column headings and prime implicants are row headings (the following table pertains to Example 4.12). Next we place a "cross" at the intersection of the row of a "covering prime implicant" with the column of the "covered minterm." The table is complete. The selection proceeds as follows:

1. *Circle each cross that is alone in its column.* Each row that has a circle corresponds to an essential prime implicant; indeed there is at least a minterm which is covered just by that prime implicant, which therefore must appear in each minimal expression. (In our running example, P_4, P_7, and P_8 are essential.)
2. *Mark each essential prime implicant with an asterisk and delete from*

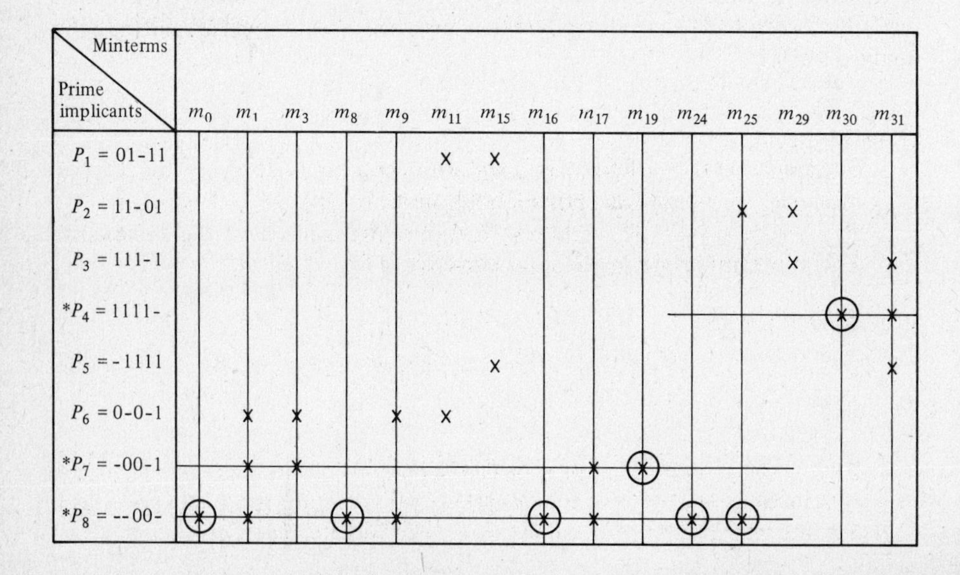

Minterms / Prime implicants	m_0	m_1	m_3	m_8	m_9	m_{11}	m_{15}	m_{16}	m_{17}	m_{19}	m_{24}	m_{25}	m_{29}	m_{30}	m_{31}
P_1 = 01-11						X	X								
P_2 = 11-01												X	X		
P_3 = 111-1													X		X
$*P_4$ = 1111-														(X)	X
P_5 = -1111							X								X
P_6 = 0-0-1		X	X		X	X									
$*P_7$ = -00-1		X	X						X	(X)					
$*P_8$ = --00-	(X)	X		(X)	X			(X)	X		(X)	(X)			

the table its row and all the columns of the minterms it covers (where the row has a cross). Indeed, since an essential prime implicant is selected for the final expression, we do not need to worry any further about the minterms it covers. The resulting table is called *reduced.* (Note that, in terms of steps 1 and 2, if m_j is covered just by essential P_k, then $C_j = P_k$, whence P_k is a factor of Φ, i.e., it is a factor of each term after the distribution.) In our example all columns but m_{11}, m_{15}, and m_{29} are deleted at this point.

3. *In the reduced table, delete all dominating columns.* (A column m_j *dominates* another column m_i if the crosses of m_j cover the crosses of m_i.) Indeed, if column m_j dominates column m_i and we select a prime implicant P_k which covers m_i, then m_j is automatically covered by P_k; thus we need not worry about the dominating column which can be removed. [In terms of steps 1 and 2, the dominated column m_i and dominating columns m_j, respectively, correspond to two terms of the form C_i and $C_j = C_i + C'_j$, whence, by absorption, $C_iC_j = C_i(C_i + C'_j) = C_i$.] In our example there is no (dominating, dominated) column pair.
4. *In the reduced table delete a dominated row only if it corresponds to a prime implicant of cost equal or larger to that of the dominating row.* (A row P_k dominates a row P_h if the crosses of P_k cover the crosses of P_h.) Indeed, the dominated row covers only a subset of the minterms covered by the dominating row, which justifies the conditions for the deletion. (In terms of steps 1 and 2, this is *not* a boolean operation but it is only dictated by our definition of minimal expression.) In our example P_5 can be deleted, because it is dominated by P_7, of the same cost; also, either P_2 or P_3 can be deleted, because they have the same cost and cover each other in the reduced table. So, we delete P_5 and P_2.
5. *If in the reduced table there is at least one cross that is alone in its column, return to step 1; else proceed.* (If steps 3 and 4 are void certainly the condition for return to step 1 is not met; in our example, we return to step 1 and find that P_3 and P_1 are essential in the reduced table and cover all the remaining minterms.)
6. *The reduced table is cyclic. We solve it by the method corresponding to relations (4.1) and (4.2).* Normally the table we get at this point is of much smaller size than the original one. Thus, in our running example $\{P_4, P_7, P_8, P_3, P_1\}$ are the prime implicants selected and the minimal expression of f is

$$f = x_5x_4x_3x_2 + \bar{x}_4\bar{x}_3x_1 + \bar{x}_2\bar{x}_3 + x_5x_4x_3x_1 + \bar{x}_5x_4x_2x_1.$$

*4.7 DESIGN OF ALL-NAND (OR ALL-NOR) COMBINATIONAL NETWORKS

We noted in Sec. 3.6 the attractiveness, from an engineering viewpoint, of NAND gates and NOR gates, since an arbitrary switching function can be realized by a network consisting of just one of these two types of gates. We also noted the boolean identity

$$ab + cd = \overline{\overline{ab} \cdot \overline{cd}} = \text{NAND}[\text{NAND}(a, b), \text{NAND}(c, d)]$$

which permits the transformation of an arbitrary SOP network into a two-level NAND network by simply replacing both the AND gates and the OR gate with NAND gates. Thus, the theory developed in the preceding sections entirely applies to the design of two-level NAND networks.

In some design situations, however, one may be willing to exchange speed for equipment, that is, to obtain a NAND network with more than two logic levels but with fewer gates. Unfortunately, there are no truly systematic methods to obtain minimal multilevel NAND networks. What is available are some good "common sense" techniques, such as the "map factoring" and the "repeated factorization" (or "transform"), which require some degree of experience to be satisfactorily used.

A thorough discussion of these techniques is well beyond the scope of this text. We shall content ourselves with a brief introduction to the transform technique. The starting point is a *minimal* SOP expression if we seek a NAND network, a *minimal* POS expression if we seek a NOR network. We shall limit our discussion to SOP expressions, and apply the following procedure.

Step 1. Examine the product terms with two or more literals, and group them so that all terms in the same group share some literals. [Two comments: (1) there is no single way to form these groups, nor is there a simple way to find the "best" grouping; (2) since the starting expression is minimal, a single literal cannot be a factor of any other term.]

EXAMPLE 4.14

Given the expression $\overline{x}_1x_2 + \overline{x}_2\overline{x}_3\overline{x}_4 + \overline{x}_1x_4 + x_2x_3\overline{x}_4$ [a minimal SOP expression of the function OR$(m_0, m_1, m_2, m_6, m_7, m_8, m_{10}, m_{12}, m_{14})$] we form the groups $\{\overline{x}_1x_2, \overline{x}_1x_4\}$ and $\{\overline{x}_2\overline{x}_3\overline{x}_4, x_2x_3\overline{x}_4\}$.

Step 2. Transform the original expression by factoring out the common literals in each of the groups formed in step 1.

EXAMPLE 4.15

$$\overline{x}_1x_2 + \overline{x}_2\overline{x}_3\overline{x}_4 + \overline{x}_1x_4 + x_2x_3\overline{x}_4 = \overline{x}_1(x_2 + x_4) + \overline{x}_4(\overline{x}_2\overline{x}_3 + x_2x_3)$$

The resulting expression is the OR of product terms, each of which is in turn the AND of some literals and of a SOP expression. In our example the factor SOP expressions are $(x_2 + x_4)$ and $(\overline{x}_2\overline{x}_3 + x_2x_3)$.

Step 3. Apply steps 1 and 2 to each of the factor SOP expressions obtained in step 2 until no more factoring is possible. At this point, the final expression can be put in NAND form in a straightforward manner, due to its structure of alternating AND and OR.

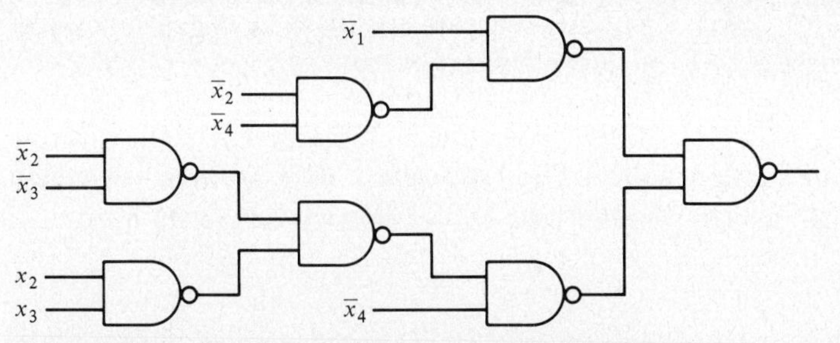

Figure 4.17 All NAND realization of a switching function.

In our example the expression

$$\bar{x}_1(x_2 + x_4) + \bar{x}_4(\bar{x}_2\bar{x}_3 + x_2x_3)$$

is already in a form that cannot be further manipulated. By applying DeMorgan's theorem, the corresponding NAND expression is

$$\overline{\overline{\bar{x}_1 \cdot \overline{\bar{x}_2\bar{x}_4}} \cdot \overline{\bar{x}_4 \cdot \overline{\overline{\bar{x}_2\bar{x}_3} \cdot \overline{x_2x_3}}}}$$

and the NAND circuit is shown in Figure 4.17.

4.8 COMBINATIONAL MODULES (MSI MODULES)

As mentioned earlier, customized logical design using discrete logic, that is, individual inverters, AND gates, and OR gates—or other modules of comparable simplicity—is only one aspect of the design practice. Indeed, current technology is making available a few complex and yet quite flexible combinational modules, by means of which most of the desirable design objectives can be achieved. Before beginning the description of specific MSI modules, it is worth gaining some motivation about them.

If we analyze an existing digital system—in particular, a digital computer—we shall find that logic gates are basically used in two main applications:

1. In *logic,* that is, in the *realization of information processing units* (which has been the focus of our attention so far).
2. In *gating,* that is, in *routing information* from register to register, in very much the same way as railroad switches are used in railroad yards.

If we further analyze logic applications, we see the frequent occurrence of some typical functions that warrant the use of specialized modules. In this chapter we shall study decoders and encoders, and then see how they can be combined to obtain some of the most versatile available modules, such as read-only memories (ROMs) and programmable logic arrays (PLAs). The latter are in a sense *universal* combinational modules, that is, "blanks" which can be "filled" to realize your specific job.

In gating applications, typical situations are those where a number of sources must be connected to a single destination, or a single source to several destinations. The modules that respond to this need are the multiplexer and the demultiplexer, respectively. It turns out that the multiplexer has such a versatile structure that it naturally lends itself to the realization of arbitrary switching functions, and has given rise to interesting logical design techniques.

4.8.1 Encoders

An *s-to-n encoder* is a module with s input lines $l_1, l_2, \ldots, l_s$ and n output lines $y_{n-1}, y_{n-2}, \ldots, y_0$. The number s is normally much larger than n. Each input line is assigned a distinct nonzero integer, that is, l_i is assigned $f(i)$, $[0 < f(i) \leq 2^n]$. The normal operation of an encoder is as follows:

(Whenever the output is to be used) only one input line carries the value 1 (active line) and all the others carry 0; the output lines display the binary equivalent of the integer $f(i)$ associated with the active line l_i.

In other words, an encoder produces the encoding of the integer associated with the active line. For instance, if l_7 is active, $n = 5$, and $f(7) = 13$, the output will be $y_4y_3y_2y_1y_0 = 01101$. Notice that for two distinct input lines l_i and l_j, the integers $f(i)$ and $f(j)$ are not necessarily distinct, that is, two distinct input lines could produce the same output; however, this is not the usual case. An encoder is simply a collection of OR gates, as shown in Example 4.16.

EXAMPLE 4.16

Suppose there are seven input lines, $l_1, l_2, \ldots, l_7$ with $f(1) = 3$, $f(2) = 5$, $f(3) = 6$, $f(4) = 9$, $f(5) = 10$, $f(6) = 12$, $f(7) = 13$. To construct the corresponding encoder we list, in a column, the binary equivalent of the integers $f(1), \ldots, f(7)$ [Figure 4.18(a)]. Clearly, at least 4 bits are needed: we call them $y_3y_2y_1y_0$. Then each function y_i will be realized by a single OR gate whose inputs are the lines l_j corresponding to 1s in the column of y_i. The network diagram is displayed in Figure 4.18(b).

A convenient display form for an encoder is based on an alternative graphic symbol for OR gates as shown below:

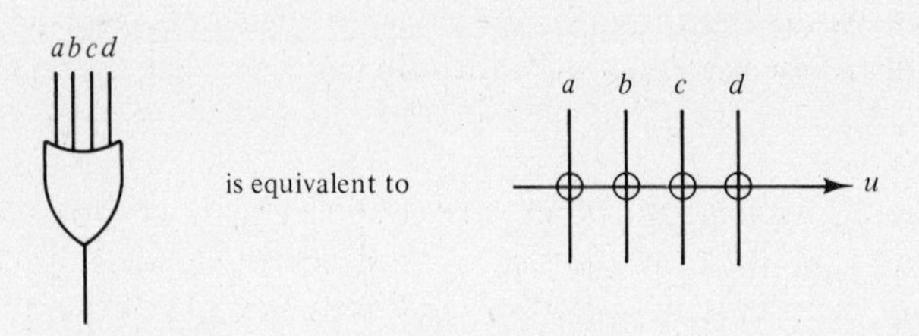

With this new convention, the scheme of the above encoder is as shown in the "grid" in Figure 4.19. Notice the striking structural identity of network and table [shown in Figure 4.18].

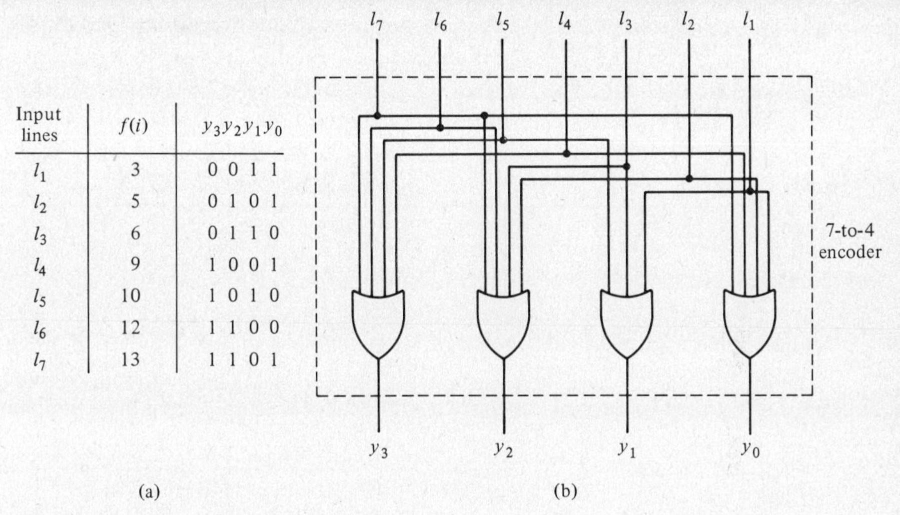

Input lines	$f(i)$	$y_3y_2y_1y_0$
l_1	3	0 0 1 1
l_2	5	0 1 0 1
l_3	6	0 1 1 0
l_4	9	1 0 0 1
l_5	10	1 0 1 0
l_6	12	1 1 0 0
l_7	13	1 1 0 1

Figure 4.18 Structure of an encoder.

An interesting variant of the previous encoder is the so-called *priority encoder,* in which the condition that only one input line be active at any one time is removed and some rule is devised for the selection of the input line among the active ones whose number is to be encoded. This rule is called *priority.* Specifically, the numbers associated with the lines are *positive* integers and are ordered increasingly; the output lines display the binary representation of the integer i, if i is the *largest* of the numbers associated with the input lines that are active. Such a type of encoder can be very useful in input/output operations of a computing system (see Chapter 8, Problems 8.7 and 8.8).

4.8.2 Decoders

An *n-to-s decoder* is a module with n input lines $x_{n-1}, x_{n-2}, \ldots, x_0$ and s output lines $u_{i_1}, u_{i_2}, \ldots, u_{i_s}$. The number s is $\leq 2^n$ and in commercially available off-the-shelf decoders is usually equal to 2^n (there exist, however, also

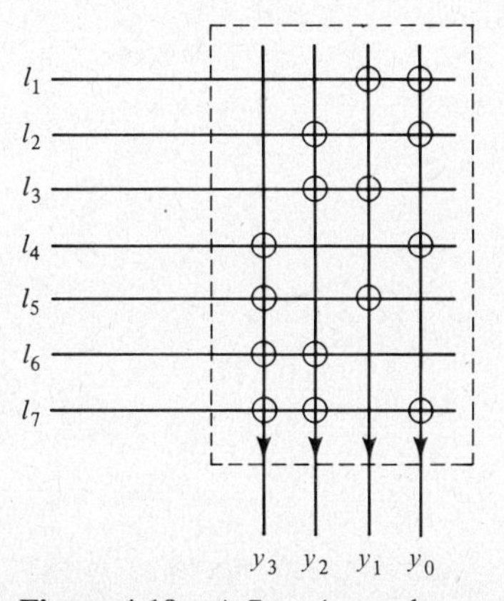

Figure 4.19 A 7-to-4 encoder.

4-to-10 decoders). A decoder is the functional inverse of an encoder. Here, each output line u_j is associated with the integer j.

> The output line u_j carries the value 1 (active line) and all the others are 0 whenever the decimal equivalent of the binary number $x_{n-1}x_{n-2} \ldots x_0$ is j.

Clearly, a decoder will have fewer than 2^n output lines only when some input combinations are known never to occur, as in the 4-to-10 decoder. Thus, from now on, we shall only consider full n-to-2^n decoders.

The simplest form of realization of a decoder is a collection of 2^n n-input AND gates when each variable is available in true and complemented form. Indeed, each output line simply decodes a minterm of the input variables, and we know that a minterm is realized by a single AND gate.

EXAMPLE 4.17

Design a 2-to-4 decoder. Given the variables x_0 and x_1 we first form $\overline{x}_0$ and $\overline{x}_1$ so that $x_0, \overline{x}_0, x_1, \overline{x}_1$ are now available. Then we combine them to form all fundamental products of two variables, as shown in Figure 4.20. In Figure 4.20 we also show an additional input, E, fed to all AND gates. This is quite common in decoders, where E, *enable,* is 1 whenever the output is to be used.

We can also adopt for AND gates a symbol analogous to that introduced earlier for OR gates, that is:

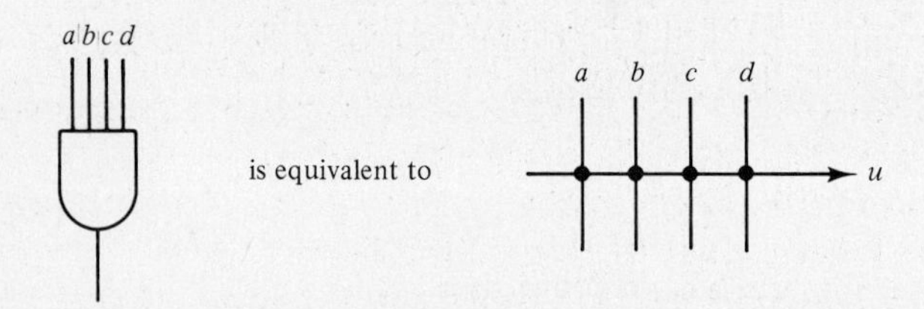

With this convention the previous decoder becomes as in Figure 4.21.

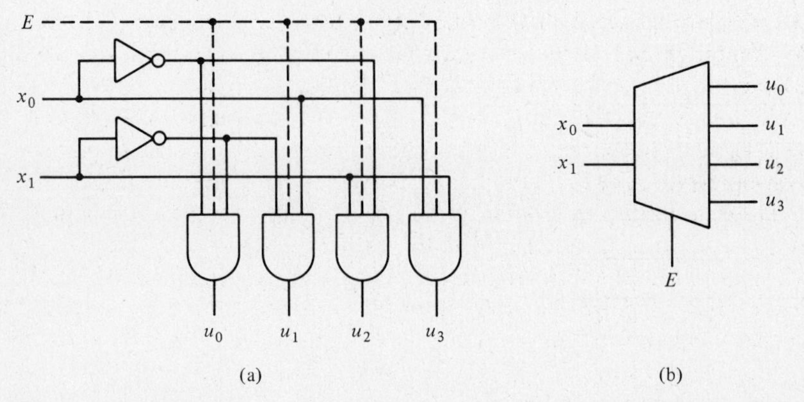

Figure 4.20 Structure of a 2-to-4 decoder (a) and its conventional graphical symbol (b).

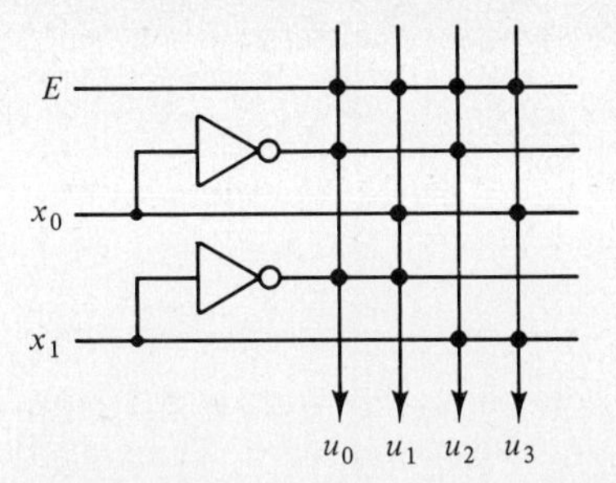

Figure 4.21 A 2-to-4 parallel decoder.

This type of realization is called *parallel decoder;* it is quite fast (only two levels of logic, including the inverters) but may run into technological difficulties because, for large n, n-input AND gates may not be easily realizable (high fan-in).

4.8.3 Transcoders, Read-Only Memories (ROMs), and Programmable Logic Arrays (PLAs)

If we cascade a decoder and an encoder we obtain a new module which can be used to translate a code into another code, that is, a code translator or *transcoder.* For example, if we wish to transform a BCD code into a 7-segment code (see Chapter 1 for a definition to both codes), we must realize the seven truth tables illustrated in Figure 4.22(a) for $S_0, S_1, \ldots, S_6$. An alternative to the direct synthesis of these functions by the general methods described earlier in this chapter, is the use of a decoder of the BCD code, whose outputs are fed to the 7-segment encoder. This realization is shown in Figure 4.22(b) and is self-explanatory. In this network we distinguish two major portions, the AND matrix and the OR matrix (also called, in jargon, AND plane and OR plane, respectively). Recalling the meaning of arrows as defined in Secs. 4.8.1 and 4.8.2, each column of the AND matrix is an AND gate while each row of the OR matrix is an OR gate.

The notion of transcoder can be readily extended to obtain two of the most versatile combinational modules available today: the read-only memories (ROMs) and the programmable logic arrays (PLAs).

An $2^n \times m$ ROM has n input lines $x_0, x_1, \ldots, x_n$ and m output lines $y_1, y_2, \ldots, y_m$[6] and it realizes an arbitrary mapping from the set of all combinations of the input variables to a subset of the combinations of the output variables. In other words, to *each* combination x of $x_0, \ldots, x_{n-1}$ we assign a combination of $y_1, \ldots, y_m$, which appears on the output lines any time x appears on the input lines. Thus, the module may be thought of as a memory, where the input lines carry an "address" and the output lines carry the memory "readout." The fact that the mapping from $(x_0, \ldots, x_{n-1})$ to $(y_1, \ldots, y_m)$ cannot be

[6]The expression "$2^n \times m$ ROM" indicates that the ROM is viewed as a matrix with 2^n rows and m columns.

BCD code				7-segment code						
b_3	b_2	b_1	b_0	S_6	S_5	S_4	S_3	S_2	S_1	S_0
0	0	0	0	0	1	1	1	1	1	1
0	0	0	1	0	0	0	0	1	1	0
0	0	1	0	1	0	1	1	0	1	1
0	0	1	1	1	0	0	1	1	1	1
0	1	0	0	1	1	0	0	1	1	0
0	1	0	1	1	1	0	1	1	0	1
0	1	1	0	1	1	1	1	1	0	1
0	1	1	1	0	0	0	0	1	1	1
1	0	0	0	1	1	1	1	1	1	1
1	0	0	1	1	1	0	1	1	1	1

(a)

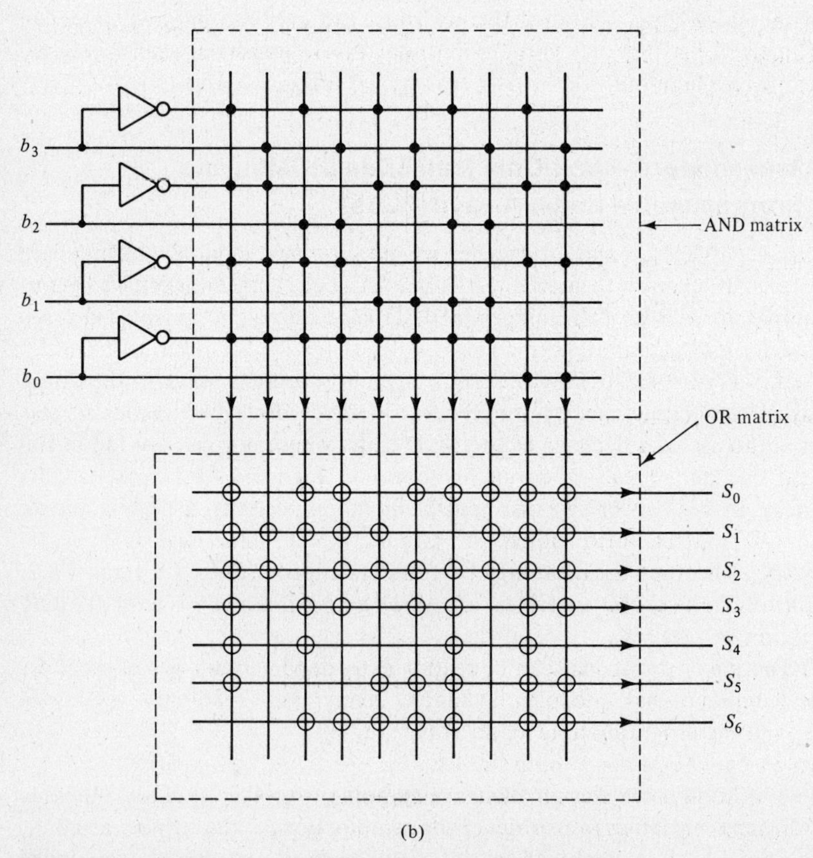

(b)

Figure 4.22 Realization of a BCD-to-7-segment transcoder.

altered once chosen, suggests that this "memory" can be "read from" but not "written into," hence the name of ROM.

Structurally, a ROM consists of a decoder for all combinations of the variables $x_0, \ldots, x_{n-1}$ followed by an OR matrix consisting of m OR gates. As in Figure 4.22, the inputs to the OR matrix are the decoder outputs, while the outputs of the OR matrix are the ROM's outputs themselves. Since the decoder is standard (an n-to-2^n decoder), we choose to represent it in Figure 4.23 with its compact module symbol, rather than by an AND matrix. Before a specific input/output map is realized, the OR matrix is full, that is, each decoder output is connected to each OR gate. In Figure 4.23(a) we have shown one such "unwritten" ROM with $n = 3$ and $m = 5$; in Figure 4.23(b) we show in tabular form a particular input/output map. Notice that there is an orderly one-to-one correspondence between the inputs to the OR gates in the OR matrix and the entries in the content portion of the table; if for every 0 in the latter we can remove the corresponding connection in the OR matrix [Figure 4.23(c)], then we obtain a ROM with the required information "written in."

From the functional standpoint, the ROM may be interpreted in two ways:

1. As an actual memory device, for information that need not be altered.
2. As a multiple-output combinational module, since each of the ROM outputs is a switching function of the ROM inputs (refer again to Figure 4.23).

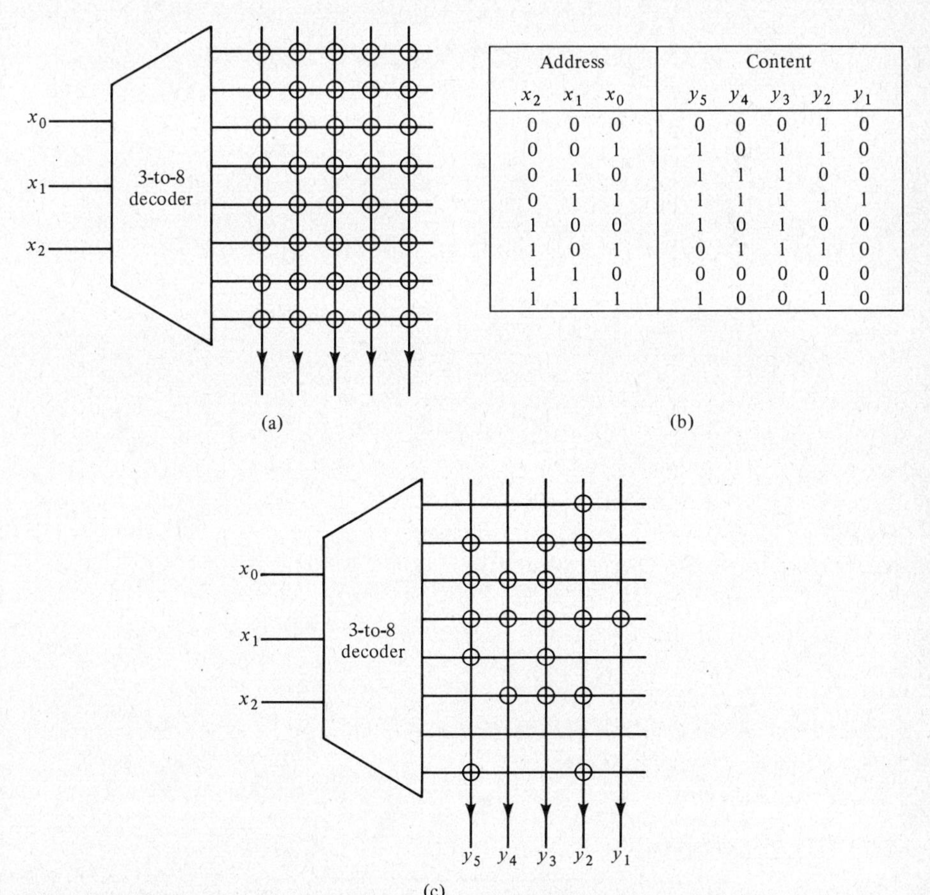

Address			Content				
x_2	x_1	x_0	y_5	y_4	y_3	y_2	y_1
0	0	0	0	0	0	1	0
0	0	1	1	0	1	1	0
0	1	0	1	1	1	0	0
0	1	1	1	1	1	1	1
1	0	0	1	0	1	0	0
1	0	1	0	1	1	1	0
1	1	0	0	0	0	0	0
1	1	1	1	0	0	1	0

Figure 4.23 An "unwritten" ROM (a), an input/output map (b), and the corresponding ROM realization (c).

From the latter viewpoint, the ROM is a "universal" combinational module, which can be specialized to realize a prescribed set of functions of the same set of inputs in an extremely straightforward manner, as is suggested by the striking structural correspondence between the pattern of 0s in the truth tables [Figure 4.23(b)] and the diode-removal pattern [Figure 4.23(c)]. We may also note that *each of the output functions of the ROM is realized in canonical SOP form* (indeed, each output of the decoder carriers a minterm of the inputs).

The last remark also points to a drawback of the ROM as a universal combinational module. We know that the minimal SOP expression of a function is generally much simpler than the corresponding canonical expression. Therefore, we may try to apply the ROM approach to minimal rather than canonical expressions: this is how the *programmable logic array* (PLA) is born. The basic difference between a ROM and a PLA resides in the AND matrix, which is a standard decoder in a ROM, while in a PLA it realizes all the product terms appearing in the expressions of the output functions. We see therefore that the switching function minimization machinery developed earlier in this chapter applies directly to PLA design, with two minor corrections:

1. There is an AND gate in the AND matrix for each product term, regardless of its number of literals. So the minimization of the number of AND gates becomes the only criterion, regardless of the number of AND gate inputs.
2. The same product term may be shared by several output functions. The minimization technique can be modified to obtain minimal number of product terms for a collection of functions. (This theory of multiple-output minimization is beyond the scope of this text, and we shall content ourselves with the SOP minimization of individual functions.)

At this point the reader may naturally ask the following question: A $2^n \times m$ ROM has 2^n product lines (the decoder outputs); how many product lines will an n-input, m-output PLA have? This very important question is answered by the manufacturers on the basis of statistical evidence. By examining a large number of actual practical examples one may obtain a reliable estimate for the number p of product lines which is adequate to cover a large fraction of the foreseeable needs. Thus a PLA is specified by three numbers, as shown below:

$$n \times p \times m$$

Inputs Products Outputs

For example, a $16 \times 48 \times 8$ PLA is typical. A significant PLA application will be discussed at the end of Chapter 5.

4.8.4 Multiplexers

An *n-to-1 multiplexer* (*in the narrow sense*) is a module with one output line w, n input lines $\{a_0, a_1, \ldots, a_{n-1}\}$, called DATA lines, and additional n input lines

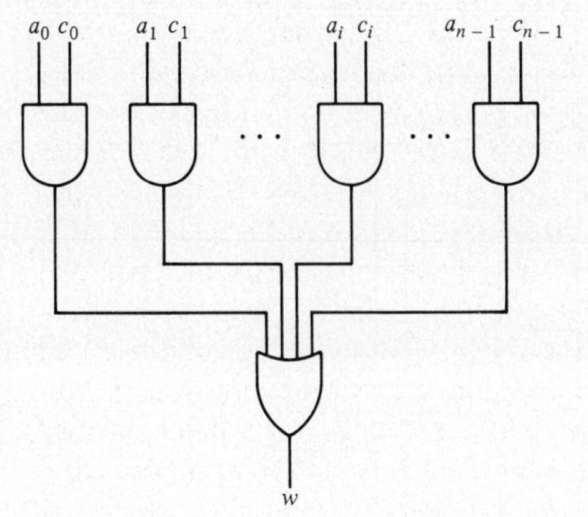

Figure 4.24 An n-to-1 narrow-sense multiplexer.

$\{c_0, c_1, \ldots, c_{n-1}\}$, called GATING lines, with the condition that only one GATING line has the value 1 (active line) at any one time. The operation of the narrow-sense multiplexer is as follows:

> If c_i is the active line then $w = a_i$, that is, line a_i is logically connected[7] to line w.

Clearly, a narrow-sense multiplexer has the following structure: Each a_i is gated by c_i ($i = 0, \ldots, n-1$), and the outputs of the corresponding n AND gates are fed to a single n-input OR gate as shown in Figure 4.24.

Very frequently, however, the term "multiplexer" is used to denote what is more adequately called *data selector/multiplexer* (we shall also use the word "multiplexer" for the latter). This module has one output line w, n input lines a_0, $a_1, \ldots, a_{n-1}$, called DATA lines, and r additional lines, called SELECT lines, $s_{r-1}, s_{r-2}, \ldots, s_0$. In the commercially available modules, n is a power of 2, that is, $n = 2^r$. The operation of the multiplexer is as follows:

> If the configuration on the SELECT lines is the representation of the integer i, then $w = a_i$.

Thus, in a multiplexer, the DATA line x_i is gated with a signal c_i which is 1 whenever the decimal equivalent of $s_{r-1}s_{r-2}, \ldots, s_0$ is the integer i; but this is exactly the function performed by a decoder of $s_{r-1}, \ldots, s_0$ (Sec. 4.8.2), where when $c_i = 1$ we also have $c_j = 0$ for all $j \neq i$. Therefore, in principle, we have a decoder for the SELECT lines, whose outputs are the GATING lines of a narrow-sense multiplexer [see Figure 4.25(a) for $n = 8$ and $r = 3$). In practice, however, decoder and multiplexer are combined into a more economical module as shown in Fig. 4.24(b) (where an "Enable" input line E is also shown).

As the preceding discussion indicates, an important use of the selector/

[7]We say that "line x is logically connected to line y" if there is a chain of gates from x to y, so that the logical value of x becomes the logical value of y.

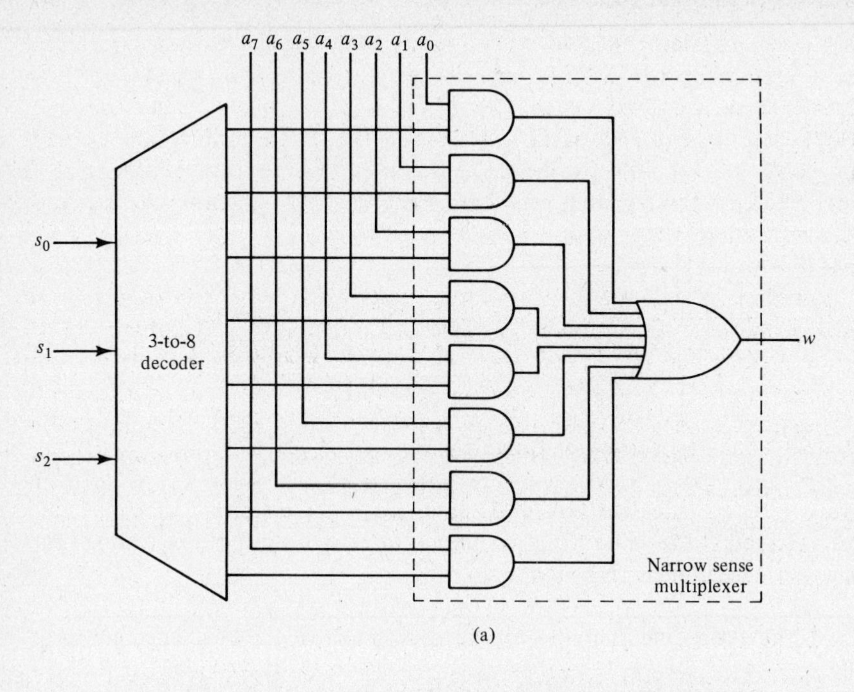

(a)

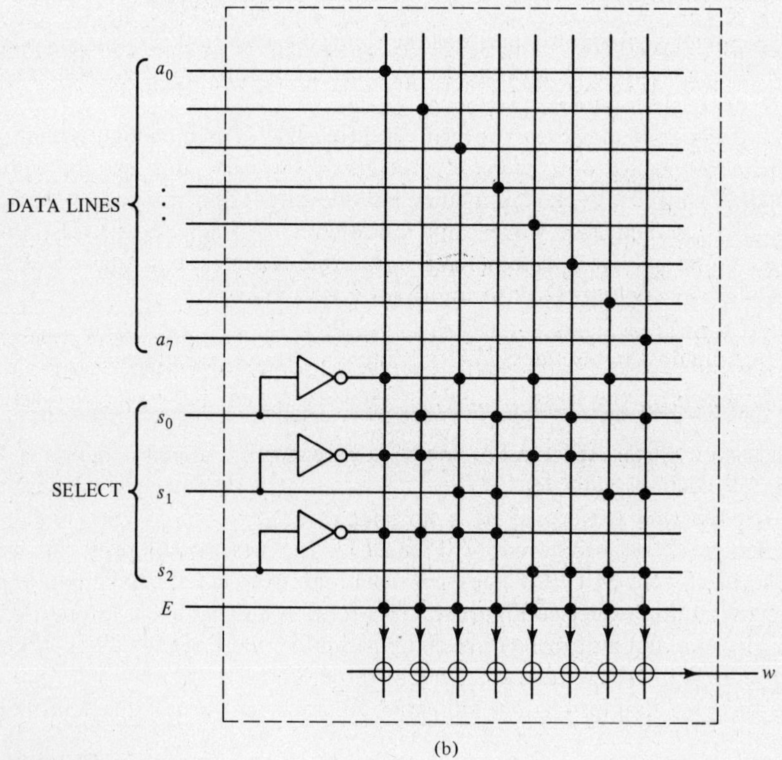

(b)

Figure 4.25 A multiplexer.

multiplexer is to select a specific input line and to "funnel" it to the single output line. In this mode the DATA lines carry independent variables while the SELECT lines are used as controls. This capability allows the use of a multiplexer as a *parallel-to-serial converter,* that is, if we assume that a binary string $x_0x_1 \ldots x_{2^r-1}$ is available *in parallel* on the input lines (and is maintained during the entire conversion), by generating an appropriate sequence of select signals we can obtain the time sequence $x_0, x_1, \ldots$ *in series* on the output line w.

An alternative mode of use for the multiplexer is the one in which the SELECT lines carry independent variables $s_{r-1}, s_{r-2}, \ldots, s_0$ and the DATA lines are used as controls. In this mode, the multiplexer can be used to generate an arbitrary switching function f of the r variables $s_{r-1}, \ldots, s_0$. We illustrate the general method with reference to the case $r = 2$, which is more readily grasped; the generalization to arbitrary r is straightforward, on the other hand. For $r = 2$, the expression of the output w of a 4-to-1 multiplexers, with DATA lines $\{a_0, a_1, a_2, a_3\}$ and SELECT lines $\{s_0, s_1\}$ [refer back to the definition and to Figure 4.25(a)] is

$$w = a_0\bar{s}_1\bar{s}_0 + a_1\bar{s}_1 s_0 + a_2 s_1\bar{s}_0 + a_3 s_1 s_0$$

Note that this expression contains *all* the minterms of the two variables s_0 and s_1, each ANDed with a boolean coefficient a_j. Now, since a given function f has a canonical SOP expression which is the OR of a subset of all minterms of its variables, this subset can be selected by setting to 1 the coefficients of the minterms which are present and to 0 those of the minterms which are absent.

EXAMPLE 4.18

For $r = 2$, realize the function $f = \text{OR}(m_1, m_2)$ with a 4-to-1 multiplexer. In Figure 4.26(a) we display the truth table of $f(s_1, s_0)$; in Figure 4.26(b) we show how to bias the multiplexer's DATA lines according to the above rule. Notice the structural identity of truth table and biasing pattern.

If, as in the previous example, the parameters a_0, a_1, a_2, a_3 can be selected only in the set of choices $\{0, 1\}$, then a 4-to-1 multiplexer can be used to generate functions of at most two variables and (by the same token) a 2^r-to-1 multiplexer

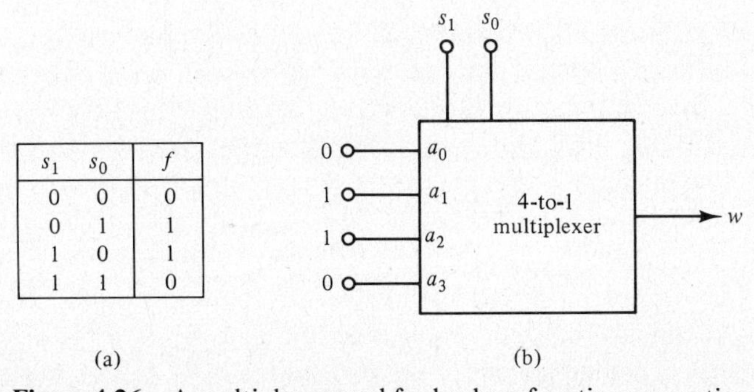

s_1	s_0	f
0	0	0
0	1	1
1	0	1
1	1	0

Figure 4.26 A multiplexer used for boolean function generation.

can be used to generate functions of at most r variables. However, there are somewhat more sophisticated ways to use a multiplexer to generate boolean functions, as soon as we allow the DATA line parameters $a_0, a_1, \ldots$ to be selected from a wider set, containing not only the constants $\{0, 1\}$ but also variables and, possibly, simple functions of the variables. For example, consider an arbitrary function f of *three* variables s_0, s_1, and s_2, and suppose that not only $\{0, 1, s_0, s_1, s_2\}$ but also $\bar{s}_0$ are available. Letting $(f_0, f_1, f_2, f_3, f_4, f_5, f_6, f_7)$ denote the truth table of the function f (each f_i is either 0 or 1), the canonical expression of f is

$$f = f_0\bar{s}_2\bar{s}_1\bar{s}_0 + f_1\bar{s}_2\bar{s}_1 s_0 + f_2\bar{s}_2 s_1\bar{s}_0 + f_3\bar{s}_2 s_1 s_0 + f_4 s_2\bar{s}_1\bar{s}_0 + f_5 s_2\bar{s}_1 s_0 + f_6 s_2 s_1\bar{s}_0 + f_7 s_2 s_1 s_0$$

which can be rewritten as

$$f = \bar{s}_2\bar{s}_1(f_0\bar{s}_0 + f_1 s_0) + \bar{s}_2 s_1(f_2\bar{s}_0 + f_3 s_0) + s_2\bar{s}_1(f_4\bar{s}_0 + f_5 s_0) + s_2 s_1(f_6\bar{s}_0 + f_7 s_0)$$

If we now are able to choose in the 4-to-1 multiplexer $a_0 = f_0\bar{s}_0 + f_1 s_0$, $a_1 = f_2\bar{s}_0 + f_3 s_0$, $a_2 = f_4\bar{s}_0 + f_5 s_0$, $a_3 = f_6\bar{s}_0 + f_7 s_0$, then obviously $w = f$, that is, the output of the multiplexer realizes f. Note that each of the parameters $\{a_0, a_1, a_2, a_3\}$ equals a function of the single variables s_0, and any such function is either 0, 1, s_0, or $\bar{s}_0$. Therefore if $\bar{s}_0$ is available, a 4-to-1 multiplexer can be used to generate an arbitrary switching function of the three variables $\{s_0, s_1, s_2\}$, and, in general, a 2^{r-1}-to-1 multiplexer can be used for an arbitrary function of the r variables $\{s_0, s_1, \ldots, s_{r-1}\}$.

EXAMPLE 4.19

For $r = 3$, use a 4-to-1 multiplexer to generate the function $f(s_2, s_1, s_0) = \text{OR}(m_1, m_4, m_5, m_6) = (0, 1, 0, 0, 1, 1, 1, 0)$. It is convenient to arrange the truth table of f [Figure 4.27(a)] as shown in Figure 4.27(b). Note that each column of the latter table is one of the forms

$$\begin{bmatrix}0\\0\end{bmatrix}, \begin{bmatrix}0\\1\end{bmatrix}, \begin{bmatrix}1\\0\end{bmatrix}, \begin{bmatrix}1\\1\end{bmatrix}$$

which respectively correspond to the functions 0, s_0, $\bar{s}_0$, 1. In addition, the correspondence between columns and parameters $a_0, \ldots, a_3$ is also shown. Therefore, by selecting $a_0 = s_0$, $a_1 = 0$, $a_2 = 1$, and $a_3 = \bar{s}_0$ we can realize the given f [Figure 4.27(c)]. In our example we have chosen $\{s_2, s_1\}$ on the SELECT lines; any one of the other two choices, $\{s_2, s_0\}$ and $\{s_1, s_0\}$, is equally admissible.

What happens, for example, if no complemented variable is available? In this case, we may still be able to realize an r-variable function with a 2^{r-1}-to-1 multiplexer. This happens, of course, if none of the parameters $a_0, a_1, \ldots$ must be selected as a complemented variable. In the preceding example, if we had

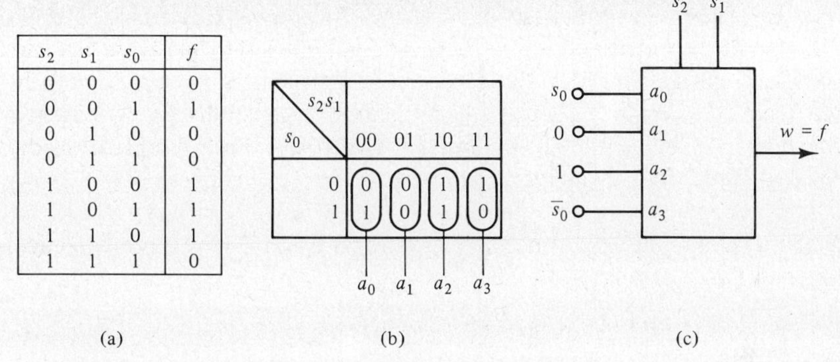

s_2	s_1	s_0	f
0	0	0	0
0	0	1	1
0	1	0	0
0	1	1	0
1	0	0	1
1	0	1	1
1	1	0	1
1	1	1	0

Figure 4.27 Generation of a switching function of three variables by means of a 4-to-1 multiplexer.

chosen $\{s_1, s_0\}$ on the SELECT lines we would have obtained the rearranged table

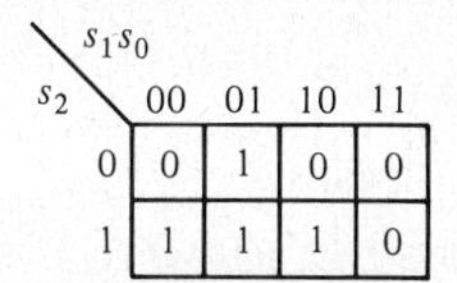

$s_2 \backslash s_1s_0$	00	01	10	11
0	0	1	0	0
1	1	1	1	0

which leads to the assignment $a_0 = s_2$, $a_1 = 1$, $a_2 = s_2$, $a_3 = 0$ (i.e., $\bar{s}_2$ is unnecessary). This is only a glance at the interesting problems arising in the generation of boolean functions by means of multiplexers. As exercises, the reader should consider the following problem:

> Any function of four variables is realizable by means of at most five 4-to-1 multiplexers.

4.8.5 Demultiplexers

A *1-to-n demultiplexer* (*in the narrow sense*) is a module with one input DATA line x, n additional input GATING lines $\{c_0, c_1, \ldots, c_{n-1}\}$, and n output lines

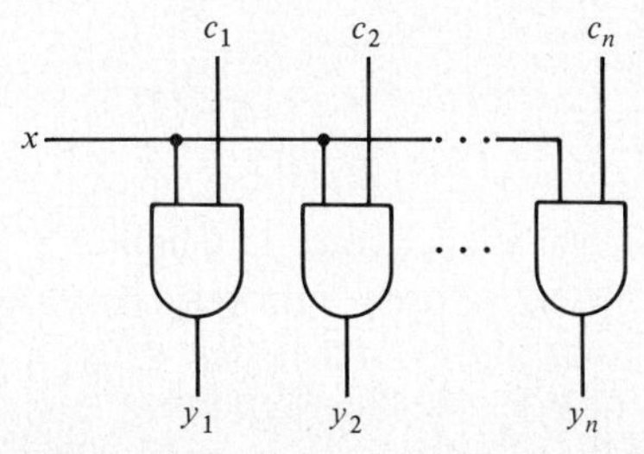

Figure 4.28 A 1-to-n narrow-sense demultiplexer.

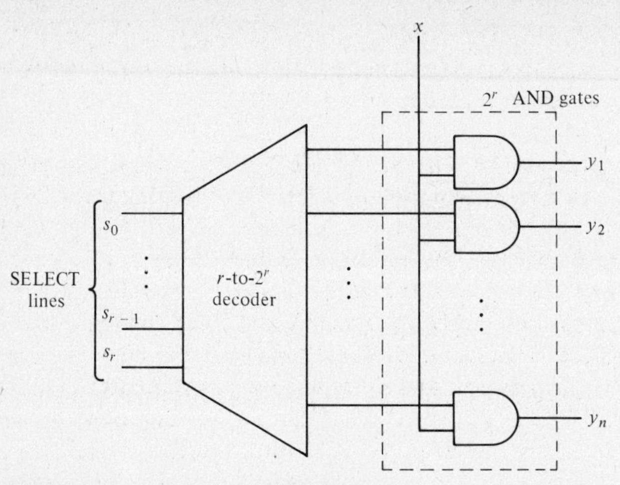

Figure 4.29 A decoder/demultiplexer.

$y_0, y_1, \ldots, y_{n-1}$. A demultiplexer is the functional inverse of a multiplexer, and its operation is as follows:

> If line c_i is active, then $y_i = x$, that is, line y_i is logically connected to the input x.

A 1-to-n narrow-sense demultiplexer is shown in Figure 4.28.

A narrow-sense demultiplexer can be extended (analogously to the narrow-sense multiplexer) to a *decoder/demultiplexer* by letting the gating lines $c_1, c_2, \ldots, c_n$ be the outputs of an r-input decoder (so that $n \leq 2^r$). The decoder inputs are normally called SELECT lines. The resulting module is shown in Figure 4.29.

There are two typical uses for the demultiplexer, both of which are the converses of corresponding uses of the multiplexer. The first is for routing one line x to one of many possible destinations. The other is as a serial-to-parallel converter: if each output of the demultiplexer is followed by a single bit storage device, and an input sequence $x_0, x_1, \ldots, x_{n-1}$ appears on the DATA line x, by appropriate control of the SELECT lines it is possible to store in parallel the given sequence in the storage devices.

NOTES AND REFERENCES

The theory of combinational networks, based on boolean algebra, had its recognizable origin in the famous paper by C. E. Shannon, quoted in Chapter 3. Soon after World War II, the interest in this topic grew immensely, and practically most of the important developments took place in the fifties. We just mention the works of Karnaugh (1953), Quine (1952), McCluskey (1956), and Muller (1954). Although some important contributions have appeared in the past 20 years, research on this topic has tapered off and the theory of combinational networks—an important part of *switching theory,* which covers also sequential networks—is now regarded as an established and mature body of knowledge.

As noted earlier in the chapter, the motivation, the objectives, and the research problems in this area have been adapting to the ever-changing technologies.

This chapter should be viewed as merely an introduction to this very elaborate field. Many important topics—such as the design of multiple-output networks, functional decomposition, all-NAND techniques, threshold logic, and fault testing—have been necessarily omitted. These topics are the domain of a course entirely devoted to logical design, and the reader interested in such topics is strongly encouraged to take a like course. Short of this, the reader may want to expand his or her knowledge in this area by consulting the many available textbooks. Among these, Kohavi (1970) presents a classical and mathematically oriented approach; Hill and Peterson (1974) is also an excellent and extensive reference, combining the mathematical fundamentals with a wealth of design examples. On the more practical side (i.e., making reference to commercially available components) is the book by Fletcher (1980). Formal rigor and practical suggestions are nicely combined in the recent text by Muroga (1979), which adds to the traditional syllabus some topics not treated elsewhere, like the two-level minimization of a switching function starting from a normal expression rather than from the canonical expression of it. This is very useful, because when the number of variables is very large, the canonical expression may be very complex, while handling of a normal expression may remain manageable throughout the minimization process.

PROBLEMS

4.1. Obtain the truth table of the output f of the following combinational network:

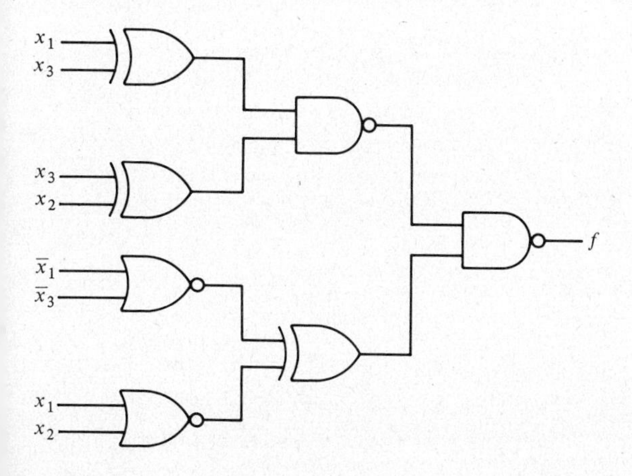

4.2. For the combinational network shown below:

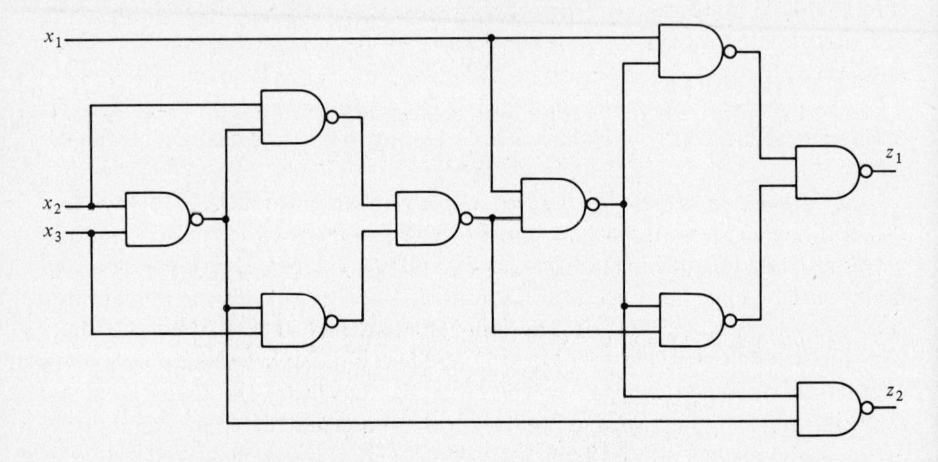

(a) Obtain the truth tables of the outputs z_1 and z_2.
(b) What does the network do?

4.3. The warning system of an elevator has the following input signals:

$G = 1$ if the gate is closed.
$L = 1$ if the elevator is loaded.
$B = 1$ if the elevator is in demand.
$U = 1$ if the elevator is moving up.
$D = 1$ if the elevator is moving down.

Give the boolean expression of an alarm signal $A(G, L, B, U, D)$ so that $A = 1$ if the gate is not closed when the elevator is moving or when the elevator is moving but is empty and not in demand.

4.4. A small corporation has 1000 shares of stock divided as follows:

A owns 125 shares.
B owns 250 shares.
C owns 375 shares.
D owns 250 shares.

At a stockholders' meeting, each share entitles its owner to one vote. A *two-thirds majority* (667 or more votes in favor) is required to pass a resolution. It is desired to design a logic network with *inputs A*, B, C, and D and *output f* to act as follows. Each shareholder casts all his or her votes in favor by setting the corresponding input to a 1 and casts all his or her votes against by setting the corresponding input to a 0. The output f is a 1 if the resolution passes and 0 otherwise. Find the truth table for $f(A, B, C, D)$.

4.5. Given $f(x_1, x_2, x_3, x_4)$ whose truth table is $\text{OR}(m_3, m_4, m_5, m_6, m_7, m_9, m_{11}, m_{13}, m_{14}, m_{15})$, obtain its *minimal* SOP and POS expressions using the K-map method.

4.6. Use the K-map method to design a minimal SOP two-stage adder, that is, a circuit to form the sum $s_1 s_0$ and the carry-out c_2 for two 2-bit addends $b_1 b_0$ and $a_1 a_0$. Note that c_0 is set to 0 and can therefore be ignored.

4.7. Derive minimal SOP expressions for each of the following functions:
(**a**) OR$(m_0, m_1, m_2, m_3, m_4, m_6, m_8, m_9, m_{10}, m_{11})$
(**b**) OR$(m_0, m_1, m_5, m_7, m_8, m_{10}, m_{14}, m_{15})$
(**c**) AND$(M_1, M_3, M_7, M_9, M_{11}, M_{13}, M_{14}, M_{15})$

4.8. Find the minimal POS expression for

$$\text{OR}(m_0, m_2, m_3, m_7, m_8, m_9, m_{10})$$

Is the answer unique?

The following four problems involve don't cares.

4.9. Let $a_3a_2a_1a_0$ be the bit string of an (8, 4, 2, 1)-BCD code. The design for a code converter from the (8, 4, 2, 1) code to the (2, 4, 2, 1) code (which is defined in Problem 1.19) is requested. Denoting by $b_3b_2b_1b_0$ the bit string of the (2, 4, 2, 1) code, give the truth tables of b_3, b_2, b_1, and b_0 as functions of a_3, a_2, a_1, and a_0, and draw the corresponding Karnaugh maps.

4.10. Find minimal SOP forms for the functions b_3, b_2, b_1, and b_0 obtained in Problem 4.9.

4.11. A (5, 4, 2, 1)-BCD code is defined exactly as the (2, 4, 2, 1)-BCD code (Problem 1.20) except that the value of the code string $a_3a_2a_1a_0$ is $a_35 + a_24 + a_12 + a_0$. Design a *display driver* which receives as input a 4-bit (5, 4, 2, 1) digit $a_3a_2a_1a_0$. The output consists of seven functions, one for each segment of the decimal digit as shown.

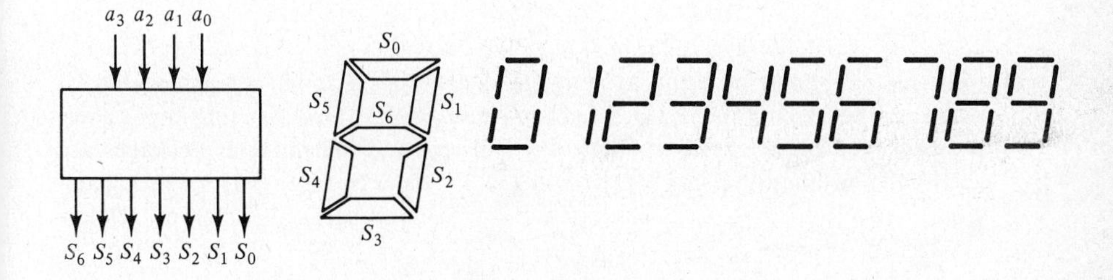

(**a**) Find minimal SOP expressions for $S_0, S_1, \ldots, S_6$.
(**b**) Find minimal POS expressions for $S_0, S_1, \ldots, S_6$.

4.12. Given the incompletely specified function of four variables OR$(m_1, m_2, m_3, m_5, m_{11})$, $\delta(m_6, m_7, m_8, m_9, m_{12}, m_{15})$:
(**a**) Find a minimal SOP expression.
(**b**) Find a minimal POS expression.
(**c**) Do the expressions obtained in (a) and (b) represent the same functions? Explain.

4.13. Design an adder cell (Figure 3.2) with five 2-to-1 multiplexers, one two-input OR gate, and one inverter.

4.14. Design an adder cell (Figure 3.2) using only 4-to-1 and 2-to-1 multiplexers. (Hint: Complementation can be realized by a 2-to-1 multiplexer.)

4.15. Consider an adder for 2-bit addends $a_1 a_0$ and $b_1 b_0$; $s_1 s_0$ is the sum and c_2 is the carry. Design the network using only 4-to-1 multiplexers. Try to use the smallest number of such multiplexers.

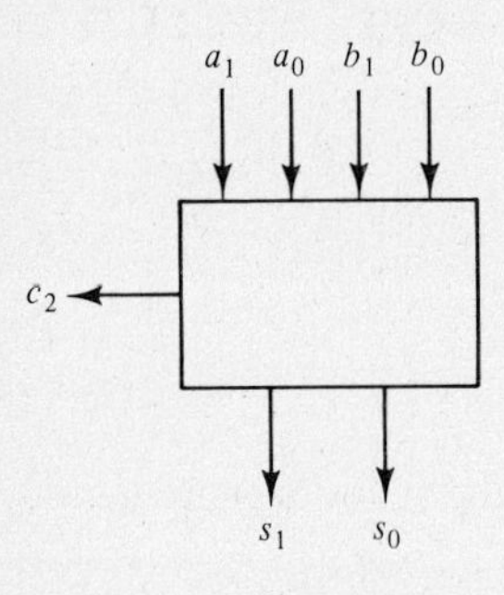

4.16. Following the method explained in the text, pp. 109 to 111, use an 8-to-1 multiplexer to realize the following function f:

x_3, x_4 (columns); x_1, x_2 (rows)

0	0	0	1
0	1	0	1
1	1	1	0
0	1	0	0

f

4.17. A 4-to-16 decoder is required. There is a choice between realizing a parallel decoder from discrete gates or using a complete decoder in a single integrated circuit package (the latter costs \$5.32). Dual four-input AND gates (two four-input gates per package) are available at \$0.48 per package, and four-inverter packages are also available (four inverters per package) at \$0.38 per package. The mounting and wiring costs are \$0.50 per package, and \$0.05 per connection (soldering). Determine the costs of the two methods.

4.18. We want to realize an 8-to-1 multiplexer with ENABLE input using *only* 2-to-1 multiplexers with ENABLE input. Draw the interconnection of a network that uses the least number of modules.

4.19. A combinational network has on its input lines the binary encoding of an integer x and on its output lines the binary encoding of an integer $y = f(x)$. Using the fewest multiplexers, design a network for $f(x) = 3x + 4$ and $0 \le x \le 7$.

4.20. Repeat Problem 4.19 for $f = x^2 + 5x + 1$ and $0 \le x \le 3$.

4.21. Repeat Problem 4.19 for $f = x^2 - 2x + 2$ and $0 \le x \le 7$.

Chapter 5

Sequential Networks

5.1 INTRODUCTION

In the two preceding chapters we have ideally assumed that the operation of combinational networks be instantaneous, that is, that no delay occurs between the time the input is applied and the time the output is available. Although physically untenable, such simplifying assumption was acceptable, due to the absence of feedback in combinational networks.

Another important element to consider is that information processing is essentially an activity that develops in time. Our introductory discussion of the operation of a digital computer in Chapter 2 is a significant illustration of this point. In other words, information processing is a sequence of actions, where the "current action" uses the results of "previous actions." This points to the necessity for *storing* data that must be available when needed. As another example, think of serial binary addition, which starts with the least-significant bits of the two operands and proceeds in order of increasing weight: obviously when adding the ith bits, we must have stored in some facility the carry bit produced when adding the $(i - 1)$st bits.

Therefore we shall begin by studying the behavior of real gates, which can be used to realize storage devices.

5.2 TIMING DIAGRAMS

From now on, each of our network variables will not be just a binary variable, but rather a *binary time waveform,* that is, a voltage function of time, which can assume at any specified instant one of the two values, which are respectively used

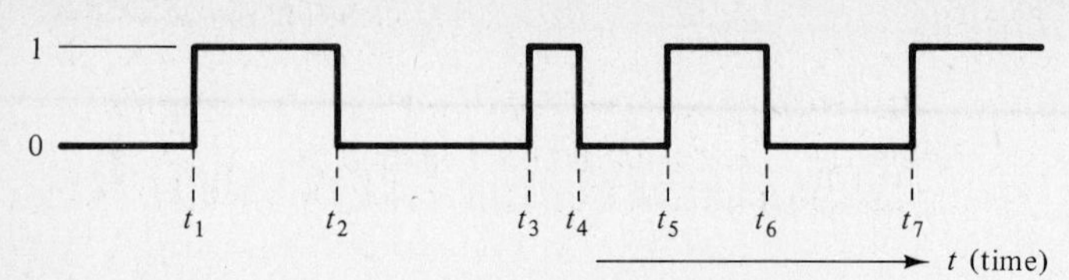

Figure 5.1 Representation of a binary waveform.

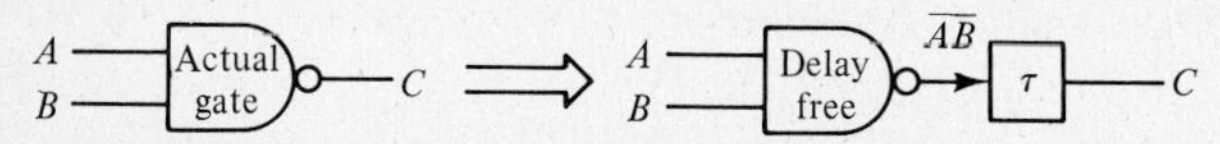

Figure 5.2 An actual gate and its model.

to represent the boolean values 1 and 0. In the following discussion, we shall assume positive logic, as defined in Sec. 4.1. With the usual convention about the representation of time (increasing from left to right), a typical binary waveform will appear as in Figure 5.1; notice that between t_i and t_{i+1} the value of the waveform is constant and that changes of value are assumed to be (ideally) instantaneous.

A circuit element—either a gate or an inverter—can no longer be assumed to have instantaneous transmission; rather, it will have a *nonzero propagation delay*. Thus we find it convenient to model an actual gate as a series connection of an ideal *delay-free gate* and a *pure delay element,* as shown in Figure 5.2, where $\tau > 0$ is the amount of delay.[1] The time analysis of a circuit element, such as the gate of Figure 5.2, is presented by means of *timing diagrams.* These are displays, *with respect to a common time reference,* of the waveforms observable at the input(s) and at the output of the circuit element. The mechanism to obtain the timing diagram is quite simple. Given the input waveforms, displayed as in Figure 5.1:

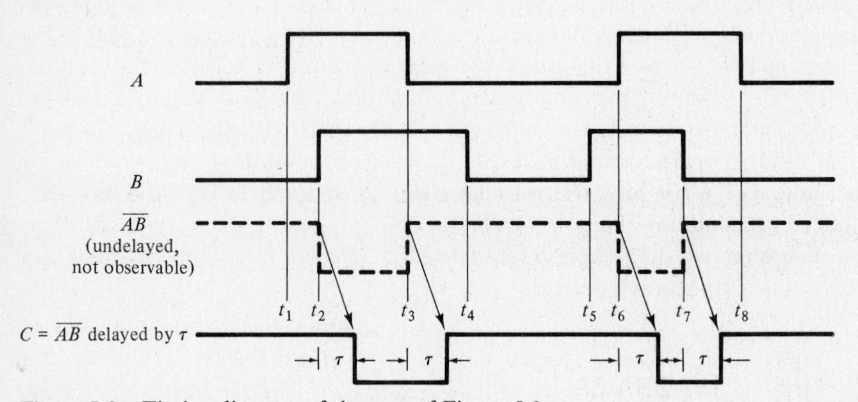

Figure 5.3 Timing diagram of the gate of Figure 5.2.

[1]Notice that this model still contains an important simplification, since it assumes that the same delay occurs for a $0 \rightarrow 1$ input change as for a $1 \rightarrow 0$ change, which may not be entirely realistic for many actual devices.

1. We construct at first its *undelayed* output waveform by computing, *instant by instant*, the specified logical function of its output values.
2. We translate to the right by the delay amount τ the undelayed waveform to obtain the actual waveform.

For the gate of Figure 5.2, application of these rules yields the waveforms of Figure 5.3. A and B are the input waveforms; note that the logical function $\overline{A \cdot B}$ is constant between t_i and t_{i+1}. Normally only the actual waveform is drawn in timing diagrams (the undelayed waveform is omitted). In dealing with networks consisting of several gates, the method described above is applied gate by gate; obviously a gate output waveform can be computed only when all of its input waveforms are known.

5.3 FEEDBACK AND MEMORY (LATCH)

We now apply the ideas developed in the previous section to show how "feedback" can be used to create storage devices.

Suppose now that we interconnect two NOR gates as shown in Figure 5.4(a), and next we add to it a *feedback* connection, which feeds the output of the downstream gate to one input of the upstream gate [Figure 5.4(b)]. We can conveniently redraw the network and bring to evidence the fact that the gates are not delay free: thus we obtain the two-input/two-output network of Figure 5.4(c), called a *latch,* which is the most elementary form of storage element and the fundamental component of all the storage elements to be studied later in this chapter. We shall now analyze the behavior of the latch, keeping in mind that, as a rule, $\tau_1 \neq \tau_2$.

The introduction of feedback may drastically change the nature of the network, when compared with those we studied in Chapters 3 and 4: in fact, if we write the switching expression of a gate output we may find it functionally dependent upon itself. [For example, if we compute the switching expression of y in Figure 5.4(c), we obtain $y = \overline{z + r} = \overline{\overline{s + y} + r} = (s + y)\bar{r}$. However, in $y = (s + y)\bar{r}$, the ys in the left and right side are *not* the same logical variable but values of the same waveform at *different* times, since gates have nonzero delays. Thus, the conventional description by switching expressions is inadequate.] Suppose, however, that we cut the feedback wire, thereby obtaining a conventional combinational network (see Figure 5.5), where one side (A) of the cut becomes an input and the other side (B) becomes an output. While keeping $r = s = 0$, we now apply a constant logical value, say 1, to input A; then $z = \overline{A + s} = \overline{1 + 0} = 0$ and $B = \overline{z + r} = \overline{0 + 0} = 1$, that is, A and B have both the value 1. This means that if we reconnect B to A, *the effect (that is, B) sustains its cause (that is, A),* that is, the value 1 on the feedback line is *stable.* Similarly, for the same condition $r = s = 0$, the value $A = 0$ is also self-sustaining or stable. In other words, while the external variables r and s are both 0, the logical value on the feedback line could be either 1 or 0, that is, the network can be in one of two distinct states for the same conditions on input lines r and s.

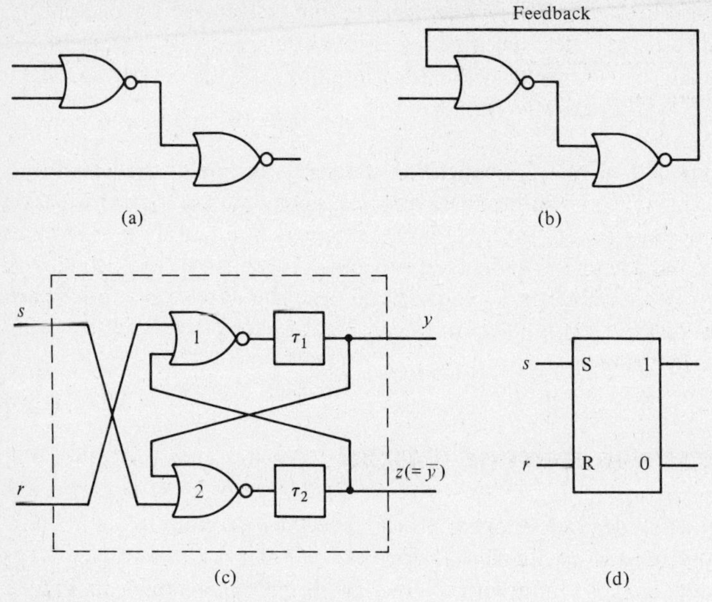

Figure 5.4 Construction of a latch.

Referring now to the network in Figure 5.4(c), we analyze what happens when the input signals r and s are not simultaneously 0. Assume that, at $t = 0$, $y = 1, z = 0$, and r and s have waveforms as shown in Figure 5.6(a). At $t = t_1$, the inputs to gate 1 are $r = 1$ and $z = 0$; then y becomes 0 at $t_1 + \tau_1$ since $\overline{r + z} = \overline{1 + 0} = 0$. At this time the inputs to gate 2 are $y = 0$ and $s = 0$, whence z becomes 1 at $t_1 + \tau_1 + \tau_2$ since $\overline{y + s} = \overline{0 + 0} = 1$. When r becomes 0 at $t_2 > t_1 + \tau_1 + \tau_2$ the inputs of gate 1 are now $r = 0$ and $z = 1$, whence the output $y = 0$ is maintained, that is, the latch has changed from $y = 1$ to $y = 0$ due to the action of r.

Next, assume that, at $t = 0$, $z = 1$, $y = 0$, and r and s have the waveforms shown in Figure 5.6(b). By an argument analogous to the preceding one, the latch changes between $t = t_1$ and $t = t_1 + \tau_1 + \tau_2$ from $y = 0$ to $y = 1$, due to the action of s.

Note that within the time interval $[t_1, t_2]$ in Figure 5.6, when either $r = 1$ or $s = 1$, the latch is changing state and there is a small interval of time $[t_1 + \tau_1, t_1 + \tau_1 + \tau_2]$ in Figure 5.6(a) and $[t_1 + \tau_2, t_1 + \tau_2 + \tau_1]$ in Figure 5.6(b) during which y and z have the same value. However, outside the time interval $[t_1, t_2]$, the

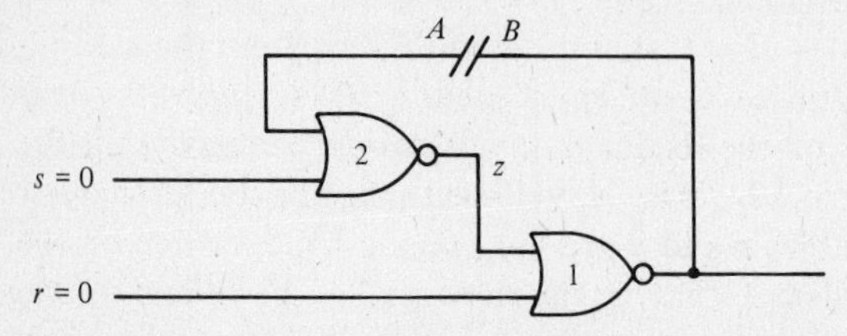

Figure 5.5 Cutting the feedback loop.

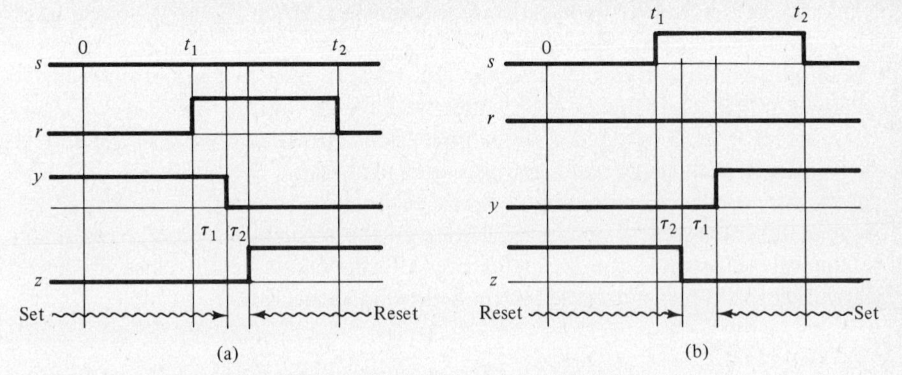

Figure 5.6 Timing diagram of the state transitions of a latch.

latch is in a stable state as previously defined and, in such condition, y and z are the complement of each other; for this reason z is usually referred to as $\bar{y}$. We also say:

> In the stable condition, the latch is said to be *set* when $y = 1$, and is said to be *reset* when $y = 0$. For this reason, s and r are respectively called the SET and RESET signals. The output y is called the *uncomplemented* output of the latch, while $\bar{y}$ is called the *complemented* output.

Normally the four terminals of the latch are labeled as follows: S for the set input, R for the reset input, 1 for the true output, and 0 for the complemented output.

Suppose now that r and s are both 1 and simultaneously change to 0. Then we shall obtain $y = 0$ or $y = 1$, depending upon whether $\tau_1 > \tau_2$ or vice versa. Clearly the operation is unreliable, so that we shall *exclude* as *illegal the configuration* $r = s = 1$. (Equivalently, we say that for the latch the condition $rs = 0$ always holds.)

In summary the latch has output $y = 0$ or $y = 1$, depending upon whether $r = 1$ or $s = 1$ occurred *last*, that is, the network remembers its input (it "latches" on it); this shows how feedback is essential for the appearance of memory in digital circuits. The circuit symbol of the latch appears in Figure 5.4(d).

*5.4 ASYNCHRONOUS SEQUENTIAL NETWORKS: A BRIEF DISCUSSION

The preceding analysis and discussion of the set-reset latch contains, in a nutshell, all the fundamental features of an extremely important class of digital networks which arise by introducing feedback in combinational networks. These

*(*Reading of this section (and all sections marked with an asterisk) can be entirely omitted, since no notion developed in later sections makes explicit reference to it; the reader who so chooses, may skip to Sec. 5.5.*)

networks are called *asynchronous sequential networks* (ASN): "sequential," because they are described by their behavior in time, "asynchronous," because the timing of network events is completely determined by the unsynchronized behavior of the input signals, over which we have no control.

As is typical of engineering applications, the study of ASNs could be conveniently subdivided into analysis and synthesis. The synthesis of ASNs makes use of rather complex and subtle techniques whose presentation is well beyond our introductory objectives. However, the procedure for analyzing ASNs adequately illustrates the basic features of this fundamental class of digital networks; this topic is discussed in the following subsection.

5.4.1 Analysis of Asynchronous Sequential Networks

The analysis of a given ASN has as its objective the interpretation of the behavior of the network. We begin from the network diagram, which is an assembly of logical gates interconnected according to the rules of combinational networks, *except the one that prohibits feedback connections.*

We shall illustrate the analysis procedure by means of a running example. Let the network of Figure 5.7 be given. The first step is to ideally cut feedback lines in order to transform the network into a feedback-free combinational one. Since each line we cut will carry an (internal) network variable, we would like to cut as few feedback lines as possible while pursuing the above objective. In some cases, this step is obvious; in other cases, it may be very complicated, and there are general procedures for this task, whose presentation, however, exceeds the scope of this text.[2] Our example is an easy case: there are two feedback lines shown cut in Figure 5.7(b).

It is important to observe the following:

Each feedback line (to be ideally cut) carries an *internal* (or *state*) *variable* y_i of the network. When we cut a line, it is convenient to assign the variable y_i to the input side of the cut, and to introduce a new variable Y_i on the output side. Y_i, which is a boolean function of the input variables $x_1, x_2, \ldots, x_n$ and of the *current* values $y_1, y_2, \ldots, y_r$ of the internal variables, represents the *future* (or *next-state*) value of y_i.

Once we have cut the feedback lines we can derive the expressions of the next state functions $Y_1, \ldots, Y_r$. In our example we have

$$Y_1 = x_1y_1 + x_2\bar{x}_1\bar{y}_2$$

$$Y_2 = x_2y_2 + x_1y_2y_1 + x_1x_2y_1$$

which we can also express in truth table form as shown in Figure 5.8(a). This truth table is conveniently repackaged in the table shown in Figure 5.8(b). (A trick that greatly aids this repackaging is to construct the truth table by

[2]See, for example, J. P. Perrin, M. Denouette, and E. Daclin, *Systèmes Logiques,* II, Dunod, Paris, pp. 253–258.

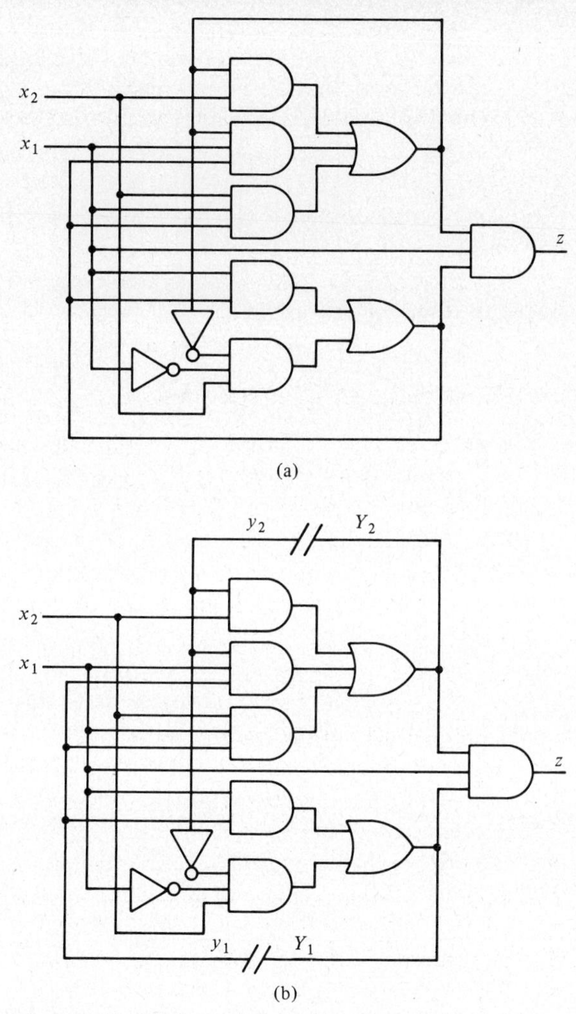

Figure 5.7 (a) A network with feedback to be analyzed. (b) The feedback lines are cut.

arranging the independent variables in the order $x_n x_{n-1} \ldots x_1 y_r y_{r-1} \ldots y_1$.) The table we have just obtained is a double-entry table, whose column headings are the input configurations and whose row headings are the configurations of the internal variables also called *internal states*. At the intersection of a row and of a column (a state-input pair or *total state*), we have as entry the values of the next state functions, that is, the configuration of the internal variables to which the network will make a transition as a result of the present state-input pair. For this reason the table is named the *transition table*. It is therefore obvious that if the values of $Y_r \ldots Y_1$ coincide with the row heading, we have a self-sustaining state (indeed the future values of $y_1, \ldots y_r$ coincide with the present values). Specifically:

A total state is *stable* when $Y_i = y_i$ for $i = 1, 2, \ldots, r$.

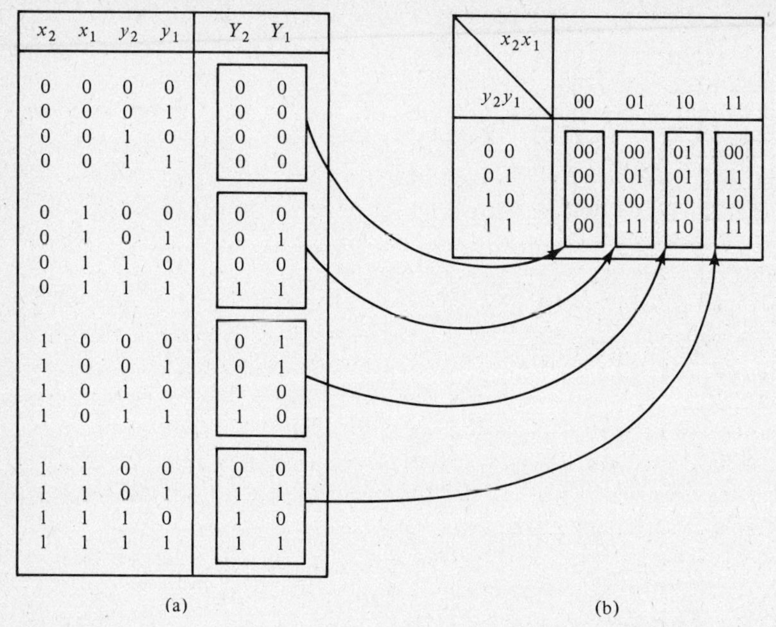

x_2	x_1	y_2	y_1	Y_2	Y_1
0	0	0	0	0	0
0	0	0	1	0	0
0	0	1	0	0	0
0	0	1	1	0	0
0	1	0	0	0	0
0	1	0	1	0	1
0	1	1	0	0	0
0	1	1	1	1	1
1	0	0	0	0	1
1	0	0	1	0	1
1	0	1	0	1	0
1	0	1	1	1	0
1	1	0	0	0	0
1	1	0	1	1	1
1	1	1	0	1	0
1	1	1	1	1	1

(a)

y_2y_1 \ x_2x_1	00	01	10	11
0 0	00	00	01	00
0 1	00	01	01	11
1 0	00	00	10	10
1 1	00	11	10	11

(b)

Figure 5.8 Truth tables of the next-state functions (a), repackaged as a "transition table" in (b).

Stable total states are normally shown encircled in the transition table which is reproduced, with this feature, in Figure 5.9, where we have also added the values of the output [i.e., each entry is a pair (next-state, present output)].

At this point, we have extracted from the network all the relevant information. Before proceeding, however, with the analysis of its behavior we must specify the regime of the input signals. In particular, we say that:

> An asynchronous sequential network operates in the *fundamental mode* when:
>
> **1.** Input variables change instantaneously.
> **2.** No more than one input variable changes at any given instant.

y_2y_1 \ x_2x_1 (Y_2Y_1, z)	00	01	10	11
0 0	00 , 0	00 , 0	01 , 0	00 , 0
0 1	00 , 0	01 , 0	01 , 0	11 , 0
1 0	00 , 0	00 , 0	10 , 0	10 , 0
1 1	00 , 0	11 , 1	10 , 0	11 , 1

Race

Figure 5.9 The transition-output table with encircled stable total states.

In the rest of Sec. 5.4, we shall only consider ASN in the fundamental mode. Assume that we are currently in a stable state (say, total state $x_2x_1y_2y_1 = 1010$); if an input variable changes (say x_1 changes from 0 to 1), property (1) of instantaneous input changes means that the total state of the network instantaneously becomes $x_2x_1y_2y_1 = 1110$, that is, we move *horizontally* in the table. If the new total state is stable, then the transition is completed (indeed in this case, the values of the next-state functions coincide with the corresponding present values of the internal variables, so no internal variable change takes place). Assume instead that the network is in stable total state $x_2x_1y_2y_1 = 1001$, and that x_1 changes from 0 to 1. In this case we move to total state $x_2x_1y_2y_1 = 1101$, which is unstable; indeed $Y_2Y_1 = 11 \neq 01$. At this point, it is essential to observe that if *only one internal variable is called to change* (in this case y_2), then the specified configuration $y_2y_1 = 11$ *will be attained* at the completion of the propagation of the signals through the network combinational logic. If, however, more than one internal variable is called to change, such as from total state $x_2x_1y_2y_1 = 0111$ when x_1 changes from 1 to 0 ($Y_2Y_1 = 00$), then, depending upon which internal variable attains first its intended value, the network will evolve differently. Specifically, in changing from $y_2y_1 = 11$ to $y_2y_1 = 00$, if y_2 changes first we temporarily obtain $y_2y_1 = 01$, otherwise we have $y_2y_1 = 10$. In other words, the two internal variables, which are called to change, enter a race. In such situation, the transition table is not fully adequate to analyze the network behavior and a more careful examination must be based on actually tracing the waveforms through the network logic. In general a race condition represents unreliable and hazardous behavior and should be absolutely avoided in design practice (unless a more careful analysis proves it to be harmless).

The reader may verify in Figure 5.9 that the only transition from a stable total state which causes a race is the one we have just mentioned; this race, however, is not serious (*noncritical race*), because it takes place in a column (heading 00) where all entries are $Y_2Y_1 = 00$, whence the network will eventually settle in total state $x_2x_1y_2y_1 = 0000$.

The analysis is almost complete, except that it may not be so easy to say "What the network does." Suppose, however, that we are told that x_1 is a periodic waveform which is equal to 1 for a small fraction of the period [Figure 5.10(a)] and that x_2 is a random input waveform, whose changes are spaced by a duration longer than the period of x_1. Then, we can sketch a typical sample of x_1 and x_2 [Figure 5.10(a)] and plot, with the aid of the transition-output table of Figure 5.9, the time sequence of internal states and outputs. (Note that new stable states are realized with a small delay after the input change.)

As an illustration, the sequence of the first six total stable states is shown directly on the transition table for the given example in Figure 5.10(b). Observing the output waveform z, we conclude that the network generates for each input "pulse" of arbitrary timing and duration, a single "pulse" of fixed duration synchronized with the waveform x_1. The network is a "pulse synchronizer," a very important module at the interface between the outside world (unsynchronized) and complex digital systems (normally synchronized).

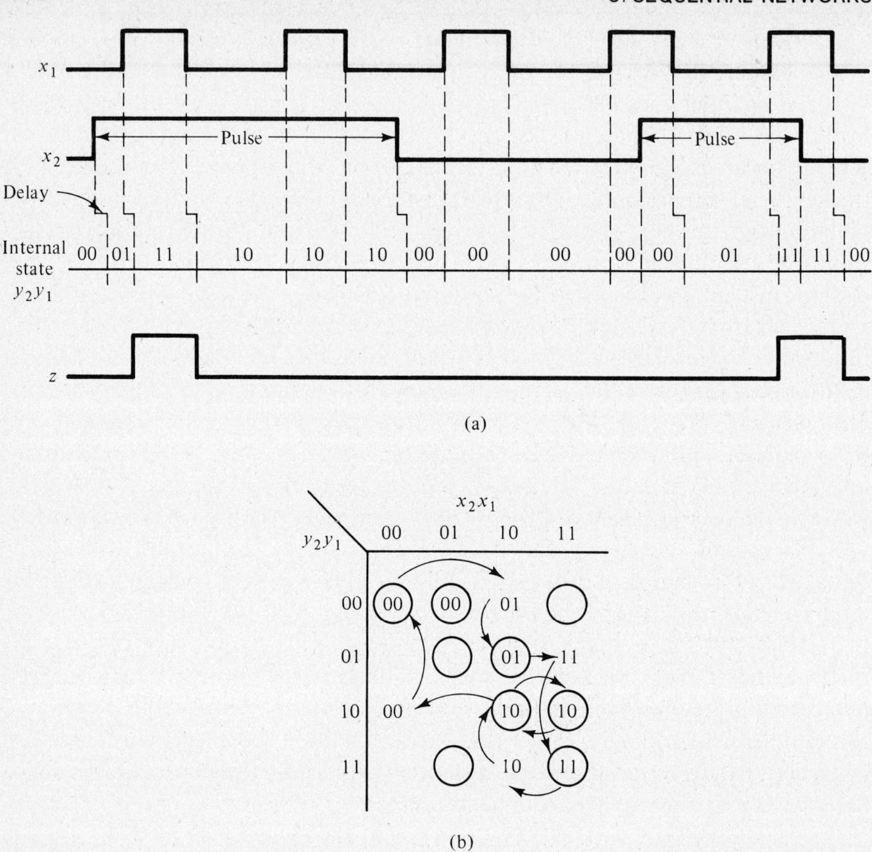

Figure 5.10 (a) Typical sample of input waveforms and corresponding output. (b) Transition sequence on transition table.

In summary, the analysis proceeds through the chain of steps illustrated in Figure 5.11.

We conclude this section with two more interesting examples. The first is the network of Figure 5.12(a). This network contains a pushbutton, which is a device frequently used to generate a boolean variable. This boolean variable is meant to be 1 when the button is pushed, and it is 0 otherwise. We must be alerted, however, to the fact that most pushbuttons exhibit the phenomenon of "contact bouncing," that is, due to microscopic imperfections of contact points, the switch, after making initial contact, bounces several times before establishing a permanent contact. Letting V_1 and 0 denoting, respectively, logical 1 and 0,

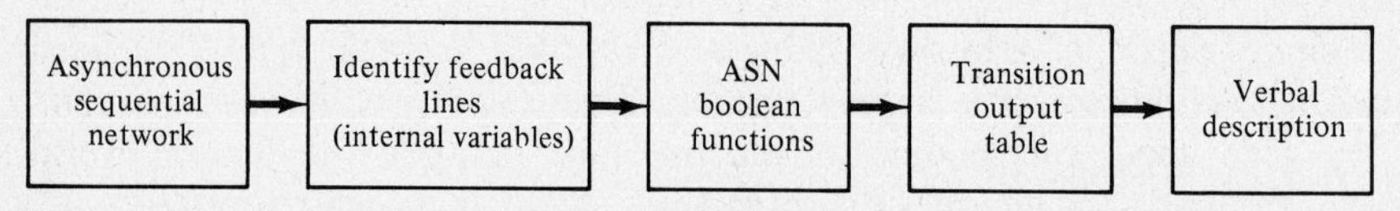

Figure 5.11 Steps in the analysis of an asynchronous sequential network.

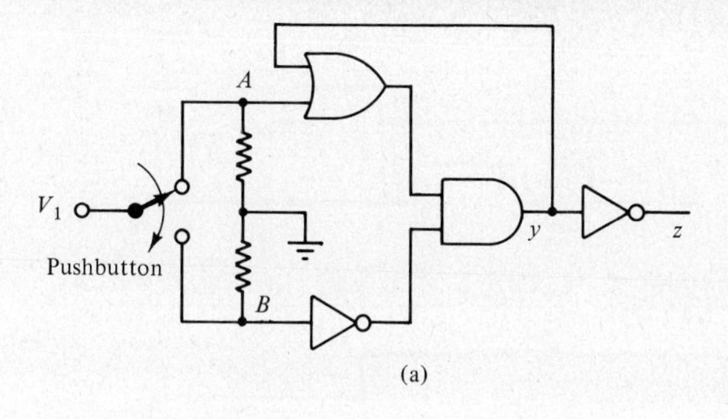

(a)

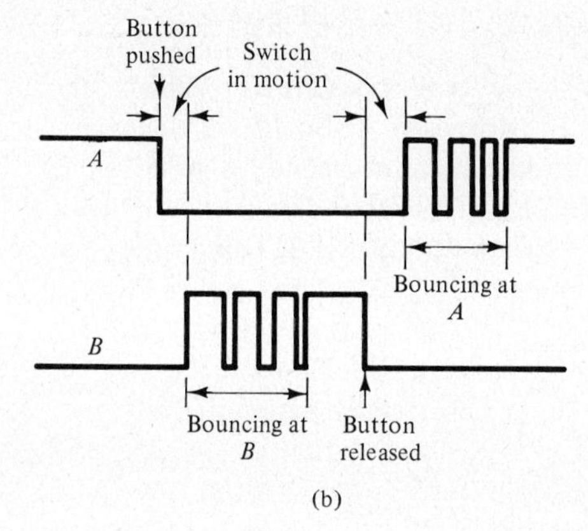

(b)

Figure 5.12 Example of ASN (contact debouncer).

typical waveforms at A and B (which are the input waveforms to our ASN) are as shown in Figure 5.12(b). Let us analyze the behavior of the ASN.

Obviously there is just one feedback line, named y. Next we obtain the network boolean equations:

$$Y = \overline{B}(A + y)$$
$$z = \overline{y}$$

The truth tables of Y and z, as well as the transition-output table are readily obtained:

A	B	y	Y	z
0	0	0	0	1
0	0	1	1	0
0	1	0	0	1
0	1	1	0	0
1	0	0	1	1
1	0	1	1	0
1	1	0	0	1
1	1	1	0	0

y \ AB	00	01	10	11
0	⓪ , 1	⓪ , 1	1 , 1	⓪ , 1
1	① , 0	0 , 0	① , 0	0 , 0

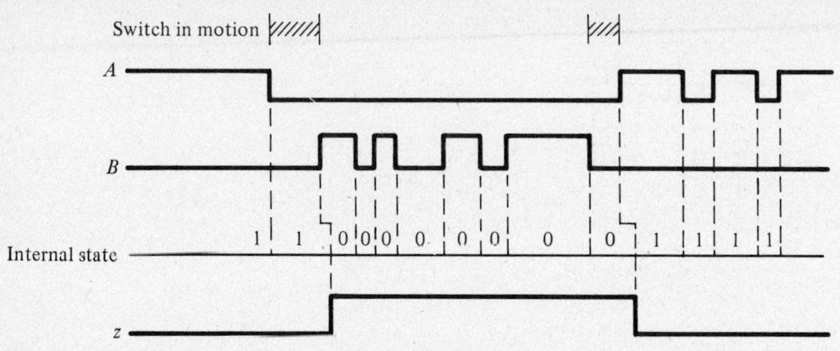

Figure 5.13 Timing analysis of the sequential network of Figure 5.12.

In the transition-output table, the stable total states are also shown. We are now ready to interpret the behavior of the proposed ASN. This is done, as usual, by deriving the behavior (state and output) corresponding to a typical input sequence. This is shown in Figure 5.13. (Note that total states ABy = 110,111 never occur since A and B can not be simultaneously 1, being the terminals of a transfer switch.) Clearly the output of the network is—apart from negligible delays at the start and at the end—a "clean" version of the waveform at input B, that is, the ASN acts as a "contact debouncer." Such network is very important and is normally deployed when switches—specially manually operated switches such as pushbuttons—provide inputs to complex digital systems.

The second example is the network of Figure 5.14. This network has two inputs, c and x, two outputs, s and r, and contains two feedback lines y_1 and y_2, respectively, coincident with s and r. By inspection of the network, we obtain the network boolean equations:

$$s = Y_1 = \overline{\bar{c} + y_1 + \overline{\overline{\bar{x} + y_2}}} = c(y_1 + \overline{\bar{x} + y_2}) = c(y_1 + x\bar{y}_2) = cy_1 + cx\bar{y}_2$$
$$r = Y_2 = \overline{\bar{c} + y_1 + \overline{\bar{x} + y_2}} = c\bar{y}_1(\bar{x} + y_2) = c\bar{y}_1 y_2 + c\bar{x}\bar{y}_1$$

From these we readily derive the following truth tables (left) of Y_1 and Y_2, and the transition-output table (right) of the network (the output is not shown since,

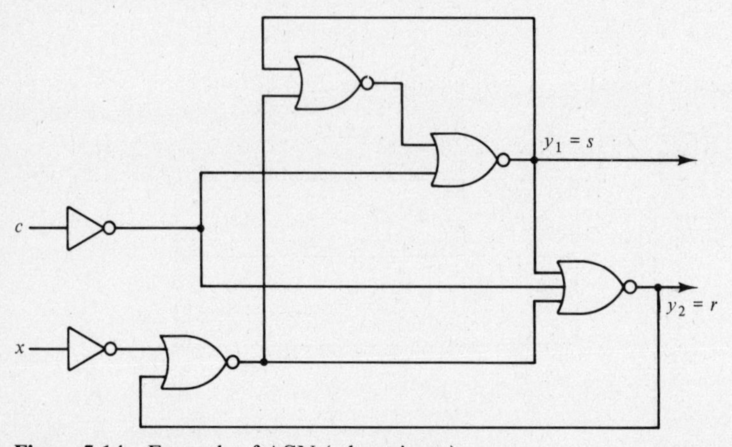

Figure 5.14 Example of ASN (edge trigger).

by hypothesis, $s = y_1$ and $r = y_2$):

x	c	y_2	y_1	Y_2	Y_1
0	0	0	0	0	0
0	0	0	1	0	0
0	0	1	0	0	0
0	0	1	1	0	0
0	1	0	0	1	0
0	1	0	1	0	1
0	1	1	0	1	0
0	1	1	1	0	1
1	0	0	0	0	0
1	0	0	1	0	0
1	0	1	0	0	0
1	0	1	1	0	0
1	1	0	0	0	1
1	1	0	1	0	1
1	1	1	0	1	0
1	1	1	1	0	1

y_2y_1 \ xc	00	01	10	11
00	(00)	10	(00)	01
01	00	(01)	00	(01)
10	00	(10)	00	(10)
11	00	01	00	01

The stable total states are shown encircled in the transition table.

To interpret the behavior of the network it is convenient to think of x as of a random (data) waveform, while c is a periodic (synchronizing) waveform. A total state is denoted by the quadruple (xc, y_2y_1). If $c = 0$ initially, the network will always make a transition to $y_2y_1 = 00$; then, since $y_2y_1 = 11$ is unreachable from any of the other states, we may ignore altogether state $y_2y_1 = 11$. Next, referring to Figure 5.15, we note that, as long as $c = 0$, the network may oscillate between total states (00, 00) and (10, 00) depending upon the value of x, but no output signal is produced. Consider now a 0 → 1 change of c (see again Figure 5.15) and assume at first that $x = 0$ (time t_1). The network then makes a transition to total

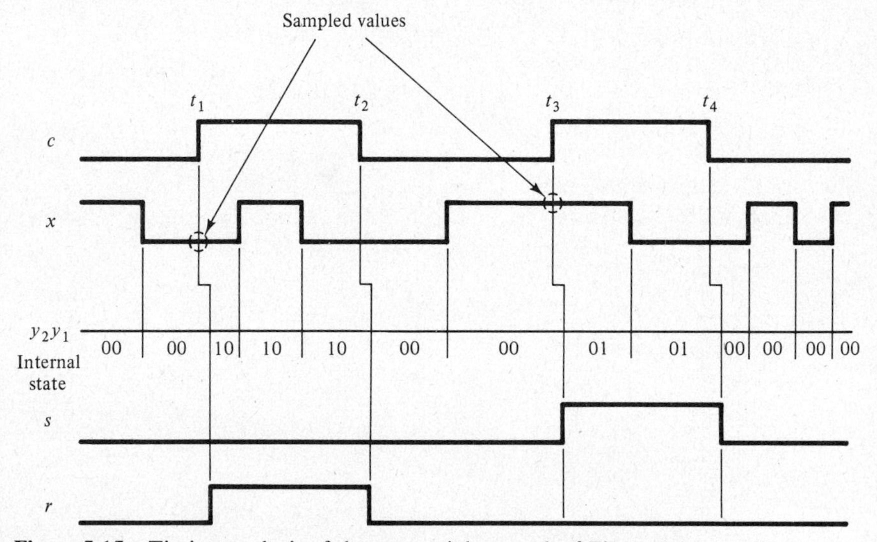

Figure 5.15 Timing analysis of the sequential network of Figure 5.14.

state (01, 10), where the output r becomes 1; while c remains equal to 1, if x oscillates between 0 and 1, the total state correspondingly oscillates between (01, 10) and (11, 10), but the output r remains at 1; the output returns to 0 only when c becomes 0 (time t_2). Next, assume that $x = 1$ when c makes a $0 \rightarrow 1$ transition (time t_3). In this case the total state changes from (10, 00) to (11, 01), and the output s becomes 1; as before, the output s remains constant at 1 while x may oscillate, and returns to 0 only when c also returns to 0 (at time t_4). From this analysis we deduce that the output is completely determined by the synchronizing waveform c and by the value of x (shown encircled in Figure 5.15) at the instant that c makes a $0 \rightarrow 1$ transition: in other words, *the* $0 \rightarrow 1$ *edge of* c *samples the value of* x. For this reason this circuit is known as an "edge trigger" and is extremely important in the realization of synchronous systems.

Note: Asynchronous sequential networks are rather delicate systems, in general. Indeed, their operation rests exclusively on the propagation of signals on feedback paths. Therefore, when the size of the problem grows (large numbers of input lines, internal variables and output lines), not only does the design become exorbitantly complicated, but the reliability of the operation is very much in doubt. Thus, very large sequential systems—such as a digital computer—are not designed as asynchronous networks, but rather as synchronous networks, where a synchronizing waveform is used to time all state transitions (as we shall discuss in the rest of this chapter). However, for small systems, not only asynchronous design is feasible and reliable, but in many cases it is the only possibility; think, for example, of a vending machine or an elevator control, where the input events are governed by the manual intervention of a human user. For such cases synchronous design would require not only a synchronizing waveform but also additional asynchronous pulse synchronizer of the type described above. Such approach is totally unrealistic and uneconomical, and should be dismissed.

5.5 GATED LATCHES AND FLIP-FLOPS (MASTER–SLAVE AND EDGE TRIGGERED)

We reconsider the basic storage element, the latch, introduced in Sec. 5.3 and modify it by gating the two input signals, SET and RESET, by means of a new signal called CLOCK. [Refer to Figure 5.16(a), where the gating is clearly

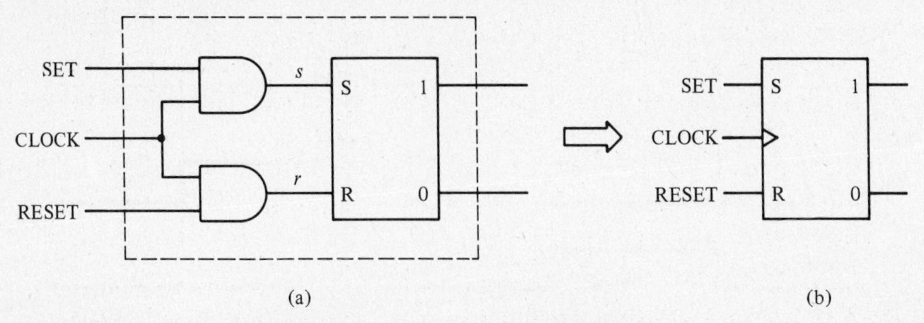

Figure 5.16 Structure and circuit symbol of clocked SR F/F.

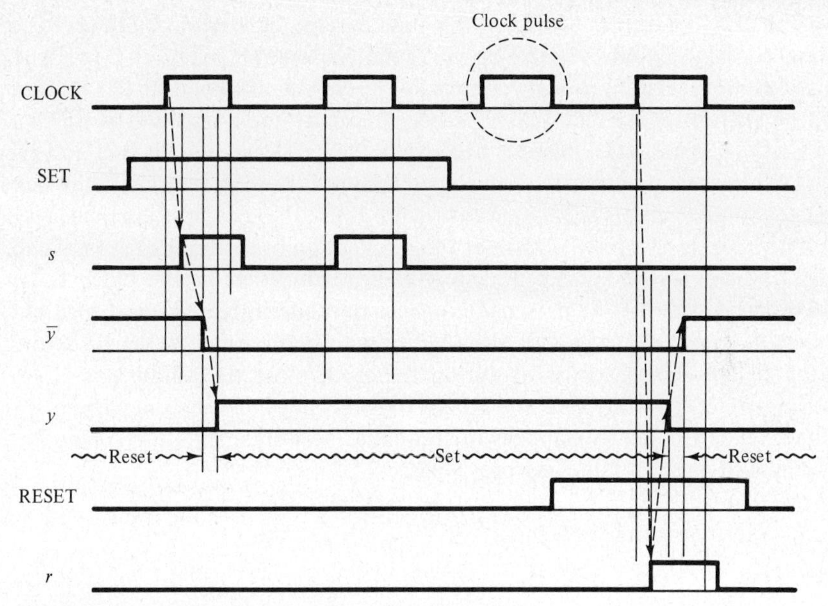

Figure 5.17 Timing diagram of the gated SR F/F.

accomplished by means of two AND gates.] CLOCK is a timing waveform whose function is described as follows:

A clock is a waveform that controls the instants at which a storage element can change its state.

Normally, CLOCK is a periodic waveform, although there is no essential need that it be so. A transition $0 \rightarrow 1 \rightarrow 0$ of CLOCK is called a *clock pulse,* and clearly the input AND gates in Figure 5.16(a) are permissive only during a clock pulse. The network shown in Figure 5.16(a) is called a *gated set–reset flip-flop*[3] (SR flip-flop, or SR F/F, for short) and is an extremely important building block of digital systems. Its usual circuit symbol is drawn in Figure 5.16(b).

We now illustrate the behavior of the clocked SR F/F, by means of a timing diagram (refer to Figure 5.17). The flip-flop is initially assumed in its reset state. We observe that, at the leading edge of the clock pulse (i.e., when the CLOCK changes from 0 to 1) the input AND gates are made permissive. This generates a rapid sequence of changes of the signals s, $\bar{y}$, and y (in this order). Thus, except for a very small delay, the *leading edge* of the clock pulse controls the timing of change of state of the flip-flop. An analogous sequence of events occurs when the flip-flop is to be reset. Notice, incidentally, that the second pulse on the s line does not affect the state of the flip-flop which is already in its set state.

The gated latch, as a type of flip-flop, has a shortcoming that, in many realizations, may be fatal to the correct functioning of the system. Indeed, let us

[3]Frequently, this type of component is referred to as *gated set–reset latch.*

revisit the timing diagram of Figure 5.17 and assume that, while CLOCK = 1, the input signals SET and RESET *do not* remain constant (see Figure 5.18). The initial transition, yielding $y = 1$ at time t_1, proceeds exactly as in the left portion of Figure 5.17. Suppose then that SET becomes 0 at time t_2 and that at time t_3, while CLOCK is still at 1, RESET becomes 1. This unleashes a chain of actions leading to a *reset* state at time t_4 (which persists until the termination of the clock pulse).

This reveals a crucial feature of the clocked latch: the latch is sensitive to the SET and RESET signals during the entire duration of the clock pulse (*window of sensitivity*). This sensitivity has no undesirable effect if one can guarantee—from an analysis of the entire digital system—that the SET and RESET signals remain constant during the clock pulse. However, this is not always possible when the SET and RESET signals depend upon the output y of the flip-flop, as frequently happens in sequential networks. This can sometimes be obviated by the two following provisions:

1. Shortening the clock pulse.
2. Artificially delaying the ensuing changes of the SET and RESET signals.

However, (as we shall discuss later in Sec. 5.8), this is not always possible or reliable, and some more robust solutions must be adopted. These solutions are the master–slave flip-flops and the edge-triggered flip-flops. Of these we shall describe in some detail below the master–slave flip-flops. For the edge-triggered SR F/Fs suffice it to say the s and r signals fed to the latch are determined by the

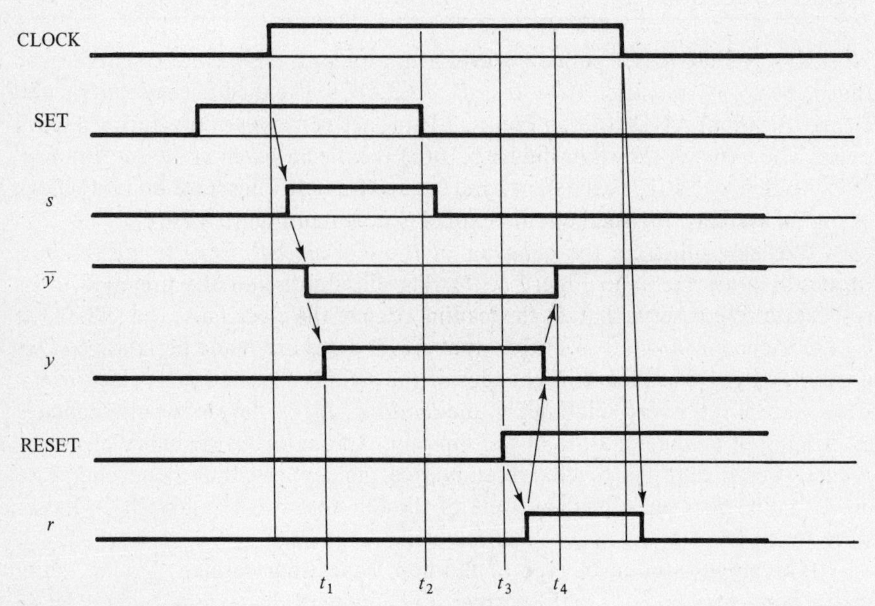

Figure 5.18 Timing behavior of the gated SR F/F if s and r vary during the clock pulse.

instantaneous sampling of the SET and RESET signals; this sampling is effected by the *leading edge* of the clock pulse, thereby effectively reducing to zero the window of sensitivity of the flip-flop.

5.5.1 Master–Slave Flip-Flops

In the master–slave interconnection, the flip-flop changes its state *after the trailing edge* of the clock pulse, so that here again the window of sensitivity is effectively reduced to zero. The master–slave SR F/F consists of two gated SR F/Fs interconnected as in Figure 5.19(a). Referring to the timing diagram of Figure 5.19(b), the clock C' is the (minutely delayed) complement of CLOCK (the slight delay is caused by the inverter). Notice that the master F/F will make a transition during the interval $[t_0, t_1]$, whereas the slave F/F will reproduce this transition only when $C' = 1$, that is, during the interval $[t_1, t_2]$. The total effect is that y changes after the trailing edges of CLOCK.

In summary, we have three major types of 1-bit storage elements:

1. The *latch,* which is unsynchronized and permanently sensitive to the input signals.

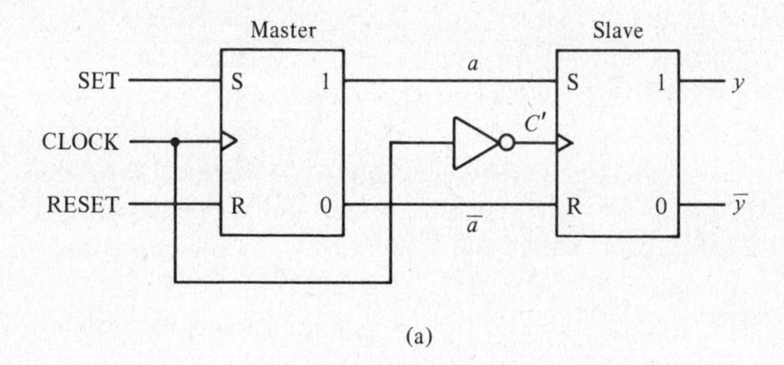

(a)

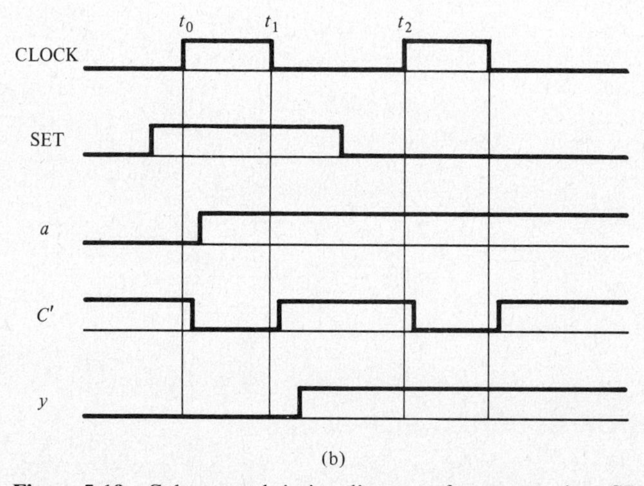

(b)

Figure 5.19 Scheme and timing diagram of a master–slave SR F/F.

2. The *gated flip-flop* (*gated latch*), where a clock waveform controls the width of the window of sensitivity to the input signals.
3. The *master–slave* and *edge-triggered flip-flops,* where additional circuitry reduces to zero the width of the window of sensitivity. The master–slave F/F switches at the trailing edge of the clock pulse, whereas the edge-triggered F/F switches at its leading edge.

With regard to circuit symbols we shall differentiate between the latch [Figure 5.4(d)] and the other two types [Figure 5.16(b)]. Indeed the gated flip-flop and the master–slave or edge-triggered flip-flops are not *functionally* different: the only differences among them are the timing reliability and the timing of switching with respect to the clock pulse (leading-edge switching for the gated and edge-triggered flip-flop, trailing-edge switching for the master–slave flip-flop).

It is now appropriate to introduce a notational convention that will be used throughout this chapter for all clocked flip-flops:

> Given a clocked flip-flop, y denotes the *present state* of the flip-flop, that is, the state *before* the next clock pulse, while Y denotes the *next state* of the flip-flop, that is, the value that y will have *after* the occurrence of the next clock pulse.

Notice that Y is not a variable observable on any flip-flop wire.

We recognize that the behavior of the SR F/F is completely described by specifying how Y (next state) depends upon y (present state) and the input signals (present input), from now on referred to as s and r. This dependence is shown in Figure 5.20, as a truth table called the *Y-table,* and by the corresponding boolean expression, referred to as the *next-state equations.* Note that since the condition $(r, s) = (1, 1)$ is ruled out, the corresponding entries in the truth table of Y are don't cares. As a consequence we obtain

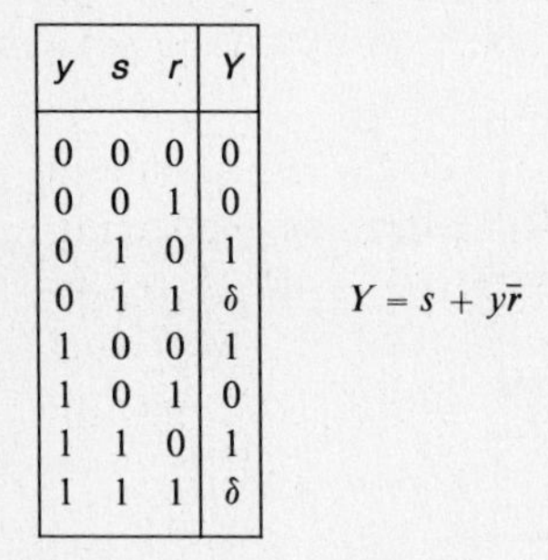

y	s	r	Y
0	0	0	0
0	0	1	0
0	1	0	1
0	1	1	δ
1	0	0	1
1	0	1	0
1	1	0	1
1	1	1	δ

$$Y = s + y\bar{r}$$

Figure 5.20 Y-table and next-state equation for the SR F/F.

5.6 OTHER TYPES OF CLOCKED FLIP-FLOPS

The SR F/F introduced in Sec. 5.5 is the prototype of clocked flip-flops. Its input logic can be further modified, thereby obtaining other types of clocked flip-flops, each of which has several attractive features. The flip-flops discussed in this section are all of the gated type; with trivial modification one can obtain their master–slave versions.

The first to be considered is the *delay flip-flop* (D F/F). It is obtained from a conventional SR F/F by adding an appropriate input steering logic as shown in Figure 5.21(a). The complementary nature of the s and r inputs guarantees the automatic satisfaction of the condition $sr = 0$. The compact circuit symbol is given in Figure 5.21(b), where the input is labeled d (mnemonic for "delay"). Notice that, since $s = d$ and $r = \bar{d}$, the next-state equation of the SR F/F, $Y = s + y\bar{r}$, is transformed into

$$Y = d + y\bar{\bar{d}} = d + yd = d$$

Thus,

$$Y = d$$

is the next-state equation of the D F/F, which shows that y is the replica, one clock period *later,* of the *present* value of y. This explains the denotation "delay flip-flop."

Next, we shall modify the SR F/F, whose Y-table is given in Figure 5.20, in a way that puts the "forbidden" input configurations to work. The resulting flip-flop, called JK flip-flop, is a versatile generalization of the SR F/F. The first modification is shown in Figure 5.22(a). From this diagram, we obtain $s = j\bar{y}$ and $r = ky$; recalling the next-state equation $Y = s + y\bar{r}$ of the SR F/F, we obtain the corresponding equation of the JK F/F, as follows (the Y-table of the JK F/F appears in Figure 5.23):

$$Y = s + y\bar{r} = j\bar{y} + y\overline{ky} = j\bar{y} + y(\bar{k} + \bar{y})$$

that is,

$$Y = j\bar{y} + \bar{k}y$$

After explicitly showing the structure of the SR F/F in Figure 5.22(b), we realize that AND gates can be combined as in Figure 5.22(c). The circuit symbol of the JK F/F is shown in Figure 5.22(d).

Finally, we observe that the JK F/F can be specialized to a storage device that changes its state any time it receives an input pulse. Indeed by connecting together the J and K inputs we obtain the *toggle flip-flop* (T F/F), shown in Figure 5.24. To obtain the next-state equation of the T F/F, we make substitutions in the next-state equation $Y = \bar{y}j + y\bar{k}$ of the JK F/F according to

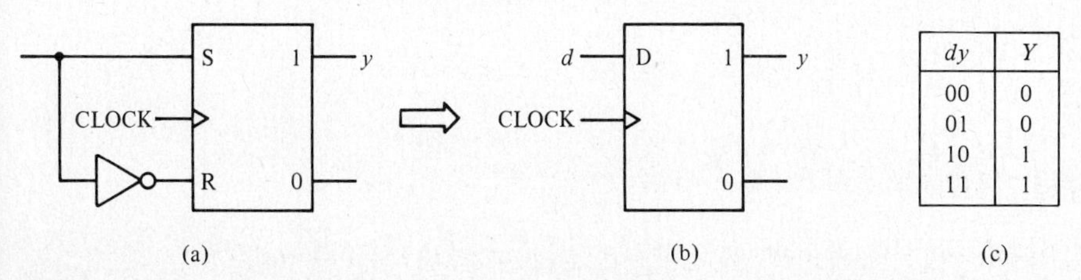

Figure 5.21 Diagram of a D F/F, its conventional circuit symbol, and its Y-table.

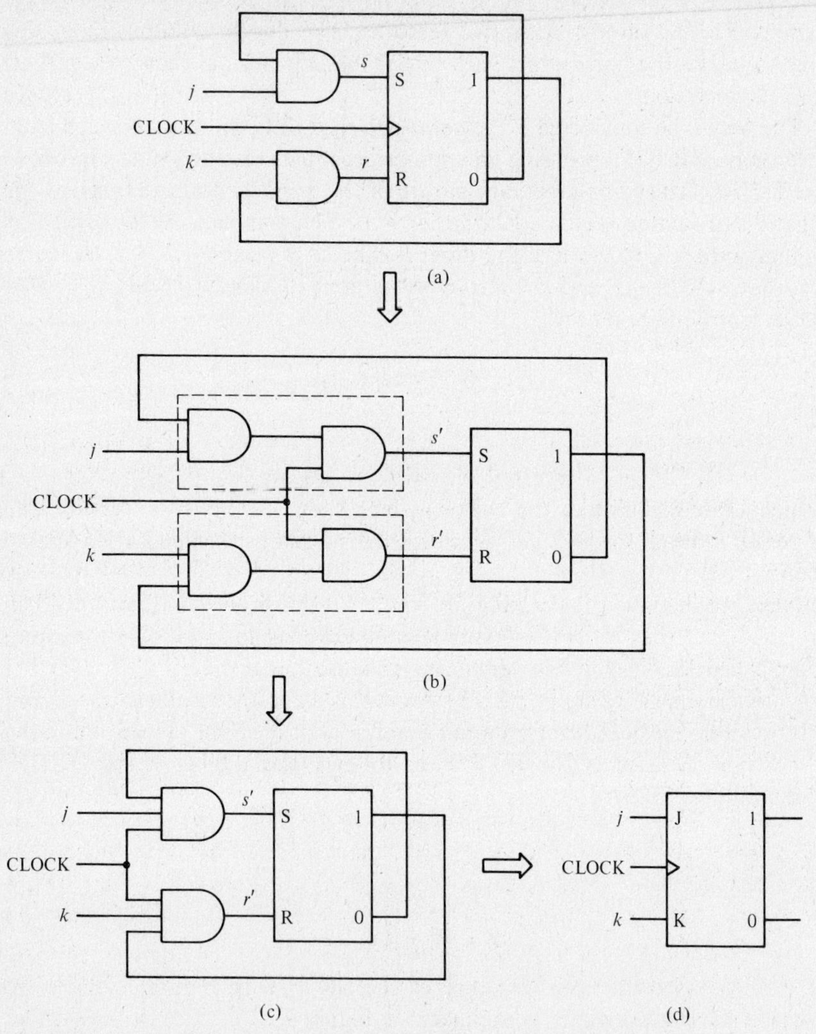

Figure 5.22 Construction of the JK F/F and its compact circuit symbol.

y	j	k	Y
0	0	0	0
0	0	1	0
0	1	0	1
0	1	1	1
1	0	0	1
1	0	1	0
1	1	0	1
1	1	1	0

$$Y = \overline{y}j + y\overline{k}$$

Figure 5.23 The Y-table and next-state equation of the JK F/F.

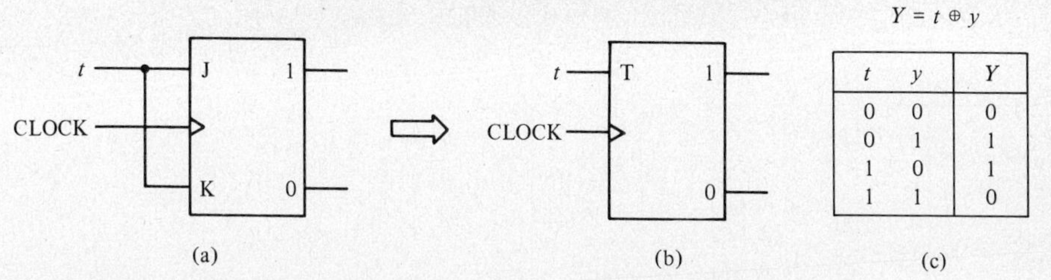

Figure 5.24 Construction of the T F/F, its circuit symbol, and its Y-table.

$j = k = t$, and obtain

$$Y = \bar{y}j + y\bar{k} = \bar{y}t + y\bar{t} = t \oplus y$$

For convenience, in Table 5.1 we have a summary of the next-state behaviors (Y-tables and next-state equations) of the four types of flip-flops.

The specific inputs of a type of flip-flop [either (s, r), or d, or (j, k), or t] are called the *excitation inputs,* or briefly the *excitation.* The next-state equation provides a means to obtain Y once y and the excitation are known. Quite important is the converse of this problem, which we need to solve in a design situation, that is: given the present state y and the next state Y, which is the excitation that will produce Y? Clearly, we need to express the excitation as a switching function of the pair (y, Y).

Perhaps the most expedient way to construct the truth tables of the excitation (excitation table) from the Y-table is the following. We view the pair of independent variables (y, Y) as a "transition" $y \rightarrow Y$ from present-state y to next-state Y. Thus, referring, for example, to the SR F/F, $(y, Y) = (0, 0)$ corresponds to the transition $0 \rightarrow 0$, which requires $s = 0$ and for which r can assume either value, that is, $(s, r) = (0, \delta)$. Analogously, $(y, Y) = (0, 1)$ corresponds to the transition $0 \rightarrow 1$ for which we must have $(s, r) = (1, 0)$; $(y, Y) = (1, 0)$ corresponds to the transition $1 \rightarrow 0$ for which we must have $(s, r) = (0, 1)$; $(y, Y) = (1, 1)$ corresponds to the transition $1 \rightarrow 1$ for which

TABLE 5.1 SYNOPSIS OF THE Y-TABLES OF THE SR, JK, D, AND T F/Fs

	SR F/F				JK F/F				D F/F			T F/F		
Row	y	s	r	Y	y	j	k	Y	y	d	Y	y	t	Y
0	0	0	0	0	0	0	0	0	0	0	0	0	0	0
1	0	0	1	0	0	0	1	0	0	1	1	0	1	1
2	0	1	0	1	0	1	0	1	1	0	0	1	0	1
3	0	1	1	δ	0	1	1	1	1	1	1	1	1	0
4	1	0	0	1	1	0	0	1						
5	1	0	1	0	1	0	1	0						
6	1	1	0	1	1	1	0	1						
7	1	1	1	δ	1	1	1	0						
	$Y = s + y\bar{r}$				$Y = \bar{y}j + y\bar{k}$				$Y = d$			$Y = t \oplus y$		

TABLE 5.2 EXCITATION TABLES OF THE SR, JK, D, AND T F/Fs

SR F/F				JK F/F				D F/F			T F/F		
y	Y	s	r	y	Y	j	k	y	Y	d	y	Y	t
0	0	0	δ	0	0	0	δ	0	0	0	0	0	0
0	1	1	0	0	1	1	δ	0	1	1	0	1	1
1	0	0	1	1	0	δ	1	1	0	0	1	0	1
1	1	δ	0	1	1	δ	0	1	1	1	1	1	0

we must have $s = 0$, while r can assume either value, that is, $(s, r) = (\delta, 0)$. This yields the truth tables in the leftmost part of Table 5.2.

The same process applied to the other types of F/Fs yields the corresponding excitation tables also shown in Table 5.2; their construction is left as an exercise for the reader.

5.7 PARALLEL REGISTERS

With the storage devices developed in the preceding sections, we are now in a position to assemble more complex modules, such as registers.

Registers were introduced in Chapter 2 as fundamental storage units of a digital system (and thus of our SEC machine), but were just represented as black boxes at that stage. We are now ready to describe in detail their internal structure.

A *register* is a bank of n flip-flops, usually D F/Fs, or SR F/Fs or JK F/Fs connected as D F/Fs (Figure 5.25). The flip-flops are numbered and displayed in the circuit diagram as an array with outputs $y_1, y_2, \ldots, y_n$. We saw in Chapter 2 that a register is typically used to temporarily store a computer word. Although a register can be loaded and unloaded both in parallel and in series, we shall presently restrict ourselves to the parallel mode of operation and return later (Sec. 5.12.1) to more versatile and complicated registers.

To load a register in parallel we must gate the inputs $x_1, x_2, \ldots, x_n$ to the appropriate flip-flops, that is, input x_i to flip-flop y_i. Referring to Figure 5.25, input x_i is gated to flip-flop y_i by a control signal called LOAD. The same signal is used to gate the clock to the clock input of the flip-flops, so that they become receptive to input information only when the LOAD signal is present. With regard to unloading the register in parallel, no special circuitry is needed since the outputs of the register flip-flops are permanently available.

5.8 SYNCHRONOUS SEQUENTIAL NETWORKS

In Sec. 2.6, before embarking on the detailed study of digital networks, we reached the important conclusions that *registers, information processing units,* and *sequencing of operation* are the three fundamental items of a digital

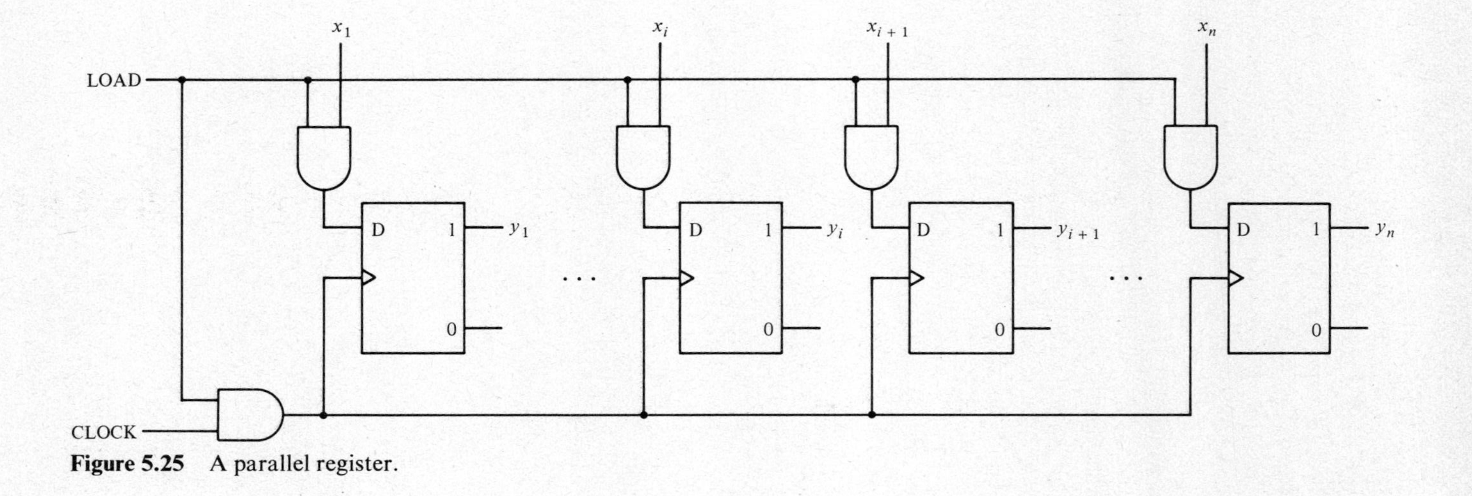

Figure 5.25 A parallel register.

computer (and, in general, of a digital system). At this point, we have gained adequate insight into the first two components of this triad: it remains to consider the third one, by far the most complex and fascinating.

A digital subsystem that controls the sequencing of various actions, such as the orderly execution of the instructions of a computer, is an important example of *sequential networks*. As we experienced already in connection with combinational networks, however, the applicability of the general analysis and design techniques we are about to study reaches beyond the somewhat restricted domain of a digital computer into the wider area of digital systems.

Basically, a sequential network is designed to process events (represented by configurations of binary variables) which occur sequentially in time.

The sequential nature of these networks implies the presence of means to store a record of the network's past history, whence the network will consist of a *memory section,* comprising a number of binary storage devices (flip-flops), and a memoryless *combinational section,* realizing a set of boolean functions (see Figure 5.27).

We shall restrict ourselves to sequential networks employing clocked flip-flops. For this class, changes of the contents of flip-flops are timed by a clock waveform, whose essential function is to sample the value of network variables (in general, outputs of the combinational portion) and to feed the sampled values to the storage devices (flip-flops).

Regardless of the type of flip-flop used, in all cases:

1. The clock pulse must be long enough to provide sufficient switching energy to the flip-flops.

This means that there is a time window of appreciable duration during which the flip-flop is receptive to whatever appears on its input. Let us examine the implications of this fact.

Normally (see again Figure 5.27) the outputs of flip-flops feed the combinational logic of the network; some outputs of this combinational logic are, in turn, the inputs to the flip-flops. Therefore, there is a combinational logic path from F/F output to F/F input, and, if the signal propagation through this path is sufficiently fast, the effects of the F/F state change occurring during the *current* clock pulse may appear at the flip-flop inputs *before* the clock pulse is terminated. Therefore, to avoid this *entirely unacceptable behavior,* one may try to reduce the duration of the clock pulse to the minimum value compatible with requirement 1. If even this remedy is inadequate, the designer must resort to a master–slave arrangement, where flip-flop state changes occur *after* the termination of the clock pulse, thereby avoiding the situation described above.[4] Note that in the master–slave mode of operation, clock pulses are rather long, thereby automatically satisfying requirement 1. Since the flip-flop is receptive to the input for the whole duration of the clock pulse, even if the input changes in this

[4]The reason why the master–slave arrangement is not automatically chosen, is its higher cost per flip-flop.

time interval, its value during a very short duration preceding the clock-pulse trailing edge (see Sec. 5.5) determines the next state of the flip-flop.

A second requirement on the clock waveform is that:

2. The spacing between consecutive clock pulses must be long enough to ensure that all network transients have died out.

Indeed, the signal changes triggered by the current clock pulse must propagate through both the flip-flops and the ensuing combinational logic. Due to large and uneven path lengths, it may take an appreciable time before flip-flop input signals attain their final intended values. Hence the requirement on clock pulse spacing.

Once the designer has chosen both the mode of operation (single or master–slave flip-flops) and the time parameters of the clock waveform, the behavior of the synchronous sequential network is described as follows: there is a very small time window (occurring after the leading edge in the single F/F arrangement and before the trailing edge in the master–slave arrangement) during which the flip-flops are receptive to the signals on their input lines. Whatever happens between consecutive time windows is therefore totally irrelevant (and could be, in reality, quite erratic, irrespective of whether master–slave or single F/Fs are used). For this reason we may dispense with the actual analysis of what happens outside the windows and, assuming that signals are stable during the window, replace the actual network with a totally adequate but greatly simplified *ideal model,* where:

1. Clock pulses are instantaneous.
2. Changes occur at clock instants and signals are stable between clock instants.

This model is referred to as a *synchronous sequential network,* and it is reached by letting the width of the flip-flop "time window" go to 0, and by making a convenient assumption about signals for time intervals during which what happens is—after all—entirely immaterial. Figure 5.26 illustrates the relation between the status of a given signal in the actual and in the idealized network.

In the following we shall omit the word "synchronous" since no other type will be considered in the remainder of this chapter. In addition, we shall always assume that the appropriate flip-flop selection has been made. Finally, the time interval between consecutive clock instants is referred to as *time unit* (not necessarily of constant duration).

An important distinction is now in order. A sequential network (SN) is an interconnection of appropriate physical functional blocks and exhibits a certain behavior, expressed as a relation between input and output sequences when the initial condition of the SN is known. It should be easy to realize that several SNs are capable of realizing the same behavior, that is, there exists an abstract model common to all the SNs that behave identically. This common abstraction is called the *sequential machine* (SM) realized by a given network, and the relation

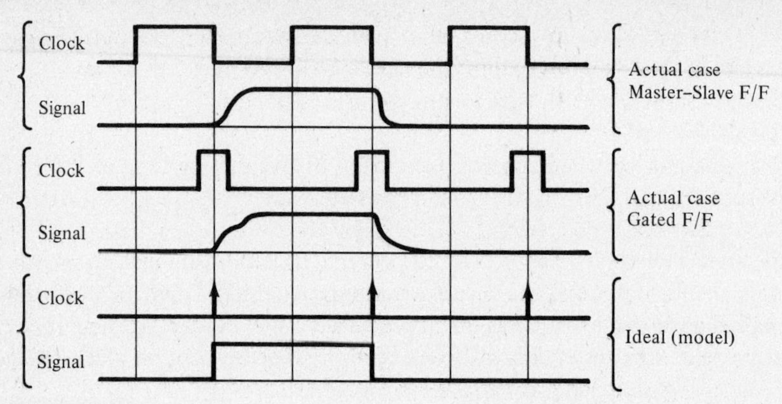

Figure 5.26 Corresponding behaviors of actual F/F input signals and idealized F/F input signals.

between SM and SN is pictorially illustrated below. The SM plays with respect to the SN very much the same role played by boolean functions with respect to combinational networks.

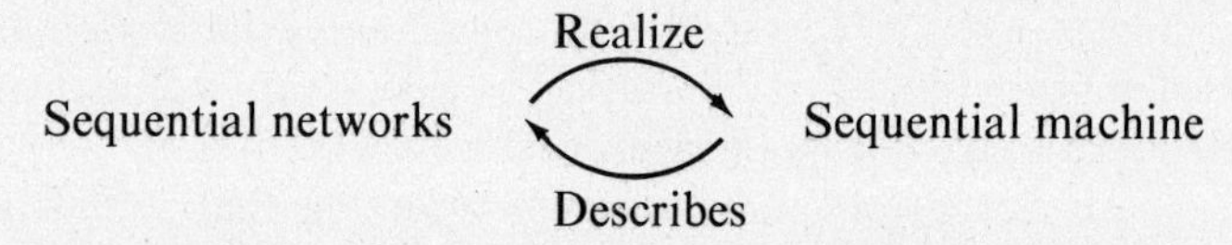

A sequential network is clearly decomposable into a collection of memory elements and a collection of combinational networks (Figure 5.27). We shall distinguish a set $x_1, x_2, \ldots, x_n$ of *input* variables, a set $y_1, y_2, \ldots, y_r$ of *internal* or *state* variables (stored in the flip-flops), a set $z_1, z_2, \ldots, z_m$ of *output* functions, and a set ϕ of *excitation* functions, to be described later; $z_1, \ldots, z_m$ and all members of ϕ are functions of $\{x_1, x_2, \ldots, x_n; y_1, y_2, \ldots, y_r\}$ and are synthesized in the combinational portion of the network.

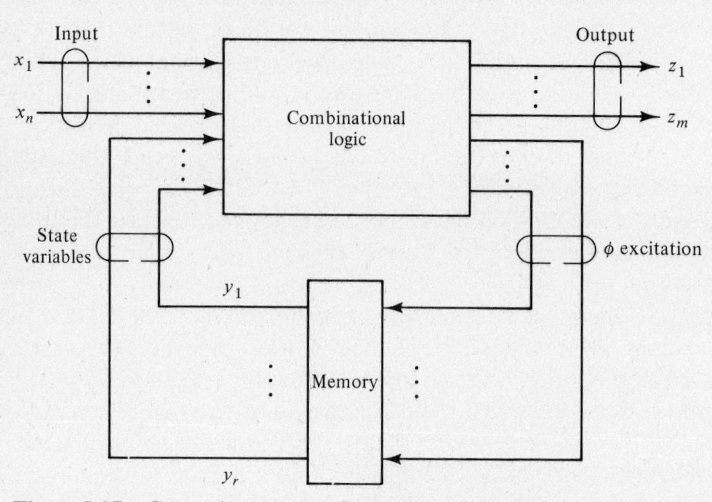

Figure 5.27 General structure of a sequential network.

For convenience of presentation, an instantaneous configuration of a specific set of binary variables will be called a *vector;* thus we shall have input vectors $(x_1, \ldots, x_n)$, state vectors $(y_1, \ldots, y_r)$, output vectors $(z_1, \ldots, z_m)$, and excitation vectors.

5.9 ANALYSIS OF SEQUENTIAL NETWORKS

The analysis of an SN can be broadly characterized as obtaining the SM describing the given SN. The procedure one may follow in carrying out this task can be conveniently organized as a sequence of steps as shown in Figure 5.28.

We shall parallel with an example the explanation of the individual steps of the analysis procedure. For simplicity, we shall assume that the memory elements are delay flip-flops, discussed in Sec. 5.6. Sequential networks using flip-flops of different types are only marginally more complicated, and we shall return to this point later in our study.

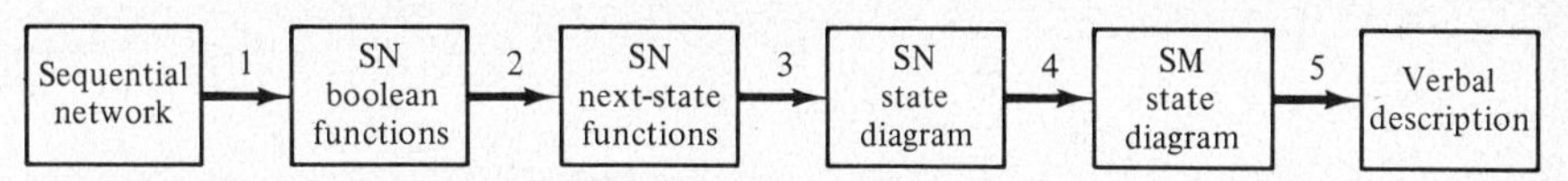

Figure 5.28 Steps in the analysis of a sequential network.

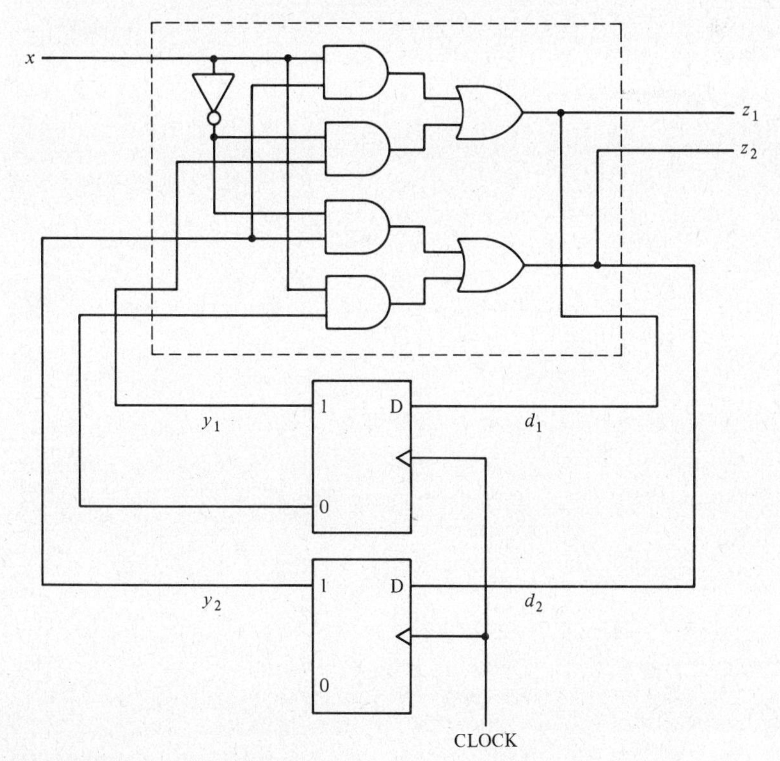

Figure 5.29 An SN to be analyzed.

Let the SN in Figure 5.29 be given for analysis. The logic enclosed by the broken lines is clearly the combinational logic of the network.

5.9.1 SN Boolean Functions

We inspect the combinational logic and write the boolean expressions being realized. We then obtain the boolean functions corresponding to those expressions (see Sec. 3.3). In this case the independent variables are x, y_1, and y_2, while z_1, z_2, d_1, and d_2 are the functions. Since $z_1 = d_1$ and $z_2 = d_2$, we need only analyze, say, d_1 and d_2. We must remark at this point that by obtaining the boolean functions we have already departed from the given network, which is characterized by *specific* realizations of those boolean functions: this is the first step in the process of abstraction which will take us from the concrete assemblage of hardware to the abstract description of its behavior. In our case, the boolean expressions are

$$z_1 = d_1 = xy_2 + \bar{x}y_1$$
$$z_2 = d_2 = x\bar{y}_1 + \bar{x}y_2$$

and the corresponding truth tables are

x	y_2	y_1	d_2	d_1	z_2	z_1
0	0	0	0	0	0	0
0	0	1	0	1	0	1
0	1	0	1	0	1	0
0	1	1	1	1	1	1
1	0	0	1	0	1	0
1	0	1	0	0	0	0
1	1	0	1	1	1	1
1	1	1	0	1	0	1

At this point the analysis of the output is completed; we shall then proceed with the analysis of the state-transition behavior.

5.9.2 SN Next-State Functions (Transition Table)

The purpose of this step is to express the state vector at the next time unit as a function of the input and state vectors at the present time unit. Letting Y_j denote the value of y_j at the next time unit, clearly when using D F/Fs we have[5]

$$(Y_r Y_{r-1} \; . \; . \; . \; Y_1) = (d_r d_{r-1} \; . \; . \; . \; d_1)$$

The *next-state table* is a double entry table, where the row headings are the present-state vectors (i.e., the combinations of the state variables at the present time unit), the column headings are the present input vectors, and the entries are the next-state and output vectors. Thus, the next-state table is simply a way to

[5] *Warning*! The situation is moderately more complex when other types of flip-flops are used, as suggested by the next-state equations in Table 5.1.

reorganize the truth tables obtained in step 1. To further simplify this inherently simple task, it is convenient when constructing the truth-tables to list the independent arguments in the following order: $x_n x_{n-1} \ldots x_1 y_r y_{r-1} \ldots y_1$. This trick—already mentioned in Sec. 5.4—permits a ready, mechanical repackaging of the truth tables into the next-state table. It must be pointed out however, that this step involves no further abstraction with respect to the result of step 1. We write below side by side the truth tables (left) and the next-state table (right) for our running example in order to evidence the natural relationship between the two:

x	y_2	y_1	d_2	d_1	z_2	z_1
0	0	0	0	0	0	0
0	0	1	0	1	0	1
0	1	0	1	0	1	0
0	1	1	1	1	1	1
1	0	0	1	0	1	0
1	0	1	0	0	0	0
1	1	0	1	1	1	1
1	1	1	0	1	0	1

		$x = 0$				$x = 1$			
y_2	y_1	Y_2	Y_1	z_2	z_1	Y_2	Y_1	z_2	z_1
0	0	0	0,	0	0	1	0,	1	0
0	1	0	1,	0	1	0	0,	0	0
1	0	1	0,	1	0	1	1,	1	1
1	1	1	1,	1	1	0	1,	0	1

5.9.3 SN State Diagram

A state diagram of a sequential network is another form of description that, being pictorial in nature, greatly appeals to intuition when the machine behavior is to be described. A state diagram is a graph whose nodes correspond in a one-to-one fashion to the state vectors and whose arcs correspond in a one-to-one fashion to the entries of the next-state table. The rule for the construction of each arc is illustrated below:

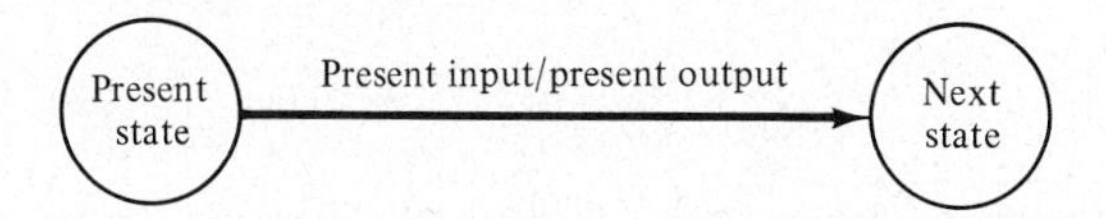

Again, the correspondence between the various items of the next-state table and the state diagram is shown in more detail in Figure 5.30. With these conventions, for our current example we obtain the state diagram shown in Figure 5.31.

5.9.4 SM State Diagram

In this step the state codes are replaced with arbitrary symbols. By doing so we ignore the specific network structure but preserve the input/output behavior of the SN: in other words, we obtain the SM describing the SN. This is the second abstracting step of the analysis procedure.

The state diagram provides us with a pictorial description of the behavior of the machine, which is far more effective than the next-state table in carrying out the next and final step of the analysis.

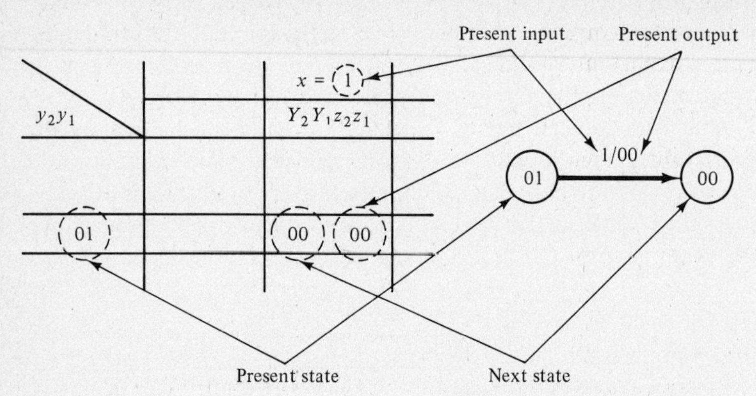

Figure 5.30 Correspondence between next-state table and state diagram.

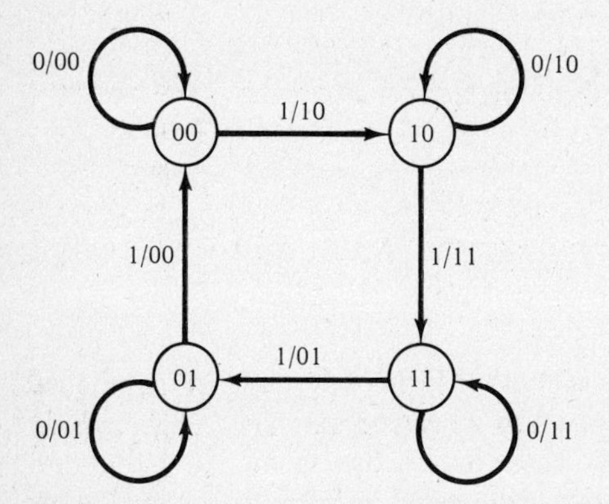

Figure 5.31 The state diagram of the given network.

5.9.5 Word Description

This is the last analysis step and the one that is least formally definable. We must interpret the SM state diagram in order to be able to describe its functions. Unfortunately, there are no general infallible recipes for this task. Typically, one considers a typical input sequence to the network and observes the corresponding output sequence; this frequently provides very useful clues. Other important clues may be as follows:

1. Look for a marked singularity in the output function of the SM. In other words, if there are very few (possibly just one) input-state pairs with a given output condition (say, 1), while all others have another output (say, 0), then most likely the network is a "sequence detector." In this case one can trace backward the input sequence(s) leading to the "singular output."
2. If the output simply monitors the internal state, then one identifies the conditions that determine state changes. This case is typical of "a counter," that is, a sequential network that counts the number of occurrences of a certain event, represented by signals on its input lines.

However useful, the impression should be dispelled that the above clues be an infallible recipe for SM interpretation.

Our running example is interpreted by using clue 1: it is a modulo-4 counter of inputs "1" and the count is being expressed in the binary Gray code (00-10-11-01) (for an explanation of the Gray code see Sec. 1.6).

5.10 SYNTHESIS OF SEQUENTIAL NETWORKS

The synthesis of a sequential network can be broadly characterized as obtaining an SN realizing the given SM. In this broad sense it can be viewed as the reverse procedure of the one outlined in Figure 5.28. However, while the analysis is deterministic in nature, several crucial choices are open to the designer at various steps of the synthesis procedure. The step sequence is outlined for convenience in Figure 5.32.

Before embarking on a detailed discussion of the steps of the design procedure, it is useful to attempt to draw a sketchy outline of the various design situations, so that different approaches and techniques are placed in the proper perspective.

As a useful guideline (and with absolutely no claim to rigor!), sequential networks can be broadly categorized in three classes:

1. Sequence detectors.
2. Counters.
3. Sequencers.

The functions of sequence detectors and counters were briefly discussed at the end of the preceding section. Normally, sequence detectors and counters are rather "small" networks, involving relatively few states. Their verbal descriptions (as the specifications of the desired behavior) are normally given in natural language and are therefore quite informal.

Sequencers on the other hand are designed to generate a complex sequence of outputs, where the output sequence is determined both by the initial condition of the sequencer and by the input sequence. Typical sizes of sequencers (in terms of states) vary over a wide range: from a relatively small control of an elevator to the control unit of a digital computer. Indeed, we come to realize at this point that the control unit of SEC, whose function was outlined in Sec. 2.6, is exactly a sequencer of the type we are about to study (we shall return extensively to this connection in Chapter 9). Although sequencers are usually large sequential networks, their verbal descriptions are normally rather close to their final formalization, which greatly simplifies their design.

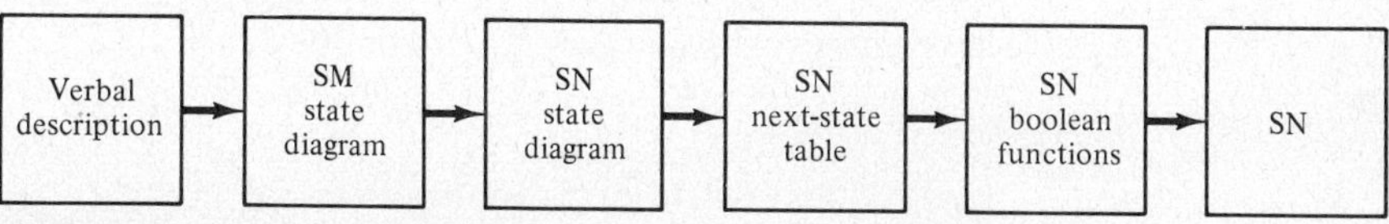

Figure 5.32 Steps in the synthesis of a sequential network.

The broad differences between detectors and counters on one hand, and sequencers on the other, are reflected in different typical solutions at various steps of the design process. The following discussion is basically geared toward detectors and counters; in Sec. 5.11 we shall consider general design techniques which are best suited to sequencers, while a detailed discussion of the latter will be undertaken in Chapter 9 in the context of the SEC system.

5.10.1 SM State Diagram

The conversion of a verbal description into a state diagram is the most important, most difficult, and least formalizable step of the procedure. In fact its objective is to set up a model of the assigned problem from the given verbal description, that is, a formal structure of previously defined elements (in our case, states and transitions) which will exhibit the prescribed behavior. It is clear that this task is the more difficult, the more imprecise is the initial word description. It is therefore recommendable, as a first preliminary step, to couch the problem statement in as formal a way as possible. Another important question, which must be answered after that, is whether the given behavior can be realized with a machine having a finite number of states. Usually this answer will emerge from the actual construction of the state diagram.

As a notational convention, let $a^{(j)}$ denote the value of some network variable a at time unit j, where j is an integer. Consider now the following problem (design of a sequence recognizer):

> Design a single input (x), single output (z) SM such that $z^{(j)} = 1$ if and only if $x^{(j-2)}x^{(j-1)}x^{(j)}$ coincides with either one of the possibly overlapping sequences 110 or 101.

For example, the two sequences given below are consistent with the prescribed behavior (the time index *increases* from left to right)

x 0 0 1 0 0 1 1 0 0 1 0 1 1 0 0 1 0 1 1 0 0

z 0 0 0 0 0 0 0 1 0 0 0 1 0 1 0 0 0 1 0 1 0

We construct the state diagram step by step, one state at a time, one arc at a time. In this process it is convenient to associate with each state we introduce a description of what it does. We begin, in this case, with a "reset" state (see Figure 5.33):

$$q_0{:} = \text{recognition of 110 or 101 not yet begun}$$

If from state q_0 we receive $x = 0$, clearly we make no progress in the recognition, so we make a transition back to q_0: if we receive $x = 1$, however, we go to state

$$q_1{:} = \text{received first symbol 1 of either 110 or 101}$$

If from state q_1 we receive $x = 0$ we go to state

$$q_2{:} = \text{received first two symbols (10) of 101}$$

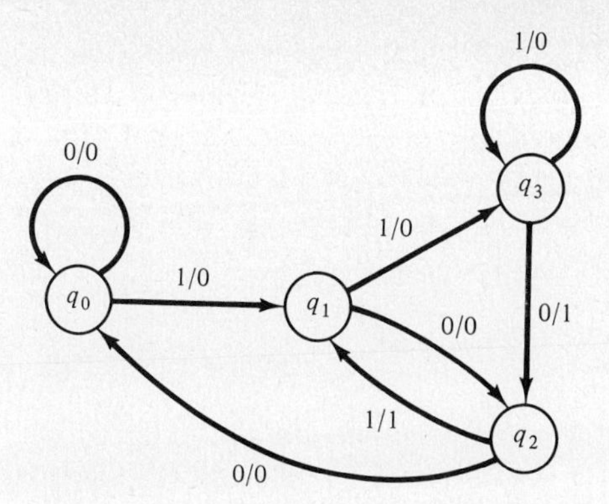

Figure 5.33 State diagram of a sequence recognizer.

If we receive $x = 1$ we go to state

$$q_3: = \text{received first two symbols (11) of 110}$$

Now, suppose we are in q_2 and receive $x = 1$; in this case we complete the recognition of 101 and generate an output $z = 1$; however, the next state will not be q_0, since the current input symbol is 1 and we have already made progress in the recognition process; thus we shall make a transition to q_1. On the other hand, $x = 0$ disrupts the recognition process and forces a transition back to q_0.

Similarly, suppose we are in state q_3 and receive $x = 0$; in this case we complete the recognition of 110 and generate $z = 1$; since our last two symbols, however, were 10 we have made progress toward the recognition of 101 and will make a transition to q_2. If in q_3 we have $x = 1$, then notice that the last two symbols were 11, and therefore we will loop back to q_3.

This completes the construction of our state diagram (Figure 5.33).

**Remark:* Frequently, when constructing an SM state diagram from a verbal description, one does not immediately recognize that a previously introduced state could be used as the destination of the current transition, and therefore redundant states are introduced.

For example, in the problem discussed above, one may obtain the state diagram shown in Figure 5.34 (rather than that of Figure 5.33), where the transition from q_3 under $x = 0$ has been directed to the "new" state q_4 rather than to the "old" state q_2.

Note that both state diagrams are correct; however, we may wish to use the one with fewer states. The minimization of the number of states has been—perhaps unjustifiably[6]—a very popular problem in the theory of sequential machines. Its solution is based on the notion of "equivalent states" and in the merging of classes of equivalent states into a single state. Although the treatment of this topic is well beyond our scope, we shall content ourselves with a weak

**The asterisk indicates that the material in this remark will not be referred to later and may be skipped.*

[6]This remark is due to the fact that there is no solid relation between the number of states and the complexity of the sequential network.

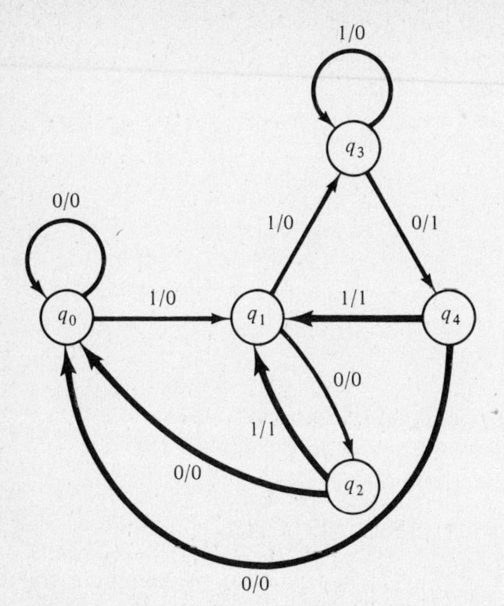

Figure 5.34 An SM state diagram with two equivalent states (q_2 and q_4).

version of equivalent states. Specifically we say that two states of a sequential machine are equivalent when for each input both states make a transition to the same state and produce the same output. This is precisely the condition of states q_2 and q_4 in Figure 5.34 (the identical transitions are given special graphical evidence). Therefore, we shall merge a pair of equivalent states q_i and q_j into a single state q_h, having the next-state transitions common to q_i and q_j; in addition q_h is reached by each and every transition reaching either q_i or q_j. Effecting this merging, we transform the diagram of Figure 5.34 into that of Figure 5.33.

5.10.2 SN State Diagram

This step entails the assignment of codes to the symbolic designations of states in the state diagram obtained in Sec. 5.10.1. This step, usually referred to as the *state assignment,* is not as trivial a task as its converse in the analysis procedure (SN state diagram → SM state diagram); rather, not only the design complexity, but the very type of design will crucially depend upon the choices made in this step.

First of all, we note an obvious constraint: the number r of state variables must be sufficient to represent the k distinct states of the diagram. Thus, we must have

$$r \geq \lceil \log_2 k \rceil$$

where $\lceil \log_2 k \rceil$ denotes "the smallest integer not less than $\log_2 k$." This only tells us that our sequential network will have at least r flip-flops, but we still have no guidance as to a desirable state assignment.

The cost of the network is conveniently broken down into the costs of the memory portion and of the combinational portion. If we choose to use a fixed (possibly minimal) number r of flip-flops, the cost of the memory portion is determined and we only have to deal with the combinational portion. At this point we have two major alternatives: to realize the combinational logic with a ROM (see Sec. 4.8.3) or with discrete gates. The two cases are significantly different. In the first case, the cost is independent of the state assignment, since the cost of the ROM depends only upon its numbers of inputs (network inputs and state variables) and of outputs (network outputs and excitation). In the second case, the cost strongly depends upon the state assignment. For fixed r, the number of distinct state assignments, although possibly astronomical, is finite, and of course some assignments will result in networks of minimal complexity. Unfortunately, no efficient method is known to find an optimal assignment, although procedures exist that are conducive to "good" assignments (not to be treated in this text, however).

An alternative to designs with minimal, or near-minimal, number of flip-flops, is one that results in a network whose structure closely resembles the state diagram itself. This design uses one flip-flop per state, and the state codings are strings of k bits only one of which is equal to 1. (For this reason, this is called the *one-hot* assignment.) This approach, which results in heavy memory and usually light logic, has found some applications in the realization of sequencers.

In brief summary, three approaches can be outlined:

1. Minimal number of flip-flops, discrete-gate logic.
2. Minimal number of flip-flops, ROM, or PLA logic.
3. One-hot assignment.[7]

Of the three, the first approach is the most intricate and the one that we shall first discuss. We assume that codes are assigned arbitrarily to the SM states, consistently with $r = \lceil \log_2 k \rceil$.

In our running example, we choose $r = 2 = \lceil \log_2 4 \rceil$, and let $q_0 = 10$, $q_1 = 00$, $q_2 = 01$, $q_3 = 11$, thus obtaining the SN state diagram shown in Figure 5.35.

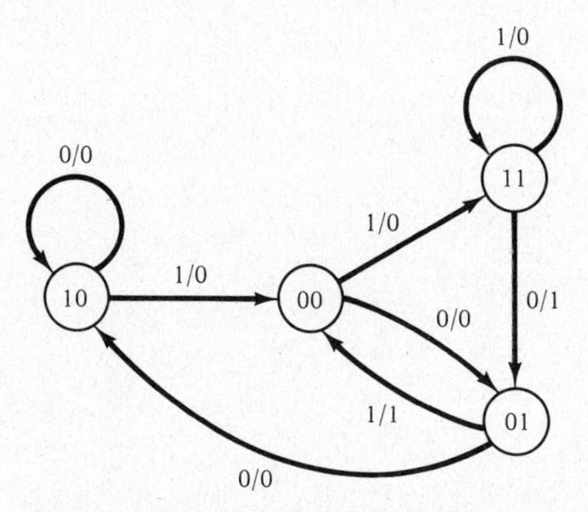

Figure 5.35 A coded state diagram.

[7]To be discussed in Chapter 9 as a possible way to realize the control unit of SEC.

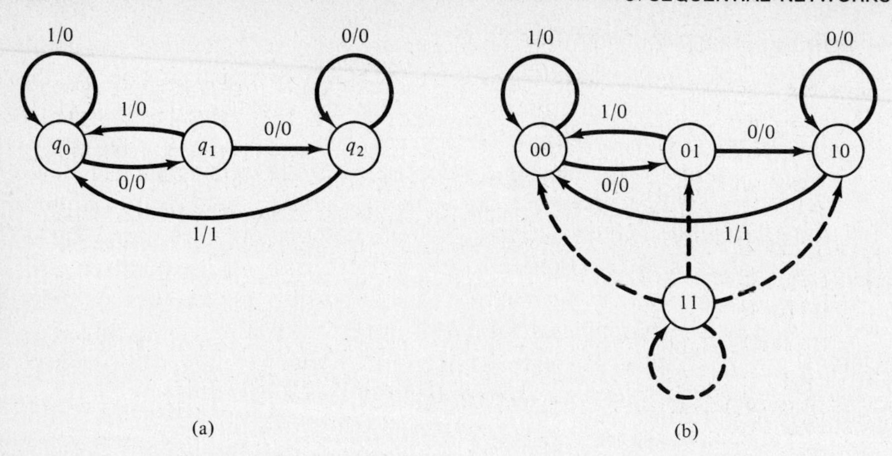

Figure 5.36 State assignment when $k \neq 2^r$.

Remark: The number of states in the SM is not necessarily a power of 2 as in the above example. However, the number of states of an SN is always a power of 2, since r internal variables yield 2^r states. In general, the SN state diagram may have some states for which there is no correspondent in the SM state diagram. This means that for these states (*extra states*) both the next state and the present output can be specified arbitrarily, and may be chosen so that a more economical realization is obtained. For example, from the SM state diagram in Figure 5.36(a) we could obtain the SN state diagram of Figure 5.36(b) where the transitions from state (11) can be specified arbitrarily.

5.10.3 SN Next-State Table

From the SN state diagram of Figure 5.35 we can readily obtain the next-state table describing it.

In our example we obtain Figure 5.37.

At this point we have all the elements to design the network of $z(x, y_2, y_1)$. We must still realize, however, the state behavior. Notice that, had we had *extra states* (see previous remark), for each one of them the corresponding line in the table is filled with don't cares.

		$x = 0$			$x = 1$		
y_2	y_1	Y_2	Y_1	z	Y_2	Y_1	z
0	0	0	1,	0	1	1,	0
0	1	1	0,	0	0	0,	1
1	0	1	0,	0	0	0,	0
1	1	0	1,	1	1	1,	0

Figure 5.37 The next-state table.

5.10.4 SN Boolean Functions

This is the last step of our synthesis procedure. Obviously, the boolean functions to be designed are the *excitation functions* which will realize the desired next-state table. At this point we must choose the type of flip-flops to be used, and proceed with the design of the corresponding excitation functions.

We shall now illustrate the construction of the truth tables of the excitation functions for the four types of flip-flop, using our running example. This will also demonstrate how to exploit the don't cares in the excitation tables of s, r, j, and k (refer to Table 5.2).

We start from the truth tables of Y_2 and Y_1 as functions of the independent variables x, y_2, y_1 (first five columns of Figure 5.38). We consider first Y_2. The truth tables of d_2 and t_2 are readily obtained, since $d_2 = Y_2$ and $t_2 = Y_2 \oplus y_2$ (sixth and seventh columns in Figure 5.38). The construction of the truth tables of s_2, r_2, j_2, and k_2 is not so straightforward but easily obtained anyway. We may either resort to the transformations displayed in Table 5.2 or, if we have no access to it and are reluctant to memorize things, we simply repeat in our current problem the process which led to the construction of Table 5.2. Specifically, we consider, one row at a time, the pair (y_2, Y_2) as a *transition from present-state y_2 to next-state Y_2* and, from our knowledge of the flip-flop, determine the excitation necessary to effect that transition. For example, if $(y_2, Y_2) = (0, 1)$ (fifth row from top in Figure 5.38), to effect the transition from $y_2 = 0$ to $Y_2 = 1$, we need $s_2 = 1$ and $r_2 = 0$ in an RS F/F, while in a JK F/F we have $j_2 = 1$ and $k_2 = \delta$ (indeed, the transition may be effected either as a setting to 1 or as a toggling to 1). Thus, in a careful but mechanical way, we may construct all of the desired excitation functions, which are displayed in Figure 5.38. Their corresponding minimal SOP expressions are

$$\text{D F/F}\begin{cases} Y_1 = \bar{y}_2\bar{y}_1 + y_2y_1 = d_1 \\ Y_2 = \bar{x}\bar{y}_2y_1 + \bar{x}y_2\bar{y}_1 + x\bar{y}_2\bar{y}_1 + xy_2y_1 = d_2 \end{cases}$$

$$\text{T F/F}\begin{cases} t_1 = \bar{y}_2 \\ t_2 = \bar{x}y_1 + x\bar{y}_1 \end{cases}$$

x	y_2	y_1	Y_2	Y_1	d_2	t_2	s_2	r_2	j_2	k_2	d_1	t_1	s_1	r_1	j_1	k_1
0	0	0	0	1	0	0	0	δ	0	δ	1	1	1	0	1	δ
0	0	1	1	0	1	1	1	0	1	δ	0	1	0	1	δ	1
0	1	0	1	0	1	0	δ	0	δ	0	0	0	0	δ	0	δ
0	1	1	0	1	0	1	0	1	δ	1	1	0	δ	0	δ	0
1	0	0	1	1	1	1	1	0	1	δ	1	1	1	0	1	δ
1	0	1	0	0	0	0	0	δ	0	δ	0	1	0	1	δ	1
1	1	0	0	0	0	1	0	1	δ	1	0	0	0	δ	0	δ
1	1	1	1	1	1	0	δ	0	δ	0	1	0	δ	0	δ	0

Figure 5.38 Truth tables of the excitation functions for the running example.

$$\text{SR F/F}\begin{cases} s_1 = \bar{y}_2\bar{y}_1, & r_1 = \bar{y}_2 y_1 \\ s_2 = \bar{x}\bar{y}_2 y_1 + x\bar{y}_2\bar{y}_1, & r_2 = \bar{x}y_2 y_1 + xy_2\bar{y}_1 \end{cases}$$

$$\text{JK F/F}\begin{cases} j_1 = \bar{y}_2 = k_1 \\ j_2 = \bar{x}y_1 + x\bar{y}_1 = k_2 \end{cases}$$

At this point we are in a position to evaluate the respective costs of the different realizations, and select the most economical one. Note that the chosen network may, in principle, deploy a mixture of different types of flip-flops.

We now discuss in detail two additional examples, which illustrate the design procedure.

EXAMPLE 5.1

We now reconsider the main example of this section, and we illustrate an alternative way of designing a sequence recognizer.

Specifically, we assume that the initial steps of the design procedure have been carried out and the SM state diagram as given in Figure 5.33 has been obtained. We now make the following state assignment:

$$q_0 = 00, \qquad q_1 = 10, \qquad q_2 = 01, \qquad q_3 = 11$$

thereby obtaining the SN state diagram of Figure 5.39(a), and the next-state table of Figure 5.39(b).

If we now choose to realize the sequential network with D F/Fs, we have the SN boolean functions:

$$d_2 = Y_2 = x$$
$$d_1 = Y_1 = y_2$$
$$x = y_1 y_2 \bar{x} + y_1 \bar{y}_2 x$$

The corresponding network is shown in Figure 5.40. The network has an extremely straightforward "next-state logic." Specifically, the flip-flops

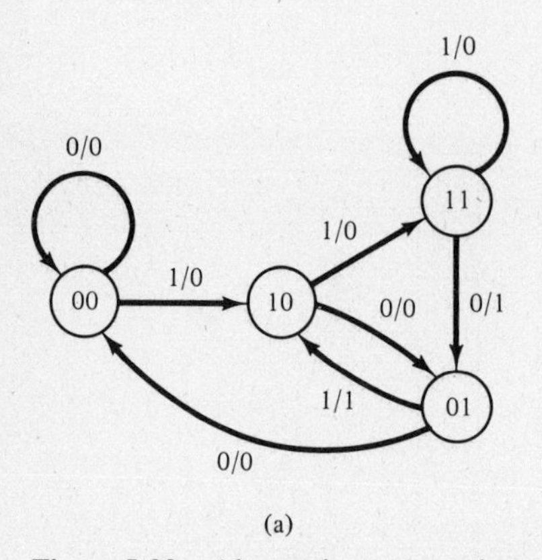

(a)

		$x = 0$			$x = 1$		
y_2	y_1	Y_2	Y_1	z	Y_2	Y_1	z
0	0	0	0,	0	1	0,	0
0	1	0	0,	0	1	0,	1
1	0	0	1,	0	1	1,	0
1	1	0	1,	1	1	1,	0

(b)

Figure 5.39 Alternative state assignment for the diagram of Figure 5.33.

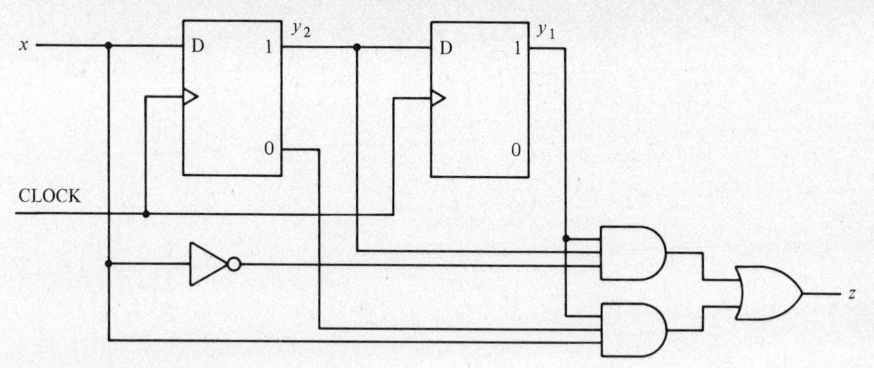

Figure 5.40 Network corresponding to the assignment of Figure 5.39(a).

are connected as a chain ($d_1 = y_2$) and the input string is simply "shifted" into this chain of flip-flops. In particular, if the input line x carries $x^{(j)}$, then $y_2 = x^{(j-1)}$ and $y_1 = x^{(j-2)}$. In other words, lines x, y_2, and y_1 carry, at any time unit, a "window" of three consecutive digits of the input string.

Then all that is needed to complete the design is to test whether this window contains any of the substrings to be detected. This test is precisely carried out by the two AND gates in the output logic of Figure 5.40, which detect the substrings $x^{(j)}x^{(j-1)}x^{(j-2)} = 011$ or 101.

This example, therefore, illustrates a general method to design sequence recognizers when we are not required *to use the least number of states or of flip-flops.* (Indeed, this method uses $n - 1$ D F/Fs if the longest string to be recognized has n bits; however, the state diagram may have fewer than 2^{n-2} states, so that $n - 2$ or fewer flip-flops are sufficient in principle for the design.) This method, referred to as *window on the input sequence,* is applicable only when the strings to be recognized may overlap.

EXAMPLE 5.2

Parity check codes, introduced in Sec. 1.7, are very simple means for error control. We recall that, in its simplest form, a parity check code is obtained by appending a single bit to a string of n bits, so that the resulting $(n + 1)$-bit string has an even number of 1s.

We are now given the problem to design a sequential checker for a parity check code whose code words are variable length strings. The checker will have two input lines x_1 and x_2 and a single output line z. Line x_1 carries binary strings of variable nonzero length, while line x_2 gives the timing of the *last* bit of each x_1 string, which is also the parity bit. The output z is the error detection signal, that is, $z = 1$ if the only if $x_2 = 1$ and the number of 1s on the just completed x_1 string is odd.

As a first step, we construct a sample sequence to get a better feeling of the problem. The 1s of the x_2 string readily determine a segmentation of the x_1 string, as shown. Then, $z = 0$ in correspondence of the nonterminal string positions; in the terminal ones, $z = 1$ as indicated. We can now proceed with the construction of the state diagram.

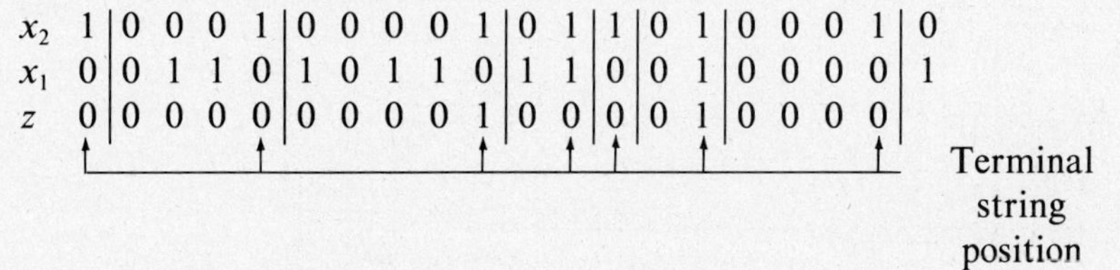

x_2 1 | 0 0 0 1 | 0 0 0 0 1 | 0 1 | 1 | 0 1 | 0 0 0 1 | 0
x_1 0 | 0 1 1 0 | 1 0 1 1 0 | 1 1 | 0 | 0 1 | 0 0 0 0 | 1
z 0 | 0 0 0 0 | 0 0 0 0 1 | 0 0 | 0 | 0 1 | 0 0 0 0 |
Terminal string position

We note first that there is a "reset state" q_0 corresponding to the beginning of an x_1 string; also, in this reset condition the parity of the 1s in the x_1 string (still to be received!) is even. We also note that x_2 acts as a reset signal to q_0, so that we will consider first inputs (x_2, x_1) of the type 00 and 01 (and defer consideration of the pairs 10 and 11). Therefore, we need an additional state q_1, corresponding to the odd parity condition. It is immediate that, for $x_2 = 0$, we obtain the partial state diagram shown in Figure 5.41(a). The condition $x_2 = 1$ resets the machine to state q_0; in addition, if the present state is q_1 and $x_1 = 0$, then $z = 1$; the same happens if the present state is q_0 and $x_1 = 1$. In all other cases, $z = 0$. This is shown in the partial diagram of Figure 5.41(b). The complete diagram appears in Figure 5.41(c).

We can now proceed with the design. There is just one internal variable, y, and, in a rather natural way, we make the state assignment $q_0 = 0$, and $q_1 = 1$. Therefore, we obtain the following SN next-state table:

y \ x_2x_1	Y, z 00	01	10	11
0	0,0	1,0	0,0	0,1
1	1,0	0,0	0,1	0,0

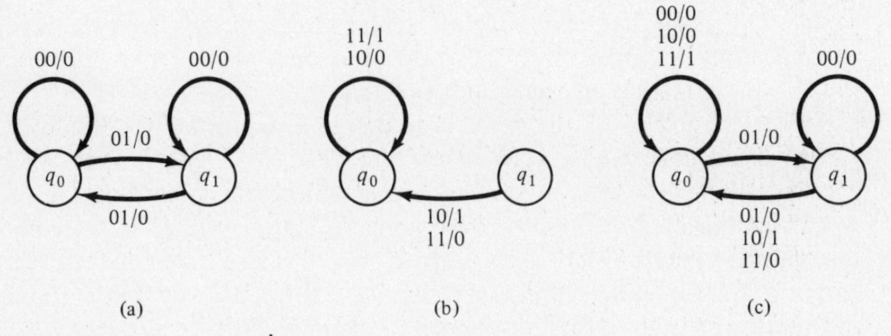

q_0 = reset state, even parity

q_1 = odd parity

Figure 5.41 State diagram of the sequential parity checker (c). Cases (a) and (b) correspond to $x_2 = 0$ and $x_2 = 1$, respectively.

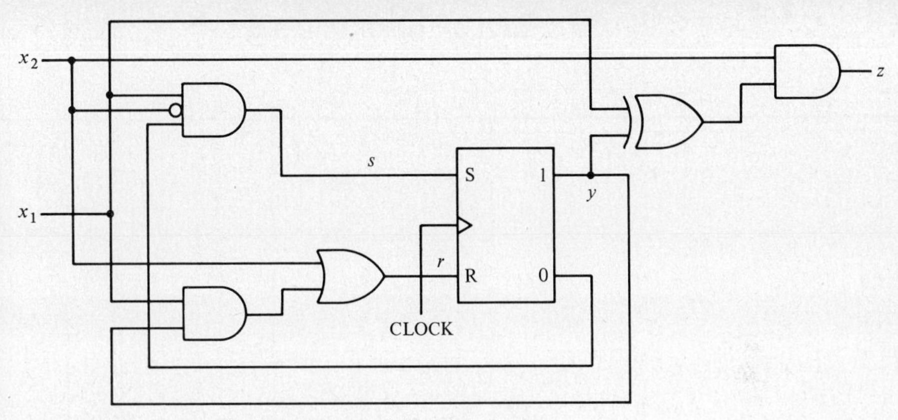

Figure 5.42 Network of the sequential parity checker.

This table is rearranged to obtain the boolean functions of Y and z (see below). In addition, choosing an SR F/F realization, we readily obtain s and r, and the boolean expressions of all three outputs of the combinational logic. The final network is illustrated in Figure 5.42.

x_2	x_1	y	Y	z	s	r
0	0	0	0	0	0	δ
0	0	1	1	0	δ	0
0	1	0	1	0	1	0
0	1	1	0	0	0	1
1	0	0	0	0	0	δ
1	0	1	0	1	0	1
1	1	0	0	1	0	δ
1	1	1	0	0	0	1

$$z = x_2(y \oplus x_1)$$
$$s = x_1\overline{x}_2\overline{y}$$
$$r = yx_1 + x_2$$

5.11 A GENERAL DESIGN TECHNIQUE BASED ON ROMs AND PLAs

One important design method of sequential networks, alluded to in Sec. 5.10.2, is based on ROMs or PLAs for the realization of the combinational logic.

Suppose that we are given a next-state table of an SN as the one shown in Figure 5.43(a). This SN has two input variables x_1 and x_2, two internal variables y_1 and y_2, and four output variables z_4, z_3, z_2, and z_1. For the curious reader, this network is a "counter" which counts up when it receives input $x_2x_1 = 01$, counts down when it receives $x_2x_1 = 10$, and does not change state for $x_2x_1 = 00$; thus the input configuration $x_2x_1 = 11$ is never used and the corresponding column does not appear in the next-state table of the network. We now choose to realize the network using T F/Fs (or, equivalently, JK F/Fs with the J and K inputs connected together), whence $t_2 = y_2 \oplus Y_2$ and $t_1 = y_1 \oplus Y_1$ (see Table 5.2). We may

State \ Input	$x_2x_1 = 00$	$x_2x_1 = 01$	$x_2x_1 = 10$
$y_2\ y_1$	$Y_2\ Y_1,\ z_4\ z_3\ z_2\ z_1$	$Y_2\ Y_1,\ z_4\ z_3\ z_2\ z_1$	$Y_1\ Y_2,\ z_4\ z_3\ z_2\ z_1$
0 0	0 0, 0 0 0 1	0 1, 0 0 0 1	1 1, 0 0 0 1
0 1	0 1, 0 0 1 0	1 0, 0 0 1 0	0 0, 0 0 1 0
1 0	1 0, 0 1 0 0	1 1, 0 1 0 0	0 1, 0 1 0 0
1 1	1 1, 1 0 0, 0	0 0, 1 0 0 0	1 0, 1 0 0 0

(a)

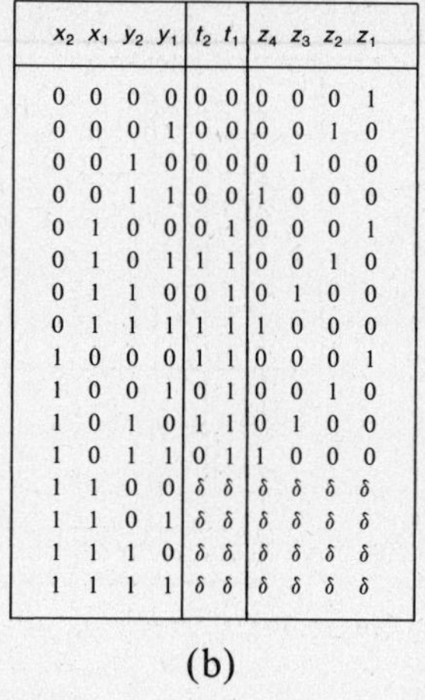

$x_2\ x_1\ y_2\ y_1$	$t_2\ t_1$	$z_4\ z_3\ z_2\ z_1$
0 0 0 0	0 0	0 0 0 1
0 0 0 1	0 0	0 0 1 0
0 0 1 0	0 0	0 1 0 0
0 0 1 1	0 0	1 0 0 0
0 1 0 0	0 1	0 0 0 1
0 1 0 1	1 1	0 0 1 0
0 1 1 0	0 1	0 1 0 0
0 1 1 1	1 1	1 0 0 0
1 0 0 0	1 1	0 0 0 1
1 0 0 1	0 1	0 0 1 0
1 0 1 0	1 1	0 1 0 0
1 0 1 1	0 1	1 0 0 0
1 1 0 0	δ δ	δ δ δ δ
1 1 0 1	δ δ	δ δ δ δ
1 1 1 0	δ δ	δ δ δ δ
1 1 1 1	δ δ	δ δ δ δ

(b)

Figure 5.43 A next-state table and an equivalent reformatting.

now redraw this table in the form shown in Figure 5.43(b), where the appropriate don't cares appear explicitly. We now apply the general two-level minimization techniques developed in Chapter 4 and obtain the boolean expressions:

$$t_1 = x_1 + x_2 \qquad z_1 = \bar{y}_1\bar{y}_2$$
$$t_2 = x_1y_1 + x_2\bar{y}_1 \qquad z_2 = y_1\bar{y}_2$$
$$z_3 = \bar{y}_1y_2$$
$$z_4 = y_1y_2$$

This combinational logic can be realized by means of a PLA, which we discussed in Sec. 4.8.3. The network is shown in Figure 5.44(a) and it consists of a 4 × 8 × 6 PLA, whose AND and OR matrices are shown in detail in Figure 5.44(b).

It is also natural to view the table shown in Figure 5.43(b) as a map from $(x_2x_1y_2y_1)$ to $(t_2t_1z_4z_3z_2z_1)$. Thus it is equally natural to interpret $(x_2x_1y_2y_1)$ as an address and $(t_2t_1z_4z_3z_2z_1)$ as a content and to use a $2^4 \times 6$ ROM to realize this map. The block diagram of the resulting network is identical to the one in Figure 5.44(a) (with the wording "$2^4 \times 6$ ROM" replacing "4 × 8 × 6 PLA"), while the AND matrix in Figure 5.44(b) is replaced by a standard 4-to-16 decoder, and the OR matrix is correspondingly modified. The ROM is shown in Figure 5.45. Here again it is worth noting how the ROM design (the OR matrix thereof) is structurally identical with the next-state table itself.

The previous example shows how the combinational portion of the same SN can be realized either with a ROM or with a PLA. This example, perhaps, is too simple to clearly contrast the two approaches. However, we can note that the PLA uses fewer product lines than the ROM, whose AND matrix is a full-blown decoder. When the difference in the numbers of product lines becomes substantial, then it may be desirable to forego the advantages of using a more standardized module (the ROM) in favor of a simpler PLA. However, when we want to retain some additional flexibility or when the network is so complex that

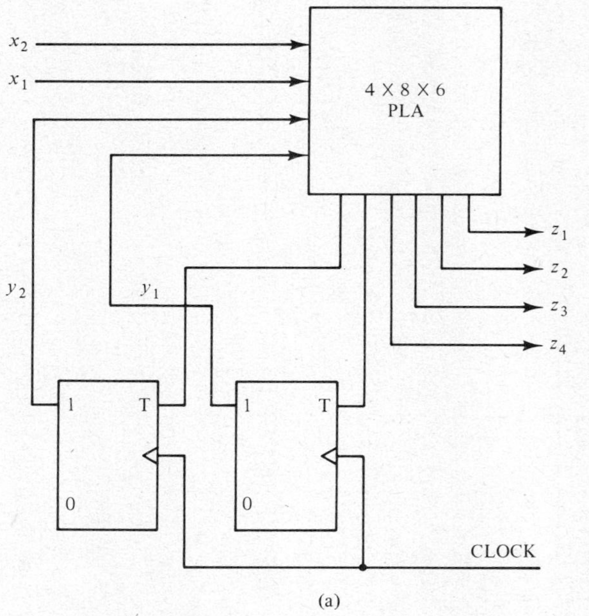

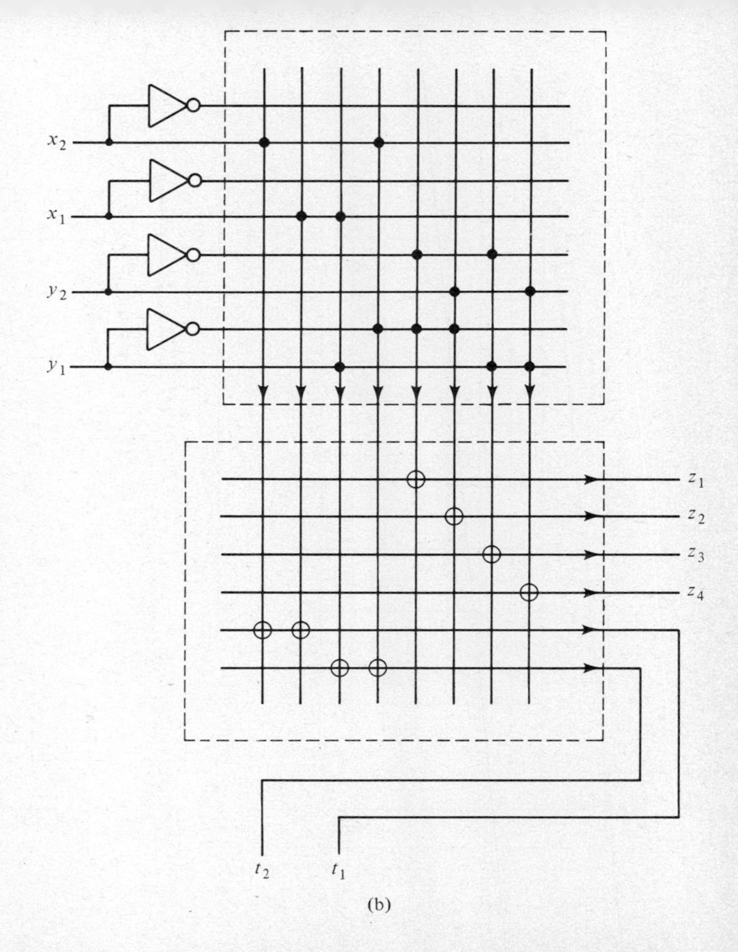

Figure 5.44 A PLA realization of a sequential network.

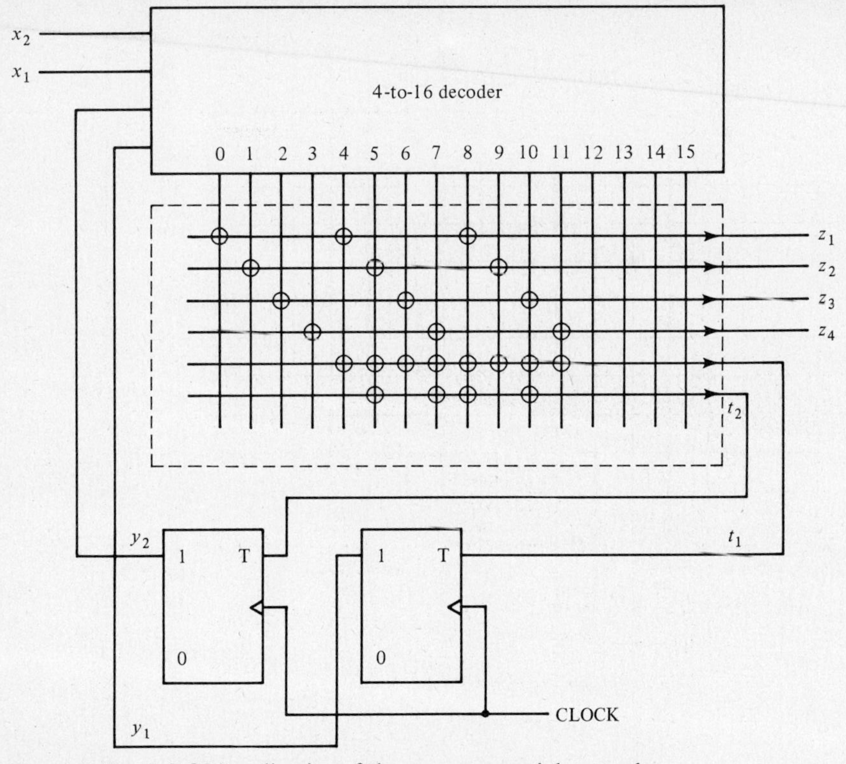

Figure 5.45 A ROM realization of the same sequential network.

the canonical form realization of the switching functions aids the observability of the network, then the ROM realization may be preferred. As we shall see, this is typically the situation that occurs in the design of the control unit of a digital computer, to be discussed in Chapter 9.

5.12 OTHER IMPORTANT SEQUENTIAL MODULES

In Sec. 5.7 we discussed parallel registers as important storage modules that are obtained by just putting together a number of flip-flops. There are other standard sequential modules, which we shall now briefly discuss: they are universal registers and counters.

5.12.1 Universal Registers

Starting from the register of Figure 5.25, which is loaded and unloaded in parallel, we may consider other modes of loading–unloading, that is: (1) parallel load–serial unload; (2) serial load–parallel unload; (3) serial load–serial unload; or any combination of these modes. A *universal register* encompasses all of the above modes, and will now be described.

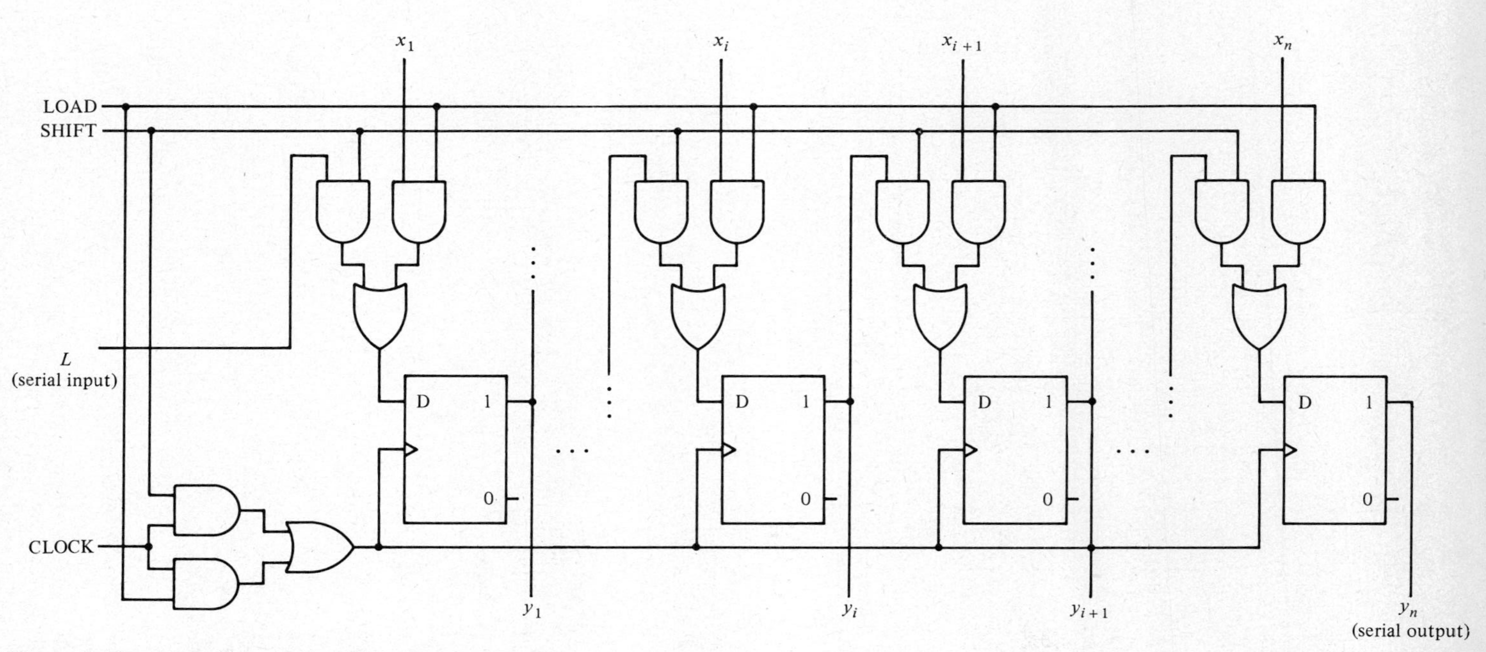

Figure 5.46 A parallel-access shift register.

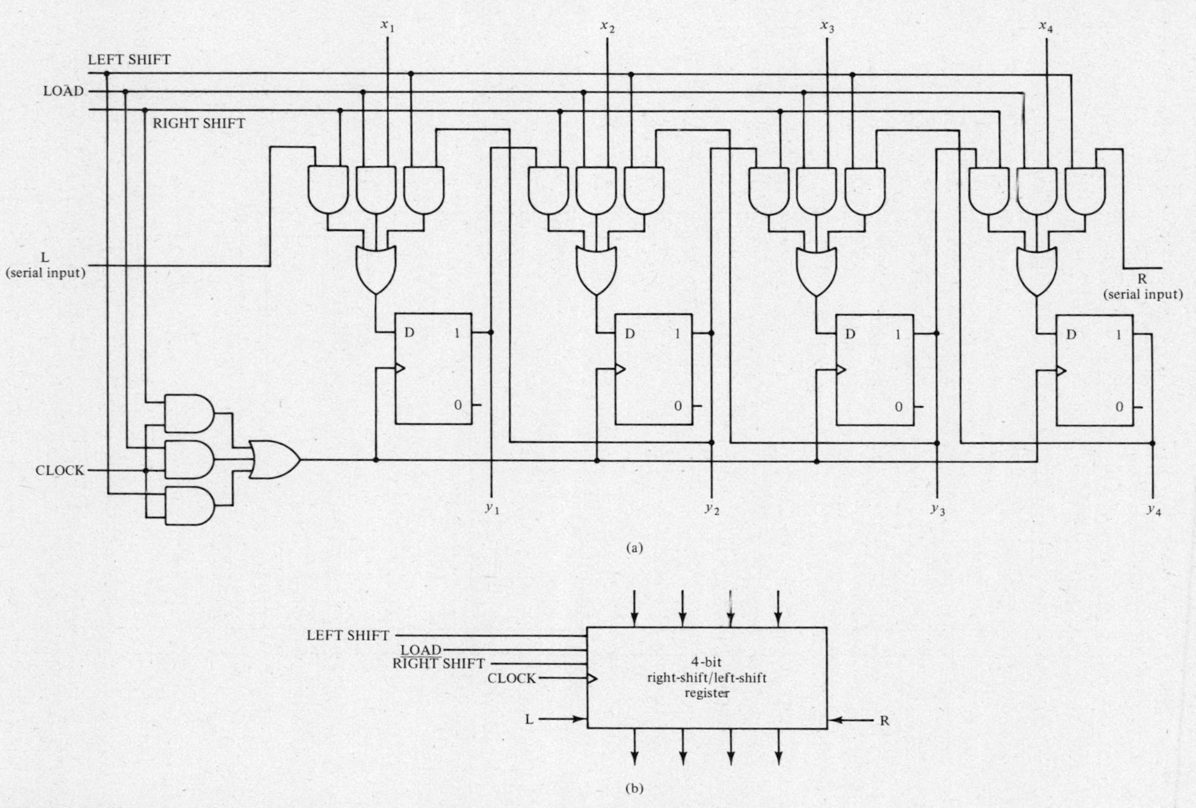

Figure 5.47 A 4-bit universal register (a) and its compact circuit symbol (b).

Since the parallel modes were described earlier, we only need consider the serial modes. To load a register serially, we feed the input sequence, one bit at a time, to a given flip-flop (say, the leftmost), while the content of the register is being shifted (to the right). Referring to Figure 5.46, the external signal L is fed to the D input of flip-flop y_1, while the shift interconnection is realized by feeding the output y_i to the D input of flip-flop y_{i+1}, for $i = 1, \ldots, n - 1$. Note however that since x_{i+1} is already feeding the input of flip-flop y_{i+1} (for the parallel load mode), the signals y_i and x_{i+1} must be multiplexed under control of the select signals SHIFT and LOAD, respectively: we are using here very simple narrow-sense multiplexers (see Sec. 4.8.4) both for the flip-flop D inputs and for the clock. Serial loading is completed by n shift operations.

Once the register has been loaded it can be serially unloaded by resorting again to shifting: at each clock time, the bit stored in the flip-flop y_i is transferred into flip-flop y_{i+1} and y_n becomes the serial output terminal; after n shift operations, the register is unloaded.

Finally, in some applications it is desirable to have a parallel-access shift register with capabilities of right shift (as described above) as well as of left shift. A self-explanatory scheme for a 4-bit universal register is shown in Figure 5.47(a), along with its compact circuit symbol [Figure 5.47(b)].

5.12.2 Counters

The *counter* is a sequential network with one input line x, called the count line, and n flip-flops so that each flip-flop output is also a network output. The set of outputs provides a coding of the number of (clock) times at which the count line x has been 1, starting from a chosen, initial instant. In some applications the count line is absent, which is equivalent to setting its value permanently to 1: in this case the counter simply steps through a sequence of states, resets itself to an initial state, and starts again. The network output is called the *count* and is normally coded in binary, although other codes are sometimes used (for example, a Gray code). In this section we shall restrict ourselves to binary codes.

The most basic type is the *binary counter,* which operates as follows:

> Let $(y_{n-1}y_{n-2} \ldots y_0)$ be the present content of the flip-flops; if $x = 0$ then $Y_j = y_j$ for $j = 0, 1, \ldots, n - 1$; if $x = 1$, then $(Y_{n-1}Y_{n-2} \ldots Y_0)$ is the binary equivalent of the integer represented by $(y_{n-1}y_{n-2} \ldots y_0) + 1$.

EXAMPLE 5.3

For $n = 4$, $(y_3y_2y_1y_0) = (1011)$, and $x = 1$, we have $(Y_3Y_2Y_1Y_0) = (1100)$.

The type of flip-flop which is best suited for use in counters is the T F/F, or, equivalently, the JK F/F with J and K connected together. By applying the general synthesis technique presented in Sec. 5.10 to the binary counter and denoting by t_i the input to F/F y_i we obtain the following results (the derivation is

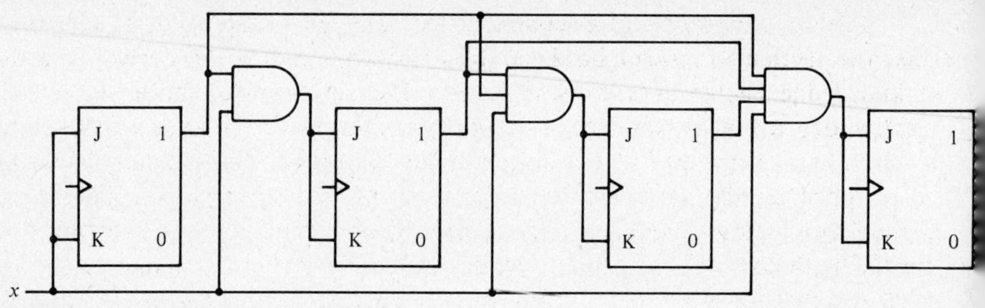

Figure 5.48 A 4-bit binary counter.

left as a problem for the reader):

$$t_0 = x$$
$$t_1 = xy_0$$
$$\vdots$$
$$t_i = xy_{i-1}y_{i-2} \cdot \cdot \cdot y_0$$

Choosing $n = 4$, and using T-connected JK flip-flops, we obtain the binary counter shown in Figure 5.48. (Of course, for safety of operation, we shall require trailing-edge switching, i.e., the use of master–slave flip-flops.)

NOTES AND REFERENCES

The concept of a state machine, that is, a machine with a set of identifiable states and with a rule to make a transition from the present state to the next state as determined by an external input, certainly goes back to the fundamental paper by Turing (1937). The notion of a sequential switching network goes back to a classical paper by McCulloch and Pitts (1943), where simple elements were connected to form time-dependent networks. Thus, the foundations of the theory of sequential networks and machines were in place in the early forties, and reached a mature state in the fifties by the contribution of several workers: we mention for their lasting significance the works of Mealy (1955) and Moore (1955) for synchronous networks, and of Huffman (1954) and Muller-Bartky (1957) for asynchronous networks.

Since then, the theory of sequential networks has evolved in several directions and has been extremely relevant to the entire computing field, both in relation to hardware and system designs (typically, for the design of control structures), and in relation to software theory and formal languages.

With reference to actual networks, sequential machine theory is now part of the background of digital network designers and is treated extensively in many textbooks already cited, such as Hill and Peterson (1974), Kohavi (1970), and Muroga (1979). It must be pointed out that this topic has attracted the attention of a large number of writers, not cited here just to avoid a lengthy list.

The abstract model of a sequential machine, as a mathematical structure (i.e., a transition structure), has been the starting point of a well-developed body of knowledge known as *automata theory*. The simplest automaton is, in our terminology, a synchronous sequential machine, also referred to as a finite-state machine (i.e., a machine with a finite number of states). This simple model can be enriched by adding new capabilities, such as "program" tapes, and so on, thereby developing a hierarchy of structures of increasing power. Contributed to mainly by theoretical computer scientists and mathematicians, automata theory reached a mature stage in the sixties. The interested reader may consult the book by Hartmanis and Stearns (1966), in addition to the text by Kohavi cited above. One important facet of automata theory is its relation to formal languages, in that for each automaton there is a family of strings of input symbols, each of which "brings" the automaton to a distinguished state (or states): this family of strings is called the language recognized by the automaton. The reader interested in the connection between automata and languages is referred to the excellent texts by Hopcroft and Ullman (1979) and the more recent one by Lewis and Papadimitriou (1981).

PROBLEMS

5.1. The following network ideally computes $z = x \oplus y$. Let the waveforms

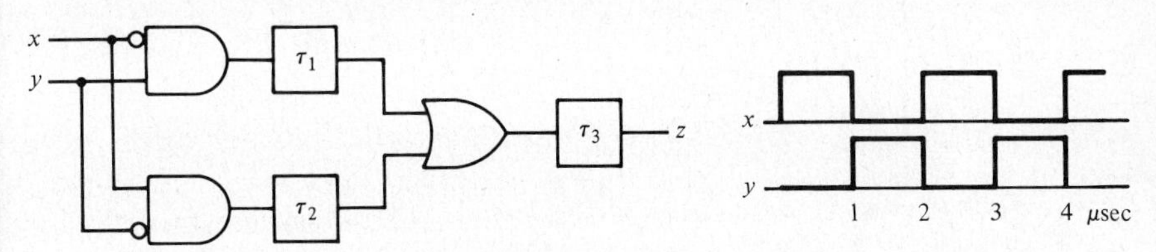

of x and y be as given.

(a) Show the waveform of z assuming $\tau_1 = \tau_2 = \tau_3 = 0$.

(b) Show the waveform of z assuming $\tau_1 = 0.1$ μsec, $\tau_2 = \tau_3 = 0.2$ μsec.

5.2. Assume $z_1 = z_2 = 0$ for $t < 0$ and that x has the waveform shown. Draw the waveforms of z_1 and z_2 assuming $\tau = 0.1$ μsec.

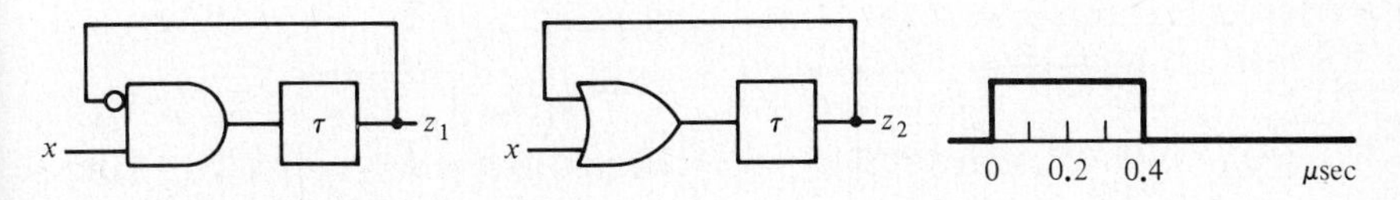

In the following problems all flip-flops are clocked.

5.3. Given the synchronous sequential network shown on page 166, analyze it and design a single finite input sequence such that, whatever the present state may be, the final state will be $(y_2, y_1) = (1, 1)$.

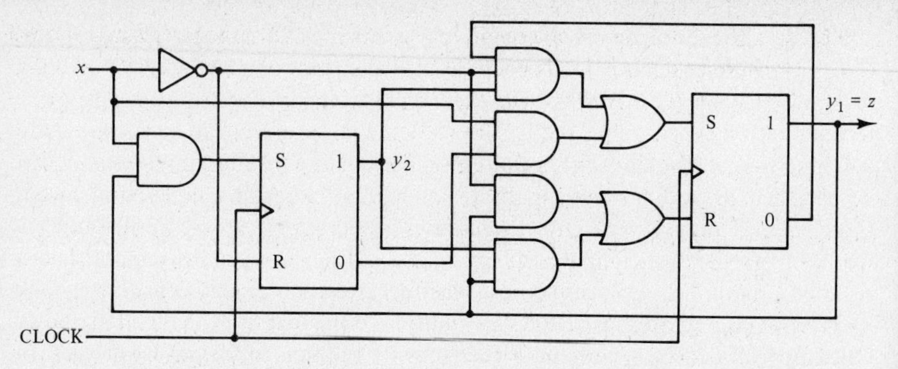

5.4. A sequential network has the following state diagram (output not shown). Assuming that the network is initially in state q_4 tabulate the sequence of states visited when the input sequence 0 1 1 1 0 1 0 is received.

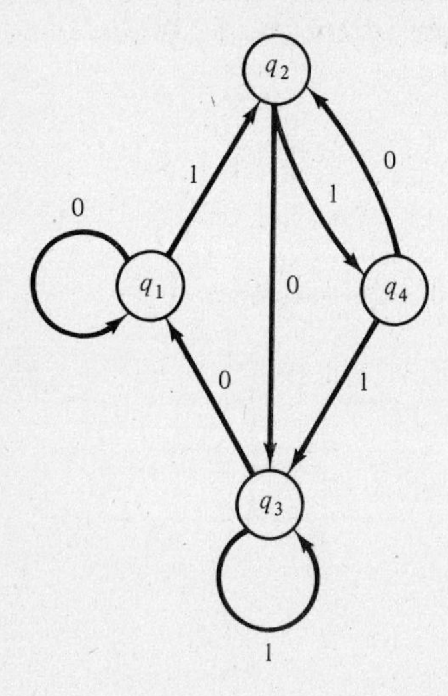

5.5. Given the following synchronous sequential network (without input lines and output lines)

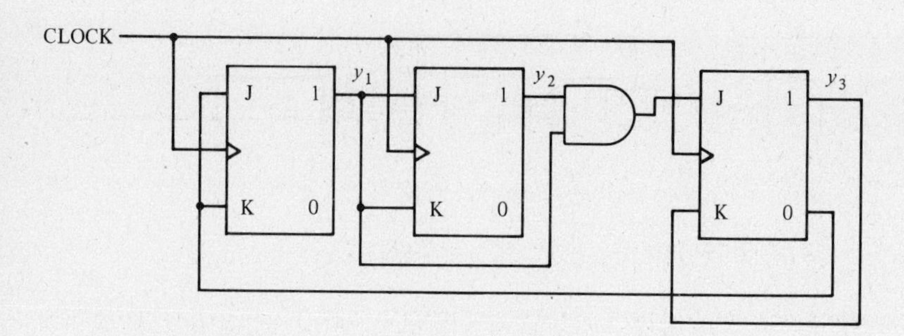

(a) Analyze it and construct the state diagram.
(b) What does the network do?

5.6. Draw the state diagram of a single input x, single output z sequence detector which produces a 1 (i.e., $z = 1$) whenever the sequence 0 1 0 1 is detected, and 0 at all other times. For example,

x . . . 0 0 1 0 1 0 1 . . .
z . . . 0 0 0 0 1 0 1 . . .

Your state diagram must have only four states.

5.7. Draw the state diagrams of the single input (x), single output (z) sequential machines, described as follows:

(a) $z = 1$ whenever the string of the last five inputs contains exactly three 1s and the string starts with two 1s.

(b) $z = 1$ whenever the sequence 1 1 1 1 occurs (overlapping sequences are accepted).

5.8. Derive a state diagram from the following word statement.

A sequential network has a binary input x and a binary output z. The output z is the same as the input x except in the following case: If two consecutive 1s appear in the input sequence, the output z is the complement of the input x for the two *input* symbols immediately following the two consecutive 1s.

A sample sequence is:

Input: 0 1 0 1 1 0 1 0 1 1 1 0 1 1 0 1 1 1 1 1 0 0 0
Output: 0 1 0 1 1 1 0 0 1 1 0 1 0 1 1 0 1 0 0 0 1 1 0

5.9. Given the following state diagram of a synchronous sequential network:

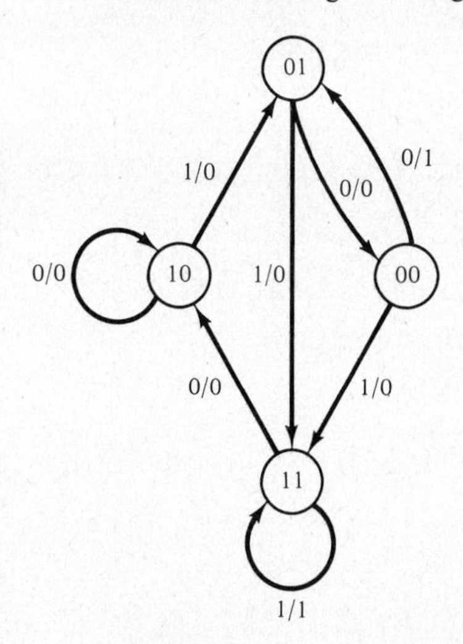

(a) Design the network using SR F/Fs.

(b) Design the network using JK F/Fs.

5.10. With reference to the network in Problem 5.3.

(a) Redesign this network (same state assignment) using D F/Fs.

(b) Same, using JK F/Fs.

5.11. Design the sequence detector of Problem 5.6 using now three D F/Fs and an appropriate state assignment (use the "window on the input sequence" method).

5.12. **(a)** Design a two-input x_1, x_2, one-output z *sequential adder,* that is, an SN which accepts two binary numbers of any length fed sequentially into x_1, x_2 from the least-significant digit, and produces the sum on z. For example:

x_1	0	0	1	0	0 . . .
x_2	1	0	1	1	0 . . .
z	1	0	0	0	1

$\longrightarrow$ From least significant to most significant

(b) Design a sequential subtractor [inputs $x_1 x_2$ as in (a); the output z gives the difference of the two numbers]. [Note: In (a) and (b) use RS F/Fs.)

5.13. Design a three-input a, b, c, one-output z SN, which accepts three binary numbers A, B, C of any length fed sequentially one-bit position at a time starting from the least-significant digit on lines a, b, c, and produces on z the value $A + B - C$. (To aid your reasoning, assume it known that at completion $A + B \geq C$). Use D F/Fs, and a ROM to realize the combinational part of the SN.

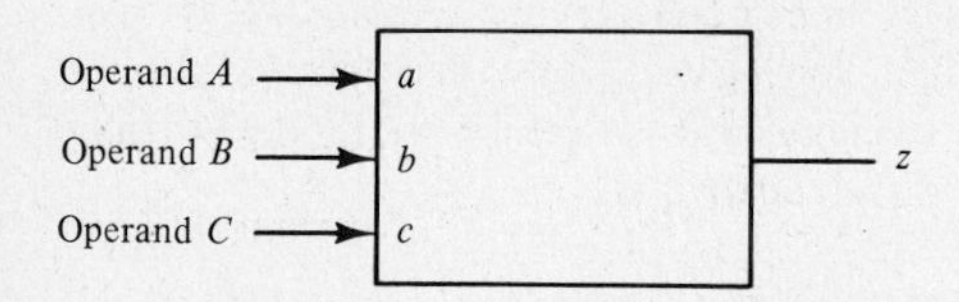

5.14. Draw the state diagram for a single input x, single output z SN, which interprets the input stream as a sequence of consecutive messages of 4 bits, and generates on z a *parity bit* at the end of each message. For example:

x	0	1	1	0	1	0	0	0	1	1	0	1	0	0	0	0
z	δ	δ	δ	0	δ	δ	δ	1	δ	δ	δ	1	δ	δ	δ	0

Parity bits

(Parity bit = 1 if the message has odd number of 1s.) Try to use the smallest number of states.

5.15. The register below is loaded in parallel with the configuration 0 1 1 0 1 0 1 1. Subsequently its content is shifted eight times. Give the content of the register after this operation.

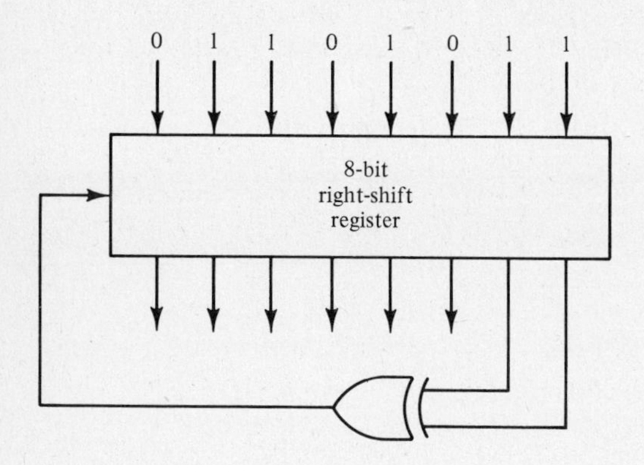

5.16. Design the combinational portion of the sequential network shown in the accompanying figure using a PLA. The network is a modulo-6 counter. [*A modulo-k counter counts in binary the number of occurrences of 1 on the input line, starting from 0 and making a transition back to 0 from (k − 1).*] The content of the counter is to be displayed on a 7-segment display.

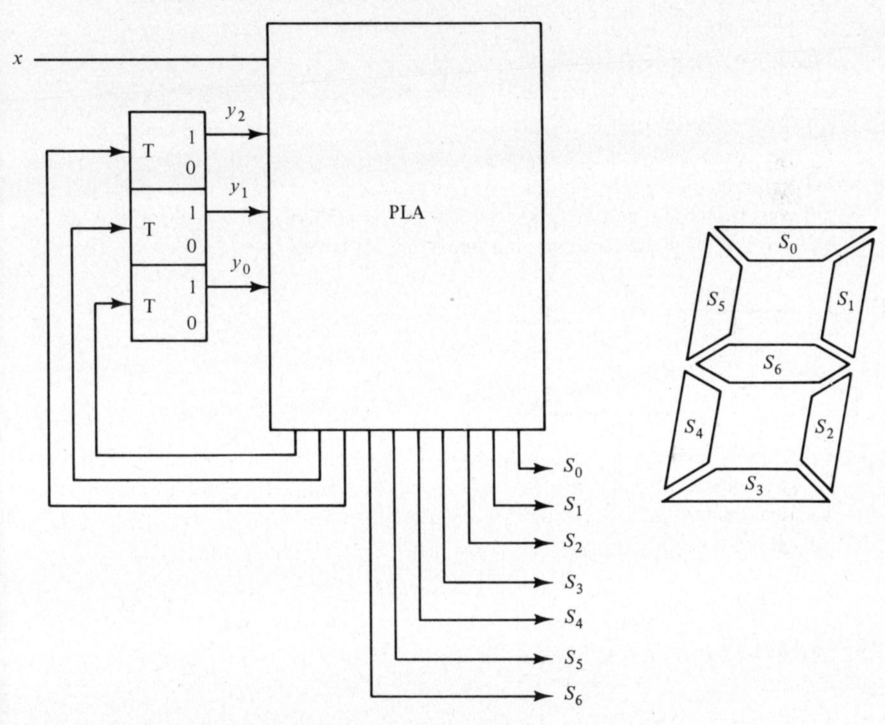

5.17. Repeat Problem 5.16 by realizing the combinational portion with a $2^4 \times 10$ ROM.

5.18. Two 7-bit right-shift registers are connected to a binary half-adder as shown below and are initially loaded as indicated. (Recall that the half-adder is described by the expressions $S = A\overline{B} + \overline{A}B$, $c = AB$.) The registers are shifted at each clock time. Give the register contents after five shifts.

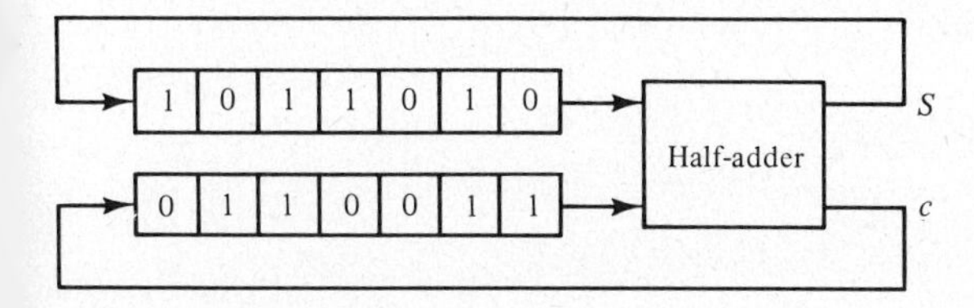

5.19. Using JK F/Fs, design a modulo-4 *Gray code counter* using two different state assignment approaches:

(**a**) Design a regular binary modulo-4 counter and transform the (y_2, y_1) signals to appropriate Gray code using combinational logic.

(**b**) Design the Gray code counter from scratch.

5.20. Design a modulo-7 counter using T F/Fs. (See Problem 5.16 for a definition of a modulo-*k* counter.)

PART THREE

SYSTEM ORGANIZATION

Chapter 6

Binary Arithmetic and the Arithmetic-Logic Unit (ALU)

6.1 ADDITION, SUBTRACTION, AND THE REPRESENTATION OF NEGATIVE NUMBERS

In Chapter 1 (Sec. 1.4) we briefly introduced the operations of addition and subtraction of binary numbers; subsequently, in Chapter 2 (Sec. 2.2) we introduced a number format in which numbers—positive and negative—are represented in *modulus and sign,* that is, by means of a sign bit (1 for negative, 0 for nonnegative numbers) and a "numerical portion" representing the absolute value of the number.

Let us now examine critically how to organize a digital system designed to perform arithmetic operations with numbers in modulus-and-sign representation. First, we would need two separate subsystems to carry out the operations of addition and subtraction, respectively, that is, an *adder* and a *subtractor*. The use of these two units is concisely described in Figure 6.1, as determined by the operation (addition or subtraction) and by the signs of the operands (same or opposite).

Instruction \ Operand signs	Same	Opposite
Addition	Adder	Subtractor
Subtraction	Subtractor	Adder

Figure 6.1 Use of adder and subtractor for given operation and operand signs.

This illustration suggests that we may restrict our attention to the row of "addition." Indeed, with regard to subtraction, the case in which minuend and subtrahend have opposite signs is equivalent to the addition of addends with identical signs; analogously, the case in which the minuend's and subtrahend's signs are identical is equivalent to the addition of two addends with opposite signs. This shows that all cases can be reduced to the operation of addition. Considering just addition, however, we see that the choice of the subsystem (adder or subtractor) depends upon the signs of the operands. In other words, which subsystem is used depends not only upon the operation code (chosen by the user) but also upon the operand signs, which are whatever they are. This situation is certainly a major drawback, because the operand signs must also be inspected before one of the two subsystems can be chosen. This reason alone makes the modulus-and-sign notation quite unsatisfactory for most arithmetic purposes. Practically all existing computers adopt a different type of representation of *relative integers* (i.e., negative and nonnegative integers) called *complement representations* or *complement notations*. Such representations not only overcome the drawback mentioned earlier, but accrue a double advantage:

1. Only one functional unit (the adder) is required.
2. The data transfers are completely determined by the instruction code.

Before entering a detailed discussion of the complement notations, we should realize that the idea is not novel to us, although we may not be fully aware of it. In Figure 6.2 we have shown the usual frame of reference for measuring polar (or central) angles, as in circle trigonometry. The angle formed by a ray R with the polar axis (the x axis in the Cartesian plane) is measured counterclockwise in *degrees,* and this measure ranges from 0° to 360°. If the angle exceeds 180° (the midrange point) sometimes we refer to it as a negative angle: indeed, the 217° angle shown in Figure 6.2 could also be referred to as a (−143°) angle.

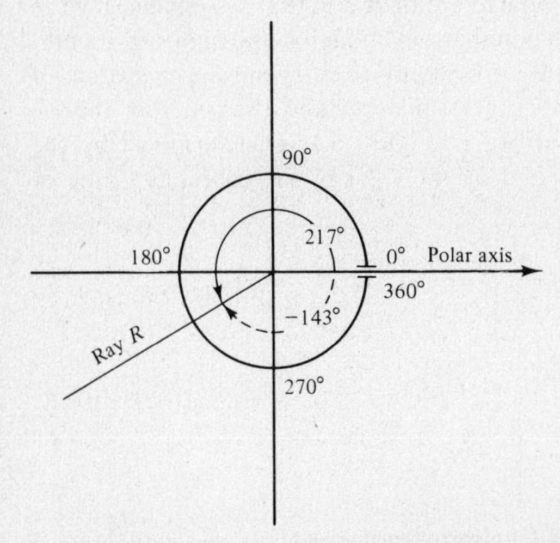

Figure 6.2 The frame of reference for polar angles.

By so doing, we are measuring the same angle either as $217° = 180° + 37°$ (no negative angles being allowed) or as $-143° = -180° + 37°$, where we note that 37 is the *complement* of 143 to 180. We could even say that all angles α in the range $0° \le \alpha < 180°$ are treated as nonnegative, while those in the range $180° \le \alpha < 360°$ are considered as negative.

This familiar notion is exactly what is involved in the complement representations of integers. Specifically, suppose that our integers are represented with n bits, so that we have the 2^n distinct binary integers $(0 \ldots 00)$, $(0 \ldots 01)$, . . . , $(1 \ldots 11)$. Suppose now that we subdivide the 360° range into 2^n intervals of identical width and that we number (in binary) each of these intervals as shown in Figure 6.3. As we did earlier for the angles in the range [0°, 360°], we now regard all binary numbers assigned to intervals in the lower semicircle as representing negative integers, as shown in Figure 6.3. This is precisely an illustration of the so-called "two's complement notation," which we now discuss in detail.

Let

$$B \equiv \underset{\substack{\uparrow \\ \text{Sign bit}}}{b_{n-1}} \underset{\substack{\uparrow \\ \text{Numerical portion}}}{\underbrace{b_{n-2}b_{n-3} \ldots b_0}} \qquad \text{Representation} \qquad (6.1)$$

be the representation of an integer B, where, as usual, b_{n-1} is the *sign bit* and $(b_{n-2}b_{n-3} \ldots b_0)$ form the "numerical portion." The value of B will be given by the formula

$$B = -b_{n-1}2^{n-1} + \sum_{j=0}^{n-2} b_j 2^j \qquad \text{Value} \qquad (6.2)$$

that is, whereas the bit positions in the numerical portion are given the same weights as in the modulus-and-sign notation, the sign bit b_{n-1} is given the (somewhat surprising) negative weight -2^{n-1}.

If $b_{n-1} = 0$, in Eq. (6.2) the first term in the right side is 0, so that

$$B = \sum_{j=0}^{n-2} b_j 2^j \ge 0$$

exactly as in the modulus-and-sign notation. All nonnegative numbers have the sign bit equal to 0; they are the numbers in the upper semicircle in Figure 6.3.

However, if $b_{n-1} = 1$, the first term in the right side of Eq. (6.2) becomes -2^{n-1}. Since the largest value of $\Sigma_{j=0}^{n-2} b_j 2^j$ is $\Sigma_{j=0}^{n-2} 2^j = 2^{n-1} - 1$, we see that the right side in Eq. (6.2) has a negative value always. Thus, negative numbers have the sign bit equal to 1; they are the numbers in the lower semicircle of Figure 6.3. Also, when $b_{n-1} = 1$, the numerical portion of $b_{n-1}b_{n-2} \ldots b_0$ no longer represents the absolute value $|B|$ of B, as it did in the modulus-and-sign notation. In fact, in this case, the right side of Eq. (6.2) must have value $-|B|$, whence

$$-|B| = -2^{n-1} + \sum_{j=0}^{n-2} b_j 2^j$$

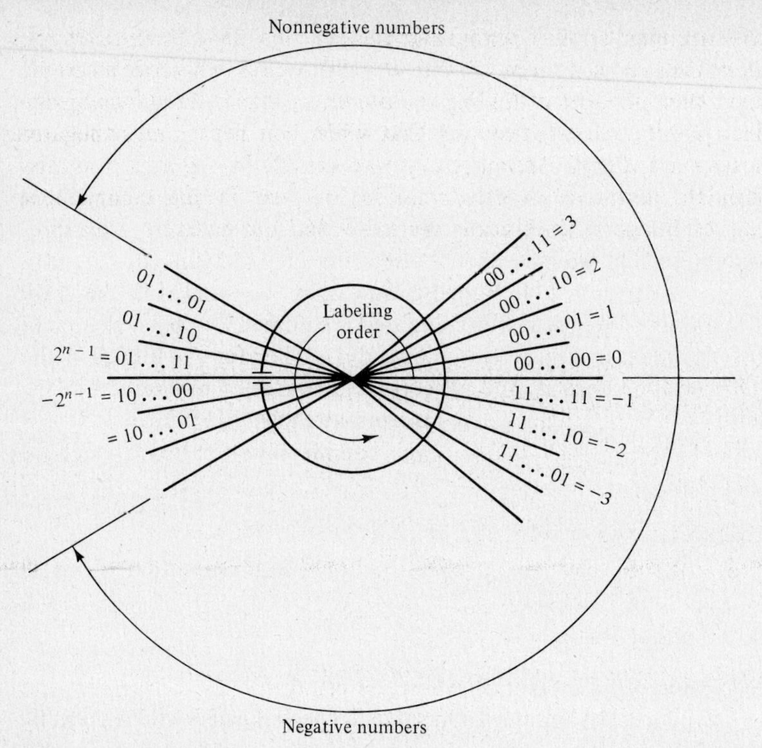

Figure 6.3 Illustration of the integers $[-2^{n-1}, 2^{n-1} - 1]$ as measures of polar angles (complement notation).

and, from this,

$$\sum_{j=0}^{n-2} b_j 2^j = 2^{n-1} - |B|$$

Clearly, $2^{n-1} - |B|$ is what must be added to $|B|$ to obtain 2^{n-1}, that is, it is the "complement of $|B|$ to 2^{n-1}." We summarize this by the following:

Two's Complement Notation with *n* Bits. For a nonnegative integer B the sign bit is 0 and the numerical portion contains the *modulus of B*; for a negative integer B, the sign bit is 1 and the numerical portion contains the *complement to* 2^{n-1} *of the modulus of B.*

Notice that the largest nonnegative integer representable with n bits is 011 . . . 1, whose value is $2^{n-1} - 1$; the smallest negative integer is 100 . . . 0, whose value is -2^{n-1}.

EXAMPLE 6.1

Let $n = 6$ (We shall deal with 6-bit numbers, the leftmost bit being the sign bit).

$N = 001101 \rightarrow b_5 = 0, (01101)_2 = 13_{10},$ whence $N = 13$

$N = 100110 \rightarrow b_5 = 1, (00110)_2 = 6_{10},$ whence $|N| = 2^5 - 6 = 26$

Before examining how to perform arithmetic, we must analyze how to obtain *the opposite* $-B$ of a given number B, represented in the two's complement notation. The operation of *finding the opposite,* that is, of *changing* sign, will be called in this text *negation*[1] (so that when you negate a nonnegative number you obtain a negative number and vice versa). In negating a number (represented in the two's complement notation) we start from a binary string $b_{n-1}b_{n-2} \ldots b_0$, whose value B, according to Eq. (6.2), is given by

$$B = -b_{n-1}2^{n-1} + \sum_{j=0}^{n-2} b_j 2^j \tag{6.3}$$

and we want to find a number $b'_{n-1}b'_{n-2} \ldots b'_0$ (the opposite of $b_{n-1}b_{n-2} \ldots b_0$) such that

$$-B = -b'_{n-1}2^{n-1} + \sum_{j=0}^{n-2} b'_j 2^j \tag{6.4}$$

If we now change the sign of both sides of Eq. (6.2) we obtain

$$-B = b_{n-1}2^{n-1} - \sum_{j=0}^{n-2} b_j 2^j \tag{6.5}$$

We now add and subtract 2^{n-1} from the right side of Eq. (6.5) and get

$$\begin{aligned} -B &= -2^{n-1} + b_{n-1}2^{n-1} + 2^{n-1} - \sum_{j=0}^{n-2} b_j 2^j \\ &= -(1 - b_{n-1})2^{n-1} + \left(\sum_{j=0}^{n-2} 2^j + 1 - \sum_{j=0}^{n-2} b_j 2^j \right) \end{aligned}$$

since $\Sigma_{j=0}^{n-2} 2^j = 2^{n-1} - 1$. Finally, we easily have

$$-B = -(1 - b_{n-1})2^{n-1} + \sum_{j=0}^{n-2} (1 - b_j)2^j + 1 \tag{6.6}$$

Equating the right sides of Eqs. (6.4) and (6.6) and noting that the identity holds for any choice of $b_{n-1}b_{n-2} \ldots b_0$, we obtain the following rule.

Rule for Two's Complement Negation. The opposite $b'_{n-1}b'_{n-2} \ldots b'_0$ of $b_{n-1} b_{n-2} \ldots b_0$ is obtained by replacing each b_j with $(1 - b_j)$ (i.e., changing each 1 to 0 and vice versa) and adding 1 to the resulting number.

Notice that $(1 - b_j)$ is "the complement to 1 of b_j," since if $b_j = 0$ then $1 - b_j = 1$ and vice versa. Thus, the string obtained from a given string by changing each of its bits to its complement to 1 (or, if you wish, to its boolean complement) is usually called the *bit-by-bit complement* of the given string. With this terminology, the above rule can be rephrased as: "The opposite of $b_{n-1}b_{n-2} \ldots b_0$ is obtained by adding 1 to its bit-by-bit complement."

[1]Frequently this operation is called *complementation.* Unfortunately, this choice of term is the origin of considerable confusion between the notions "to represent a number is two's-complement notation" and "to complement a number" (i.e., to change its sign).

Now, assume that b_k $(0 \leq k \leq n - 2)$ is the least-significant 1 bit in the string $b_{n-2}b_{n-3} \ldots b_0$, that is, $b_k b_{k-1} \ldots b_1 b_0 = 10 \ldots 00$. The bit-by-bit complement of $b_k b_{k-1} \ldots b_1 b_0$ is $01 \ldots 11$. When we add the integer 1 (represented as $00 \ldots 01$) to this string considered as a binary number, we have

2^k	2^{k-1}	2^{k-2}	. . .	2^1	2^0	weights
1	1	1	. . .	1		carries
0	1	1	. . .	1	1	+
0	0	0	. . .	0	1	
1	0	0	. . .	0	0	sum

that is, the carry propagates all the way to the position of weight 2^k, where it stops. This means that $b'_j = b_j$ for $j = 0, 1, \ldots, k$, whereas $b'_j = 1 - b_j$ for $j = k + 1, k + 2, \ldots, n - 1$. This is expressed by the following:

Alternative Rule for Two's Complement Negation. The opposite $b'_{n-1}b'_{n-2} \ldots b'_0$ of $b_{n-1}b_{n-2} \ldots b_0$ is obtained by complementing all and only the bits to the left of the least-significant 1.

EXAMPLE 6.2

Find the opposite of $B \equiv 010110100$ (i.e., "negate" $B \equiv 010110100$):

Bit-by-bit complement	101001011
+1	000000001
	101001100

Alternatively,

```
010110|100
      |^_______Least-significant 1
101001|100
```

Finally, suppose we wish to obtain the opposite B' of the number $B \equiv 10 \ldots 0$, whose value is -2^{n-1}. The value of B' is obviously 2^{n-1}. But we know that such positive number is not representable with n bits, since the largest positive n-bit number in the two's complement notation has value $2^{n-1} - 1$. If we require that the opposite of a number be always defined (that is, given any number, its opposite can also be represented), we may restrict to $[-2^{n-1} + 1, 2^{n-1} - 1]$ the interval of admissible integers.

6.2 ADDITION AND SUBTRACTION OF INTEGERS IN THE TWO'S COMPLEMENT NOTATION

As indicated at the beginning of Sec. 6.1, the fundamental advantage of the complement notation is that the mechanics of addition of signed integers is independent of the operand signs; since we noted that subtraction can be viewed as an instance of addition, then only one subsystem, the adder, is necessary. In this section we shall substantiate this claim.

We begin by viewing the operation of addition in terms of the interpretation of numbers as angles, previously introduced in Figures 6.2 and 6.3. As we saw earlier, in the complement notation both nonnegative and negative numbers are represented as counterclockwise angles, the former with angles not exceeding 180°. Let A and B be two positive numbers, whose corresponding angles are shown in Figures 6.4(a) and 6.4(b). If we wish to compute $A + B$, we just trace counterclockwise (i.e., in the positive direction) an angle equal to B starting from the upper ray of angle A: the resulting angle corresponds to $A + B$ [Figure 6.4(c)]. Suppose now we wish to compute $A - B$. Obviously, this can be done by tracing clockwise (i.e., in the negative direction) an angle equal to B starting again from the upper ray of A. Alternatively we may add to angle A an angle equal to $360° - B$, thereby obtaining the same result as subtracting B from A. This is exactly the mechanism of the complement notation.

Let us first consider the addition of identical-sign operands A and B. The basic requirement is that the available n bit positions be sufficient to represent the sum; if A and B are positive, then we must have $(A + B) < 2^{n-1}$, while if A and B are both negative then we must have $|A + B| \leq 2^{n-1}$. If either of these conditions is violated, we say that we have an *overflow*. The detection of overflows is of crucial importance. We let $A \equiv a_{n-1}a_{n-2} \ldots a_0$ and $B \equiv b_{n-1} b_{n-2} \ldots b_0$.

Case 1. $A \geq 0$, $B \geq 0$ In this case $a_{n-1} = b_{n-1} = 0$, and the numerical parts $a_{n-2} \ldots a_0$ and $b_{n-2} \ldots b_0$, respectively, represent $|A|$ and $|B|$. The sum S of the

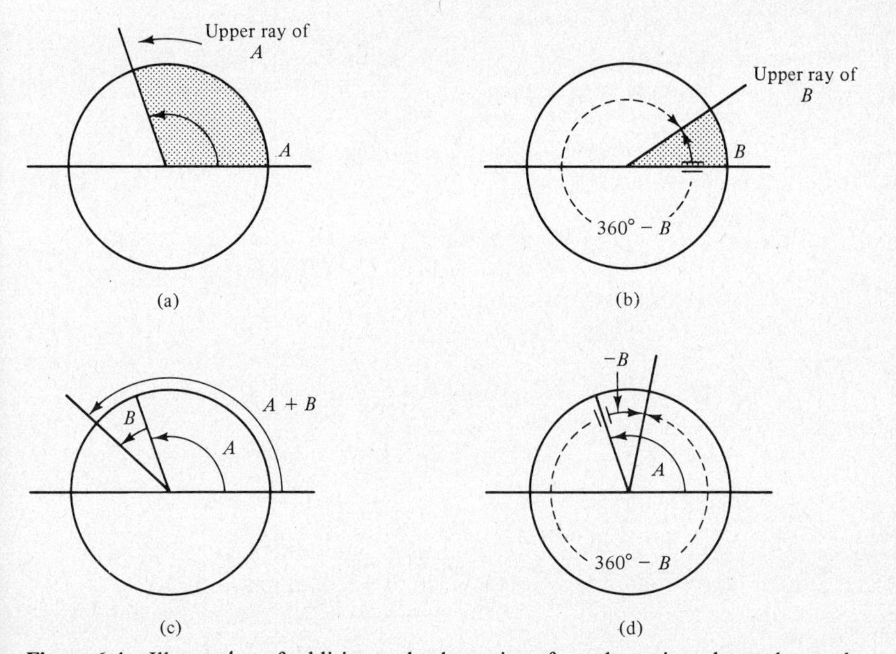

Figure 6.4 Illustration of addition and subtraction of numbers viewed as polar angles.

numerical parts is thus $|A| + |B|$. Since the largest number representable by the numerical part $s_{n-2} \ldots s_0$ of the sum is $2^{n-1} - 1$, whenever $|A| + |B| > 2^{n-1} - 1$, overflow occurs. In fact, since $S = |A| + |B| \geq 2^{n-1}$, we have a carry $c_{n-1} = 1$ (of weight 2^{n-1}) into the sign position (of weight -2^{n-1}). This yields a wrong sign bit for the sum $s_{n-1} = 1$ (the sum is a negative number?!) and a carry $c_n = 0$.

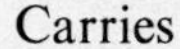

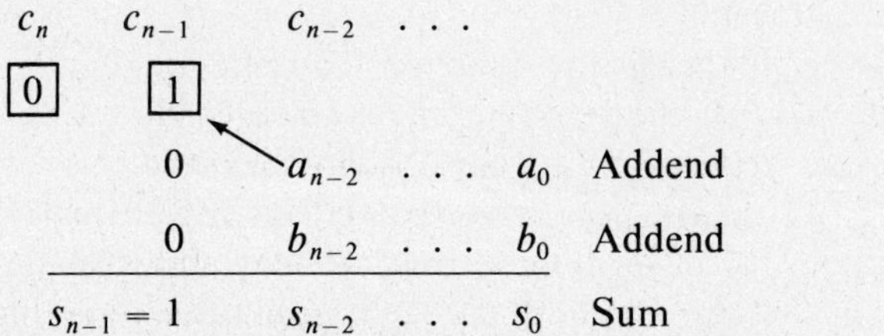

c_n c_{n-1} c_{n-2} . . .
0
1
0 a_{n-2} . . . a_0 Addend
0 b_{n-2} . . . b_0 Addend
$s_{n-1} = 1$ s_{n-2} . . . s_0 Sum

Overflow in the Addition of Positive Operands

In the next example, as well as in the subsequent ones, we shall use 6-bit operands in the two's complement representation, that is, $2^{n-1} - 1 = 31$ and the range of the numerical portion is [0, 31].

EXAMPLE 6.3

$|A| + |B| > 31.$

Notice: Overflow							
	[0	1]	1	1	1	1	Carries
		0	0	1	1	0 1	$A = 13$
		0	1	0	1	1 1	$B = 23$
		1	0	0	1	0 0	$S = -28$, wrong

However, when $|A| + |B| \leq 2^{n-1} - 1$, we have $c_{n-1} = 0$ and the numerical portion $s_{n-2} \ldots s_0$ correctly represents $|A| + |B|$. Notice also that $c_n = 0$ in this case.

EXAMPLE 6.4

$|A| + |B| \leq 31.$

0	0	1	0	0	0		Carries
	0	0	1	1	0	1	$A = 13$
	0	0	1	0	0	0	$B = 8$
	0	1	0	1	0	1	$S = 21$, correct

Case 2. $A < 0$, $B < 0$ In this case $a_{n-1} = b_{n-1} = 1$ and $a_{n-2} \ldots a_0 = 2^{n-1} - |A|$, $b_{n-2} \ldots b_0 = 2^{n-1} - |B|$. The sum S of the numerical parts is given by

$$S = 2^{n-1} - |A| + 2^{n-1} - |B| = 2^{n-1} - (|A| + |B| - 2^{n-1})$$

Now, overflow occurs whenever $|A| + |B| > 2^{n-1}$. In this case we obtain $S < 2^{n-1}$, that is, there is no 1-carry generated into the sign position (i.e., $c_{n-1} = 0$). As a consequence the sign bit s_{n-1} of the sum is 0 (which is erroneous) and $c_n = 1$.

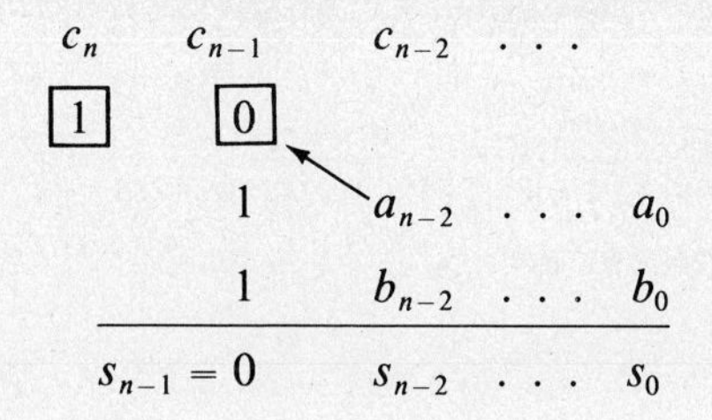

Overflow in the Addition of Negative Operands

EXAMPLE 6.5

$|A| + |B| > 32.$

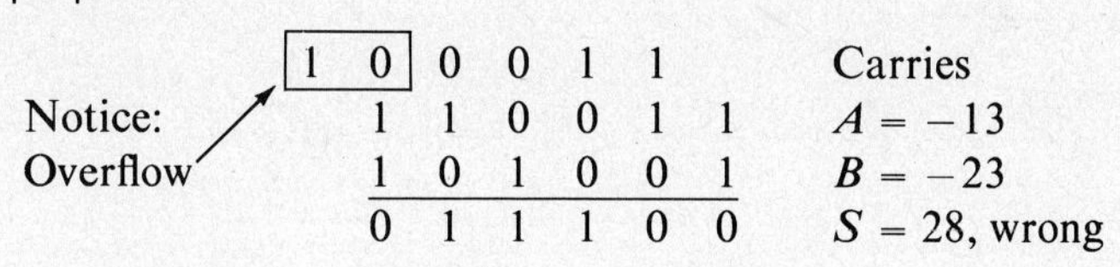

However, when $|A| + |B| \leq 2^{n-1}$, we obtain $S \geq 2^{n-1}$. Thus the numerical part $s_{n-2} \ldots s_0$ contains $S - 2^{n-1} = 2^{n-1} - (|A| + |B|)$, which is correct, s_{n-1} is equal to 1 (which is also correct), and c_n is also equal to 1.

EXAMPLE 6.6

$|A| + |B| \leq 32.$

1	1	0	0	0	0		Carries
	1	1	0	0	1	1	$A = -13$
	1	1	1	0	0	0	$B = -8$
	1	0	1	0	1	1	$S = -32 + 11 = -21$, correct

We now consider the case of opposite-sign operands. Without loss of generality, we shall assume that $A \geq 0$ and $B < 0$. We must still distinguish whether $|A| \geq |B|$ or $|A| < |B|$. Notice that no overflow will ever occur when adding two operands with opposite signs.

Case 3. $A \geq 0, B < 0, |A| \geq |B|$ In this case $a_{n-1} = 0$ and $b_{n-1} = 1$; also $a_{n-2} \ldots a_0 = |A|$ while $b_{n-2} \ldots b_0 = 2^{n-1} - |B|$. Thus, the sum of the numerical parts has value $2^{n-1} + |A| - |B| \geq 2^{n-1}$. This implies that a carry $c_{n-1} = 1$ is generated; the weight of this carry is 2^{n-1}, which correctly leaves $|A| - |B|$

$$
\begin{array}{ccccc}
c_n & c_{n-1} & c_{n-2} & \ldots & \\
\boxed{1} & \boxed{1} & & & \\
 & 0 & a_{n-2} & \ldots & a_0 \\
 & 1 & b_{n-2} & \ldots & b_0 \\
\hline
 & s_{n-1} = 0 & s_{n-2} & \ldots & s_0
\end{array}
$$

Addition of Opposite-sign operands; Non-negative Result

in the numerical part $s_{n-2} \ldots s_0$ of the sum. Moreover the carry $c_{n-1} = 1$ is added to $b_{n-1} = 1$, yielding $s_{n-1} = 0$ and $c_n = 1$. Notice that c_{n-1} and b_{n-1} cancel each other because they have weights 2^{n-1} and -2^{n-1}, respectively, so that $s_{n-1} = 0$ is the correct sign bit of the sum.

EXAMPLE 6.7

$A \geq 0, B < 0, |A| \geq |B|$.

1	1	0	0	1	1		Carries
	0	1	1	0	0	1	$A = 25$
	1	1	0	0	1	1	$B = -13$
	0	0	1	1	0	0	$S = 12$

Case 4. $A \geq 0, B < 0, |A| < |B|$ Again, $a_{n-2} \ldots a_0 = |A|$ and $b_{n-2} \ldots b_0 = 2^{n-1} - |B|$. The sum of the numerical parts has value $2^{n-1} - (|B| - |A|) < 2^{n-1}$, that is, the carry c_{n-1} has value 0. Thus $s_{n-2} \ldots s_0 = 2^{n-1} - (|B| - |A|)$, correctly because the sum is negative; moreover, the sum of c_{n-1},

$$
\begin{array}{cccccc}
c_n & c_{n-1} & c_{n-2} & \ldots & & \\
\boxed{0} & \boxed{0} & & & & \\
 & 0 & a_{n-2} & \ldots & a_0 & \\
 & 1 & b_{n-2} & \ldots & b_0 & \\
\hline
 & s_{n-1} = 1 & s_{n-2} & \ldots & s_0 &
\end{array}
$$

Addition of Opposite-Sign Operands; Negative Result

a_{n-1}, and b_{n-1} gives $s_{n-1} = 1$ (correctly, because the sum is negative) and $c_n = 0$.

EXAMPLE 6.8

$A \geq 0, B < 0, |A| < |B|$.

0	0	1	1	1	1		Carries
	0	0	1	1	0	1	$A = 13$
	1	0	0	1	1	1	$B = -25$
	1	1	0	1	0	0	$S = -32 + 20 = -12$

We summarize the preceding analysis in the table given in Figure 6.5, where the condition of $c_n c_{n-1}$ is explicitly considered.

B \ A	≥ 0	< 0
≥ 0	Overflow if $c_n c_{n-1} = 01$	Always correct result $c_n c_{n-1} = \begin{cases} 00 \\ 11 \end{cases}$
< 0	Always correct result $c_n c_{n-1} = \begin{cases} 00 \\ 11 \end{cases}$	Overflow if $c_n c_{n-1} = 10$

Figure 6.5 Summary table of addition of numbers in two's complement notation.

Inspection of this table yields the following simple rule for overflow detection:

Correct operation occurs when c_n and c_{n-1} are identical. Overflow occurs whenever c_n and c_{n-1} are different.

*6.3 THE ONE'S COMPLEMENT NOTATION

We now mention another complement notation for signed integers, the one's complement notation, which finds application in some computing systems, because some designers find it attractive. As we shall see shortly, there are pros and cons in its adoption.

The one's complement notation is defined by saying that the numerical value of the n-bit string $b_{n-1}b_{n-2} \ldots b_0$ is

$$B = -b_{n-1}(2^{n-1} - 1) + \sum_{j=0}^{n-2} b_j 2^j \tag{6.7}$$

that is, *the weight of the sign bit* b_{n-1} *is* $-(2^{n-1} - 1)$ *rather than* -2^{n-1}. Notice that when B is negative ($b_{n-1} = 1$) the numerical part $b_{n-2}b_{n-3} \ldots b_0$ represents the number $(2^{n-1} - 1 - |B|)$ that is, the *complement of* $|B|$ *to* $2^{n-1} - 1$.

To obtain the opposite $b'_{n-1}b'_{n-2} \ldots b'_0$ of $B \equiv b_{n-1}b_{n-2} \ldots b_0$ we argue as in connection with the two's complement notation, and obtain

$$\begin{aligned} -B &= b_{n-1}(2^{n-1} - 1) - \sum_{j=0}^{n-2} b_j 2^j \\ &= -(2^{n-1} - 1) + b_{n-1}(2^{n-1} - 1) + (2^{n-1} - 1) - \sum_{j=0}^{n-2} b_j 2^j \\ &= -(1 - b_{n-1})(2^{n-1} - 1) + \sum_{j=0}^{n-2} 2^j - \sum_{j=0}^{n-2} b_j 2^j \\ &= -(1 - b_{n-1})(2^{n-1} - 1) + \sum_{j=0}^{n-2} (1 - b_j)2^j \end{aligned}$$

which yields the simple identities $b'_j = 1 - b_j$ for $j = 0, 1, \ldots, n - 1$, or the following rule:

Rule for One's Complement Negation. The opposite $b'_{n-1}b'_{n-2} \ldots b'_0$ of b_{n-1} $b_{n-2} \ldots b_0$ is obtained as the bit-by-bit complement of it.

The simplicity of this rule for negation (compare it with the corresponding one for the two's complement notation) is what makes the one's complement notation attractive. On the other hand, the gain in simplicity in computing the opposite of a number is offset by some complication in performing arithmetic operations.

Again referring to the addition of signed integers A and B, the cases ($A \geq 0$, $B \geq 0$) (Case 1 of Sec. 6.2) and ($A \geq 0$, $B < 0$, $|A| \leq |B|$) (Case 4 of Sec. 6.2) behave analogously in the two notations, and the reader can easily verify this assertion. Cases 2 and 3, however, behave in a slightly different way. For example, let us consider the case corresponding to Case 3 in Sec. 6.2:

$A \geq 0, B < 0, |A| > |B|$ In this case, $a_{n-1} = 0$ and $b_{n-1} = 1$; also $a_{n-2} \ldots a_0 = |A|$, while $b_{n-2} \ldots b_0 = 2^{n-1} - 1 - |B|$. Thus the sum of the numerical parts has value $2^{n-1} + |A| - |B| - 1$. Since $|A| > |B|$, then $|A| - |B| - 1 \geq 0$, whence $2^{n-1} + |A| - |B| - 1 \geq 2^{n-1}$. This implies that a carry $c_{n-1} = 1$ is generated; the weight of this carry is 2^{n-1}, which leaves in the numerical part $s_{n-2}, \ldots, s_0$ of the sum the value $|A| - |B| - 1$, that is, the *desired value minus 1*. Moreover,

$$
\begin{array}{ccccc}
c_n & c_{n-1} & c_{n-2} & \cdots & \\
\boxed{1} & \boxed{1} & & & \\
 & 0 & a_{n-2} & \cdots & a_0 \\
 & 1 & b_{n-2} & \cdots & b_0 \\
\hline
 & s_{n-1} = 0 & s_{n-2} & \cdots & s_0
\end{array}
$$

the carry $c_{n-1} = 1$ is added to $b_{n-1} = 1$, yielding $s_{n-1} = 0$ and $c_n = 1$. Thus the result has the correct sign ($s_{n-1} = 0$) but the value represented in the numerical portion must be corrected by adding 1 to it. The characteristic condition for this correction to take place is $c_n = 1$, that is, the carry-out of the most significant position (c_n) becomes a carry into the least-significant position. For this reason, the correction is known as *end-around carry*. This two-step process in executing additions is the unattractive feature of the one's complement notation.

In contrast with the two's complement notation, the opposite of $100 \ldots 0 = -2^{n-1} + 1$ is $011 \ldots 1 = 2^{n-1} - 1$ (i.e., it is defined). However, there are *two* distinct representations of the integer 0: $00 \ldots 0$ and $11 \ldots 1$.

6.4 ADDER CELL (FULL ADDER)

There are various common realizations for the adder cell (also called *full adder*) shown as a black box in Figure 3.2(c), and each one exhibits some advantages. The truth tables for the sum S and carry-out C' outputs were given in Figure 3.2(b); the corresponding expressions are

$$
\begin{aligned}
S &= \bar{A}\bar{B}C + \bar{A}B\bar{C} + A\bar{B}\bar{C} + ABC \\
C' &= AB + AC + BC
\end{aligned}
\qquad (6.8)
$$

It is easily seen that the following are equivalent expressions for S:

$$
\begin{aligned}
S &= ABC + (A + B + C)(\bar{A}\bar{B} + \bar{A}\bar{C} + \bar{B}\bar{C}) \\
&= ABC + (A + B + C)(\overline{AB + AC + BC}) \\
&= ABC + (A + B + C)\overline{C'}
\end{aligned}
$$

the latter of which corresponds to the circuit realization shown in Figure 6.6.

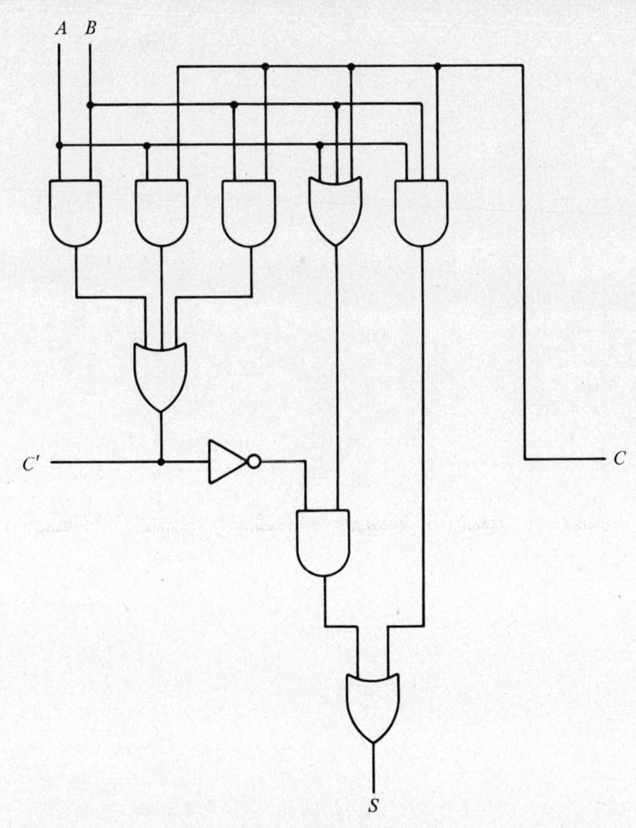

Figure 6.6 A realization of the adder cell.

An alternative realization is obtained by transforming expressions (6.8) as follows:

$$C' = AB + AC + BC = AB + C(A + B)$$

$$S = \bar{A}\bar{B}C + \bar{A}B\bar{C} + A\bar{B}\bar{C} + ABC = (\bar{A}\bar{B} + AB)C + (\bar{A}B + A\bar{B})\bar{C}$$

$$= (\overline{\bar{A}B + A\bar{B}})C + (\bar{A}B + A\bar{B})\bar{C} = (\bar{A}B + A\bar{B}) \oplus C$$

Now, $(\bar{A}B + A\bar{B})$ is equivalent to $AB \oplus (A + B)$, as shown below:

$$\begin{aligned} AB \oplus (A + B) &= \overline{AB}(A + B) + AB(\overline{A + B}) \\ &= (\bar{A} + \bar{B})(A + B) + AB \cdot \bar{A}\bar{B} \\ &= \bar{A}B + A\bar{B} \end{aligned}$$

Thus we have

$$S = AB \oplus (A + B) \oplus C$$

A realization corresponding to these expressions is shown in Figure 6.7. (Incidentally, a similar realization occurs in a very popular integrated circuit, the SN 74181.)

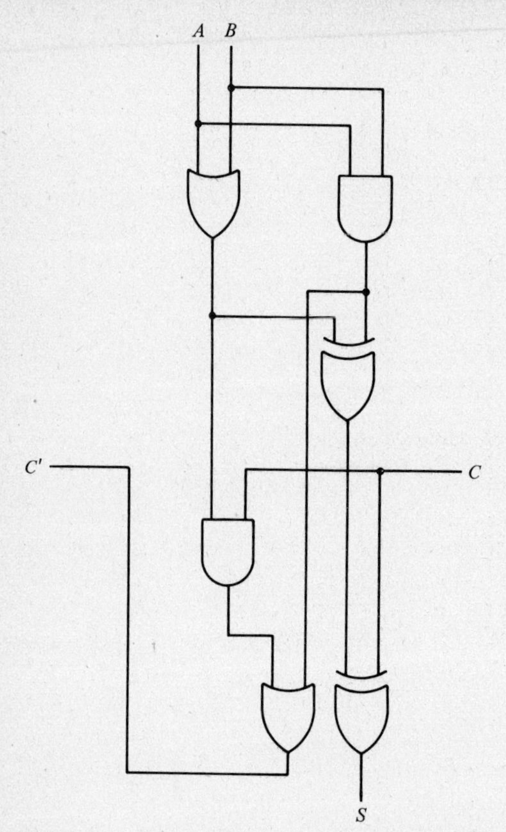

Figure 6.7 Alternative realization of the adder cell.

6.5 PARALLEL ADDER SUBSYSTEMS

We shall now study the realizations of subsystems designed to perform the parallel addition of two n-bit numbers. As we saw earlier (Sec. 3.1) the most basic realization is an array of n adder cells, interconnected so that the carry-out from the ith cell is the carry-in into the $(i + 1)$st cell. For convenience, the diagram of this type of parallel adder is repeated in Figure 6.8, where the overflow detection network is also shown.

This kind of adder is commonly referred to as "ripple through," for the reason that the carry may have to propagate through all the stages before the result $s_{n-1} \ldots s_0$ becomes stable (and may thus be used): this happens, for example, in the worst case when the two operands are 111 . . . 11 and 000 . . . 01. To appreciate the disadvantages of this long propagation, at first we shall analyze in detail the input/output delay in an individual adder cell. We assume, quite realistically, that signals experience a unit of delay τ in traversing an electronic gate. We also assume, for concreteness, that the adder cell is realized as in Figure 6.6 (repeated in Figure 6.9(a) with the gates numbered for ease of reference). Since in this analysis we are not interested in the gate types, we consider the more abstract model shown in Figure 6.9(b) (called a *graph*) where

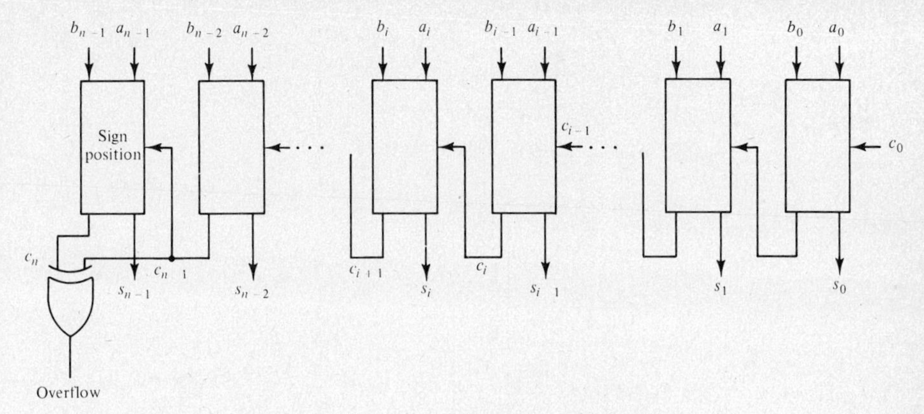

Figure 6.8 A parallel binary adder (ripple-through parallel adder).

each gate has been replaced by a *node* (a circle) and each connection is shown as an *arc* (a segment with an arrow at one end). The delay $T(I, U)$ incurred between an input I and an output U—in units τ—is given by the number of nodes of the *longest* path from input I to output U. Thus, in our case $T(A, C') = T(B, C') = T(C, C') = 2$, while $T(A, S) = T(B, S) = T(C, S) = 5$; for example, $T(A, S) = 5$ is given by the path 1–6–7–8–9. This is easily summarized by the scheme in Figure 6.9(c), which is self-explanatory.

Assume now that all inputs, $a_{n-1} \ldots a_0$ and $b_{n-1} \ldots b_0$ be simultaneously applied at time 0. Then the output s_i becomes stable at the latest at time $(5 + 2i)\tau$, since the carry may have to propagate through the i cells $0, \ldots, i-1$—experiencing a delay 2τ through each cell—and an additional delay of 5τ is experienced from input c_i to output s_i. Therefore, if we are adding two n-bit numbers, the sum bit s_{n-1} becomes available with a delay $(2n + 3)\tau$; if τ is, say, 2 nsec and $n = 60$, then we have a delay of 246 nsec! Clearly, the ripple-through adder can be deployed only for small n or in systems where speed is not of great concern: in all other cases alternate realizations must be used, as we shall now discuss.

Since a parallel adder is a combinational network, each output s_j ($j = 0, 1, \ldots, n-1$) is a switching function $s_j(a_{n-1}, \ldots, a_0; b_{n-1}, \ldots, b_0; c_0)$ and as such it is, in principle, realizable with a three-level network (one level to provide $\bar{a}_{n-1}, \ldots, \bar{a}_0, \bar{b}_{n-1}, \ldots, \bar{b}_0$ and two levels for a standard AND-to-OR realization). However, for large n, such networks would be inordinately bulky and even technologically infeasible. Clearly, suitable fast adders must represent compromises between the two given extremes.

A very interesting and fast scheme is the so-called "carry-look-ahead" parallel adder; as the name indicates, all the carries are computed separately as functions of the input variables and are then fed to the adder cells. The design is based on the following considerations. Let a_i and b_i be the $(i + 1)$st bits of the two operands, and let c_i and c_{i+1} be the corresponding carry-in and carry-out bits, respectively. As is well known,

$$c_{i+1} = a_i b_i + a_i c_i + b_i c_i$$

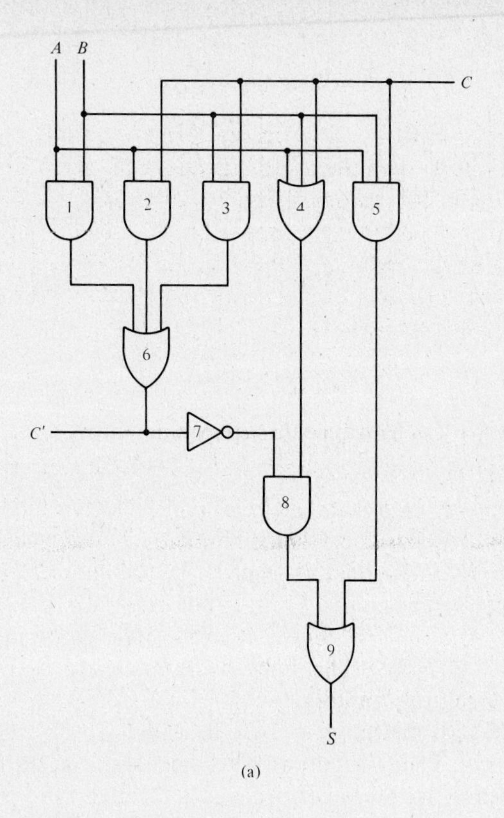

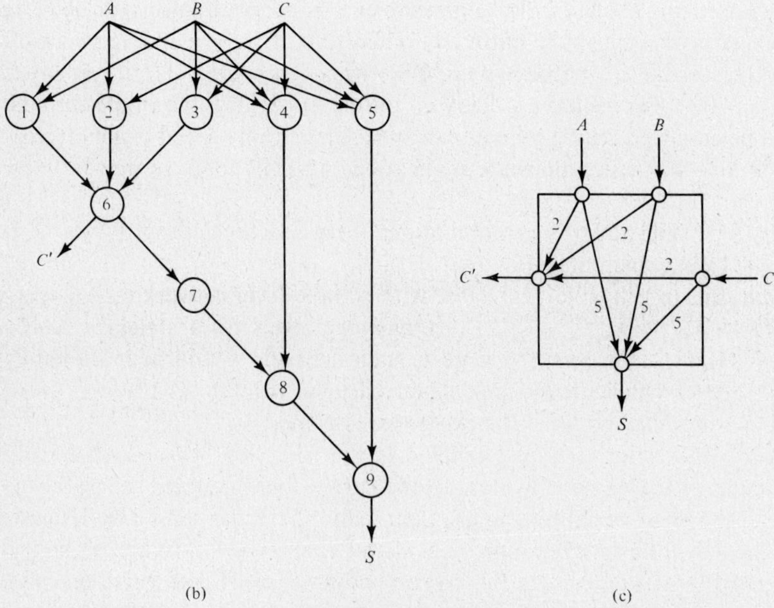

Figure 6.9 Propagation delay analysis in an adder cell.

or equivalently

$$c_{i+1} = a_i b_i + (a_i + b_i)c_i$$

So when $a_i = b_i = 1$ we have $c_{i+1} = 1$ irrespective of c_i, that is, a carry c_{i+1} is *generated;* it is therefore convenient to introduce a new function g_i, the GENERATE function for bit position i, which is given as $g_i = a_i b_i$. On the other hand, when $a_i + b_i = 1$ a carry is *propagated* from the $(i + 1)$st place to the $(i + 2)$nd place: we set $(a_i + b_i) = p_i$ and call it the PROPAGATE function for bit position i. Thus from each pair of bits a_i and b_i we create the two functions g_i and p_i. With these definitions we have

$$c_{i+1} = g_i + p_i c_i \tag{6.9}$$

Notice that the two functions g_i and p_i are available in the adder scheme given in Figure 6.7, which is thus modified as shown in Figure 6.10(a); in Figures 6.10(b) and 6.10(c) we respectively give the circuit symbol for this modified adder cell and its delay diagram (notice that an EXCLUSIVE OR gate contributes two units of delay). Now, Eq. (6.9) can be applied to c_i, that is, $c_i = g_{i-1} + p_{i-1}c_{i-1}$, and so on, thereby obtaining

$$\begin{aligned}
c_{i+1} &= g_i + p_i(g_{i-1} + p_{i-1}c_{i-1}) \\
&= g_i + p_i g_{i-1} + p_i p_{i-1}(g_{i-2} + p_{i-2}c_{i-2}) \\
&= g_i + p_i g_{i-1} + p_i p_{i-1} g_{i-2} + p_i p_{i-1} p_{i-2} c_{i-2} \\
&\vdots \\
&= g_i + (p_i)g_{i-1} + (p_i p_{i-1})g_{i-2} + \cdots \\
&\quad + (p_i p_{i-1} \cdots p_{i-k+1})g_{i-k} \\
&\quad + (p_i \cdots p_{i-k})c_{i-k}
\end{aligned} \tag{6.10}$$

The latter expression gives c_{i+1} as a function of $g_i, g_{i-1}, \ldots, g_{i-k}, p_i, p_{i-1}, \ldots, p_{i-k}$, and c_{i-k}, and corresponds to the two-level AND-to-or realization shown in Figure 6.11: obviously there is a limitation on the fan-in of the gates, so that the difference $(i - k)$ cannot be very large.

A typical compromise realization consists in subdividing the array of n adder cells into contiguous subarrays, each consisting of a small number of cells, for example, 4 or 5; within each subarray a full carry-look-ahead is realized, whereas the carry ripples from subarray to subarray as shown in Figure 6.12.

Each four-stage carry-look-ahead unit is based on the following expressions, which are obtained by specializing Eqs. (6.10):

$$\begin{aligned}
c_0 &= c_{IN} \\
c_1 &= g_0 + p_0 c_{IN} \\
c_2 &= g_1 + p_1 g_0 + p_1 p_0 c_{IN} \\
c_3 &= g_2 + p_2 g_1 + p_2 p_1 g_0 + p_2 p_1 p_0 c_{IN} \\
c_4 &= g_3 + p_3 g_2 + p_3 p_2 g_1 + p_3 p_2 p_1 g_0 + p_3 p_2 p_1 p_0 c_{IN}
\end{aligned}$$

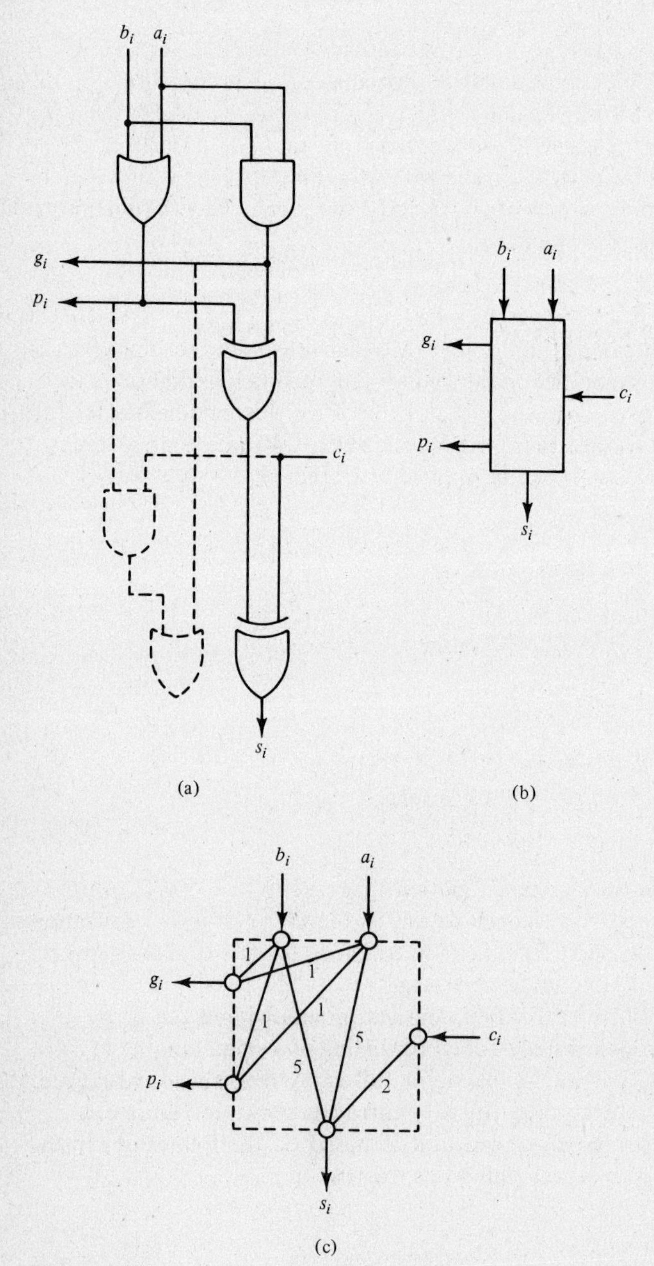

Figure 6.10 A modified adder cell suitable for a carry-look-ahead adder, its symbol, and its delay diagram.

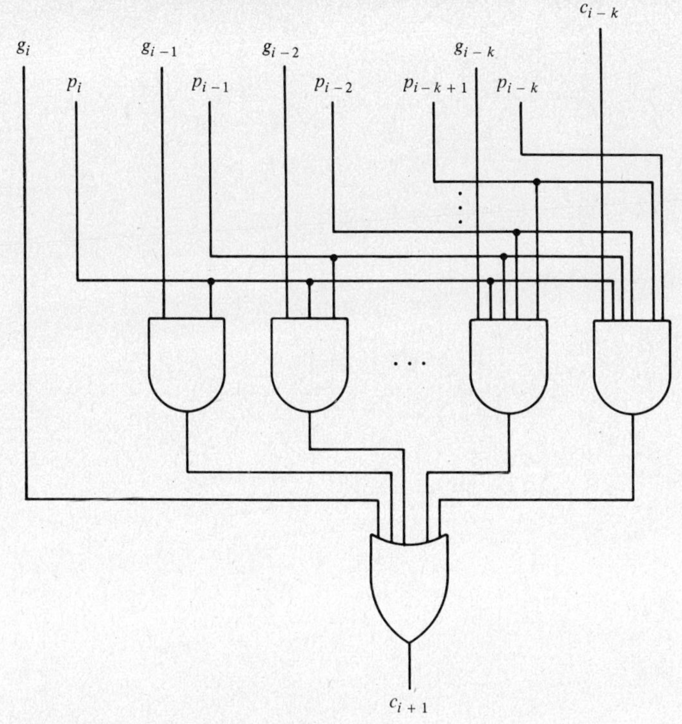

Figure 6.11 Two-level realization of c_{i+1}.

Note that maximum fan-in used in the look-ahead unit is 5. The obvious delay diagram for the look-ahead unit is shown in Figure 6.13: we point out that the path $c_{\text{IN}} \to c_0$ has delay 0, the paths $g_i \to c_{i+1}$ $(i = 0, \ldots, 3)$ have delay τ, while all other paths have delay 2τ.

Equivalent to the delay diagram of Figure 6.13 are the following equations, where we make the assumption that all functions g_i and p_i $(i = 0, 1, 2, 3)$ have a delay τ:

$$\begin{cases} & \text{delay}(c_0) = \text{delay}(c_{\text{IN}}) \\ (i = 1, 2, 3, 4) & \text{delay}(c_i) = \max[\text{delay}(c_{\text{IN}}) + 2\tau, 3\tau] \end{cases}$$

where the term 3τ is due to $\text{delay}(p_i, g_i) + 2\tau = \tau + 2\tau = 3\tau$.

With this preliminary analysis, we are now able to compute the delay of each output of a complete adder subsystem, like the one partially shown in Figure 6.12. The analysis is illustrated in Figure 6.14. We perform the following steps:

1. Determine the delays of all the inputs to the look-ahead units. (The p_is and g_is all have delay τ; the delay of the c_{IN} inputs is determined by tracing the corresponding path. See Figure 6.14.)
2. Determine the delay of all other carry outputs from the look-ahead units. (For the rightmost unit this delay is due to the g_i and p_i inputs; for the remaining units, it is due to the c_{IN} input. See Figure 6.14.)

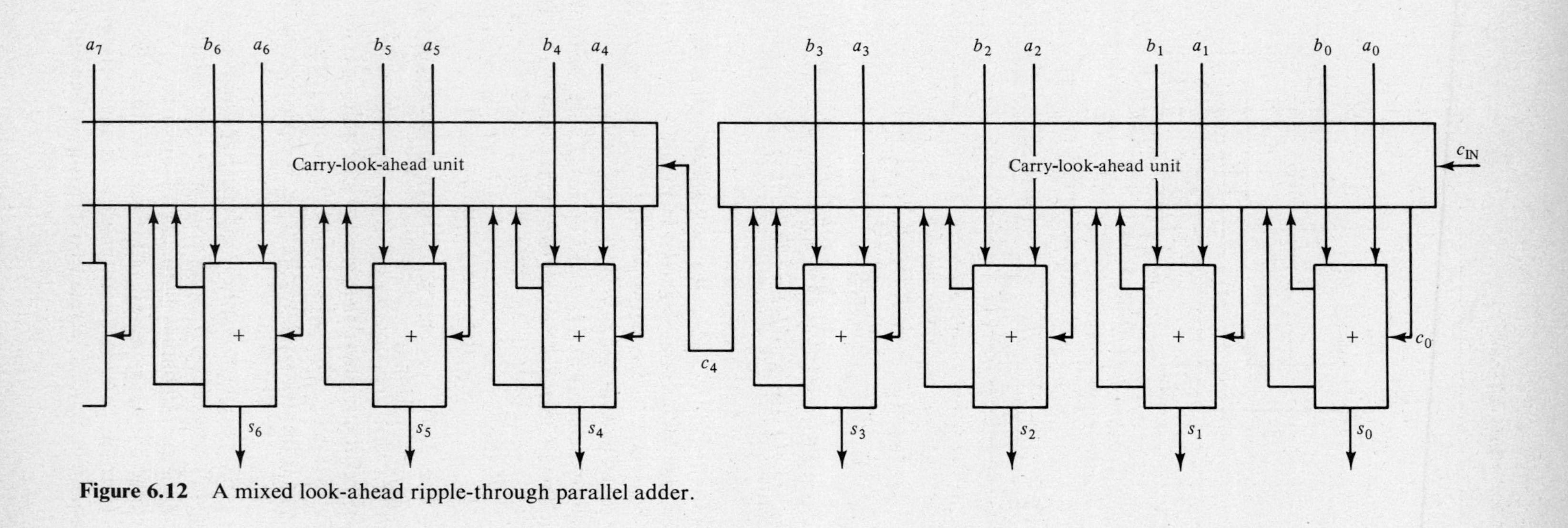

Figure 6.12 A mixed look-ahead ripple-through parallel adder.

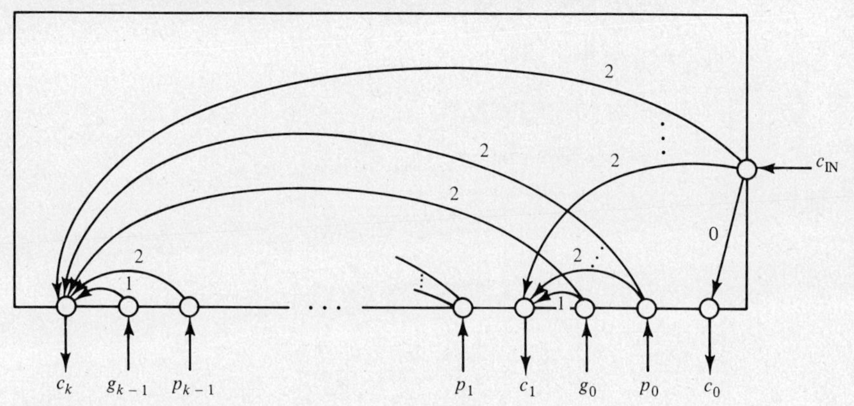

Figure 6.13 Delay diagram for the look-ahead unit. Each arc is labeled with its delay, expressed in units τ.

3. Using the delay diagram of the adder cell, shown in Figure 6.10(c), determine the delay of the s outputs of each adder cell. [For all cells, except the rightmost one, this delay is due to the delay of the carry input; for the rightmost cell, the delay is due to the operand inputs (a_0, b_0). See Figure 6.14.]

Problems 6.8, 6.9, and 6.10 at the end of this chapter concern interesting variants and extensions of the ideas presented in this section.

Remark. For the analytically inclined reader, we now describe a formal method to obtain the delay performance of a mixed adder, as the one partially shown in Figure 6.12. Suppose we have n-bit numbers and the carry-look-ahead unit spans m cells, and consider $s_j, j \geq 1$ [i.e., the output of the $(j+1)$st adder cell]. We now find unique integer quotient q and remainder r of the division of $(j-1)$ by m, that is, $j - 1 = mq + r$, with $r < m$. Clearly the carry c_j into the $(j+1)$st adder cell is generated within the $(q+1)$st carry-look-ahead unit. We denote by $T(x)$ the time at which a circuit variable x becomes available and by $\max\{x, w, \ldots, t\}$ the largest element in a set of numbers $\{x, w, \ldots, t\}$. Notice that $T(a_0) = \cdots = T(a_{n-1}) = T(b_0) = \cdots = T(b_{n-1}) = 0$; also, from Figure 6.9, $T(p_i) = T(g_i) = \tau$ for all values of i. It follows from Figure 6.13 that

$$T(c_j) = \max\{T(g_j) + 2\tau, T(c_{mq}) + 2\tau\}$$

where c_{mq} is the carry-in into the $(q+1)$st carry-look-ahead unit. It is a simple exercise to verify that $T(c_o) = 0$ and $T(c_{mq}) = (2q+1)\tau$. Since

$$T(s_j) = \max\{T(a_j) + 5\tau, T(c_j) + 2\tau\}$$

and $T(g_j) = \tau$, $T(a_j) = 0$, we obtain

$$\begin{aligned} T(s_j) &= \max\{T(a_j) + 5\tau, T(g_j) + 2\tau + 2\tau, T(c_{mq}) + 2\tau + 2\tau\} \\ &= \max\{5\tau, 5\tau, (2q+1)\tau + 4\tau\} \\ &= \max\{5\tau, (2q+5)\tau\} = (2q+5)\tau \end{aligned}$$

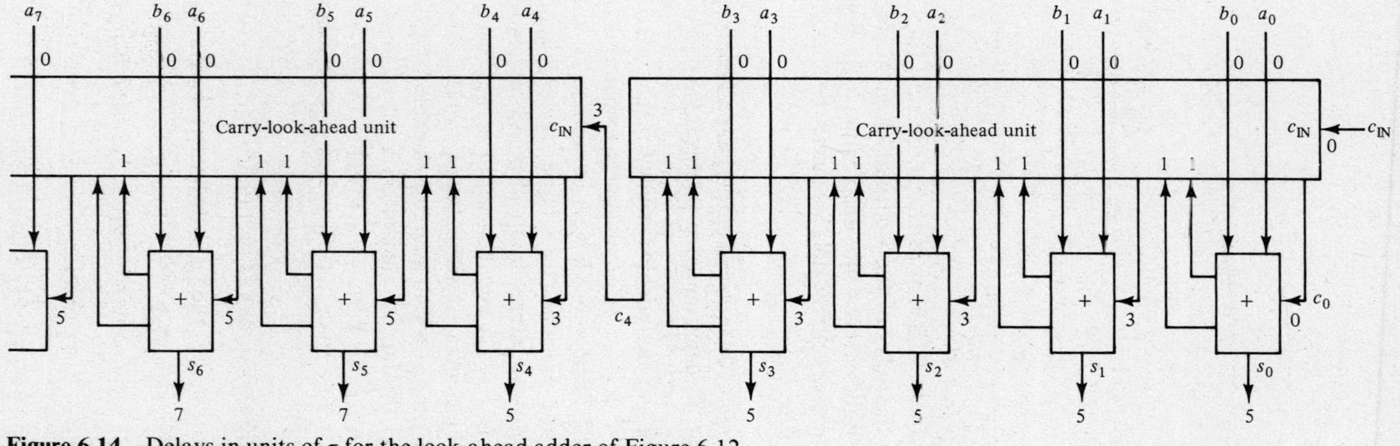

Figure 6.14 Delays in units of τ for the look-ahead adder of Figure 6.12.

Thus assuming, for example, $n = 60$ and $m = 4$, output s_{59}, the most-significant digit of the sum, becomes available at time $(2 \times 14 + 5)\tau = 33\tau$, since $(59 - 1) = 4 \times 14 + 2$. This shows that there is a considerable speedup with respect to the delay 123τ of the corresponding ripple-through adder.

6.6 ARITHMETIC-LOGIC UNIT (ALU)

Any computing system, from a calculator to a large mainframe machine, contains an adder. So does our SEC machine, for which we shall now design this important functional module.

We are now convinced of the advantages of the complement notation, and we stipulate *the adoption of the two's complement notation to represent signed integers in our machine,* thereby burying the original proposal (made in Sec. 2.2) to use the modulus-and-sign notation. Moreover, in the interest of speed of operation, we choose to deploy a parallel adder rather than a serial adder.

Our present objective, however, is to modify the parallel adder in order to transform it into a very versatile unit which can be "programmed" to perform a variety of useful functions besides binary addition: this unit is the so-called arithmetic-logic unit.

The *Arithmetic-Logic Unit* (briefly, ALU) is an extremely important subsystem of the central processing unit. It is the ALU which performs the actual transformations of data and "computes" new information from given information. We shall now describe the basic features of an ALU, with some simple and concrete exemplification, but with absolutely no claim of generality.

The general representation of a typical ALU is given in Figure 6.15. As we see, the ALU is characterized by three sets of lines, two input sets, the *input data lines* and the *control lines,* and one output set, the *output data lines*. Of course, the input data lines and the control lines can be logically connected to the outputs of some registers, and analogously the output data lines can be logically connected to the inputs of some registers. We shall not consider any of these registers, however, as part of the ALU, which therefore is a purely combinational structure and could, in principle, be completely described as a network realizing a set of boolean functions (in other words, each output data line is a boolean

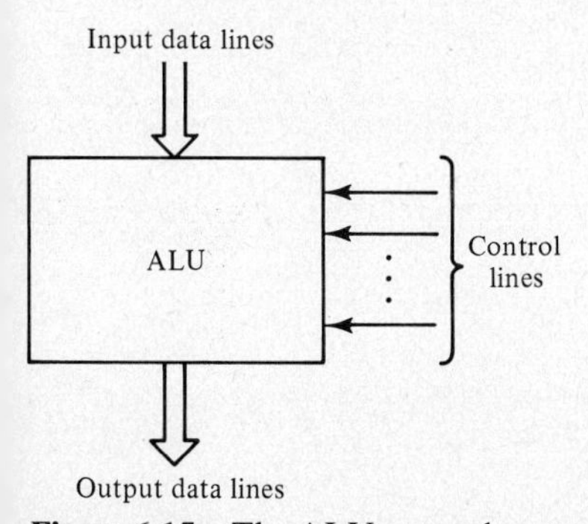

Figure 6.15 The ALU as a subsystem.

function of the input data lines and of the control lines). We shall refer to the data on the input lines as the operands; the control lines select the operation to be performed on the operands.

If the ALU were designed to perform only one operation (for example, addition), there would be no need for control lines. If more operations were required in the computing systems, one could think of designing a separate unit for each of these operations. However, there is an economic advantage in sharing circuitry. Therefore, a typical ALU is a highly versatile subsystem, capable of performing a sufficiently large repertoire of operations on the input data: as we said, the selection of the operation is assigned to the control lines.

For simplicity we assume that the parallel adder, to be modified into an ALU, is of the ripple-through type and that the adder cell is like that shown in Figure 6.7.

As a general comment, an ALU could be viewed as a bit-organized system, that is, a network consisting essentially of n identical subnetworks, one for each of the bit positions, suitably interconnected via the carry line, as shown in Figure 6.16, where $a_{n-1} \dots a_1 a_0$ and $b_{n-1} \dots b_1 b_0$ are the two n-bit operands, $c_0, c_1, \dots, c_n$ are the carry bits, and $f_0, \dots, f_{n-1}$ are the output lines (collectively denoted as the vector F). The control signals have been omitted. Our objective is the design of one of the subnetworks shown in Figure 6.16, which will be referred to as a *bit unit*. (While keeping here some generality, we remind the reader that $n = 21$ in the SEC.)

First we consider a reasonable repertoire of *arithmetic functions* for the ALU. Obviously, the ALU must be capable of performing *addition* and *subtraction* of two n-bit operands A and B. Since we have assumed that operands are represented in the two's complement notation, we realize immediately that the operation of subtraction ($A - B$) can be achieved by means of three concurrent actions:

1. *Complement the subtrahend lines* (referred to here as the B lines).
2. *Set to 1 the input carry* c_0 to the least-significant adder stage position.
3. *Perform addition.*

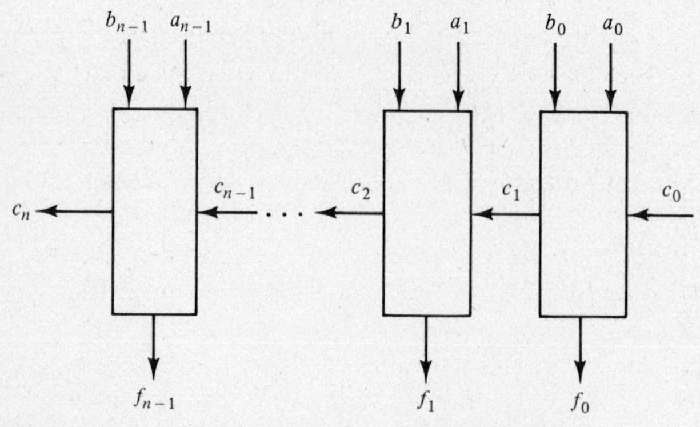

Figure 6.16 Functional organization of an ALU.

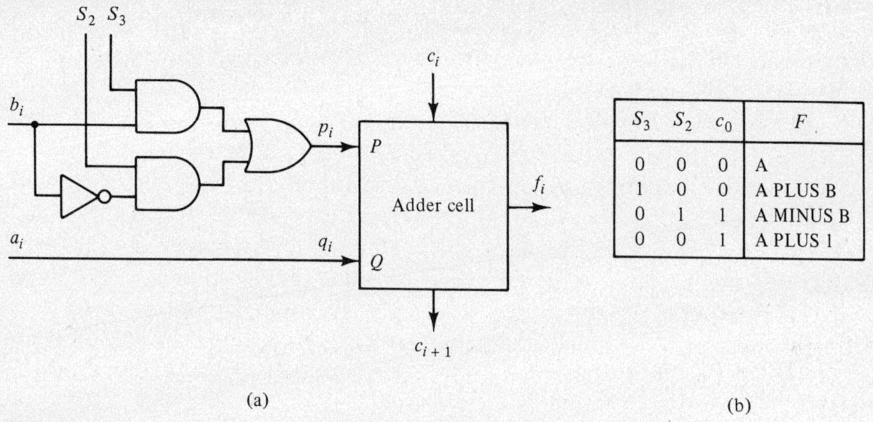

S_3	S_2	c_0	F
0	0	0	A
1	0	0	A PLUS B
0	1	1	A MINUS B
0	0	1	A PLUS 1

Figure 6.17 First modification of adder cell and resulting operations.

To implement step 1, for each adder cell [Figure 6.17(a)] we introduce a *selection circuit* (a narrow-sense multiplexer), which places on the P input of the $(i + 1)$st cell, called p_i, either b_i or $\overline{b}_i$, depending upon the values of two *control signals*, S_2 and S_3; action 2 is implemented by using c_0 itself as a control signal. Using the three control signals c_0, S_2, and S_3 we can implement the *direct transfer* $F = A$ and the *arithmetic operations* $F = A$ PLUS B, $F = A$ MINUS B, $F = A$ PLUS 1, as indicated in Figure 6.16(b). These three operations provide adequate arithmetic capabilities.

It is desirable, however, to extend the repertoire of operations executable by the ALU to include some, or possibly all, of the *logical operations* on the two operands A and B, in a bit-by-bit fashion. To achieve this objective we shall modify the cell of Figure 6.16(a); these modifications however must be compatible with the presently executable operations.

First of all, since logical operations on two strings of bits are executed bit by bit, each bit position is independent of the others; this means that the carry connections, which laterally link adjacent cells, must be effectively disabled. This can be effected by setting the carry lines c_i to a *fixed* condition (so that they do not carry any information), that is, either $c_i = 0$ or $c_i = 1$ for $i = 0, 1, \ldots, n$. We can simply obtain $c_i = 1$ by introducing a signal L into the adder cell as in Figure 6.18; obviously $L = 0$ leaves unaffected the previous behavior of the cell, while $L = 1$ forces $c_{i+1} = 1$. Thus, by setting $c_0 = L = 1$ we obtain $c_0 = c_1 = c_2 = \ldots = c_{n-1} = c_n = 1$ and each bit position is independent of the others. Clearly, $L = 0$ for arithmetic operations and $L = 1$ for logical operations (L stands for Logical).

When $c_0 = L = 1$ (logical operations), the adder cell computes the function $f_i = p_i \oplus q_i \oplus 1 = p_i q_i + \overline{p}_i \overline{q}_i$. We must now see how to modify the input functions p_i and q_i in order to realize the desired logical capabilities.

Consider the two networks shown in Figures 6.19(a) and 6.19(b): since $AB \cdot (A + B) = AB$ and $AB + (A + B) = A + B$, the two networks are equivalent. So we can add the AND gate (with output p_i^*) and the OR gate (with output q_i^*) as

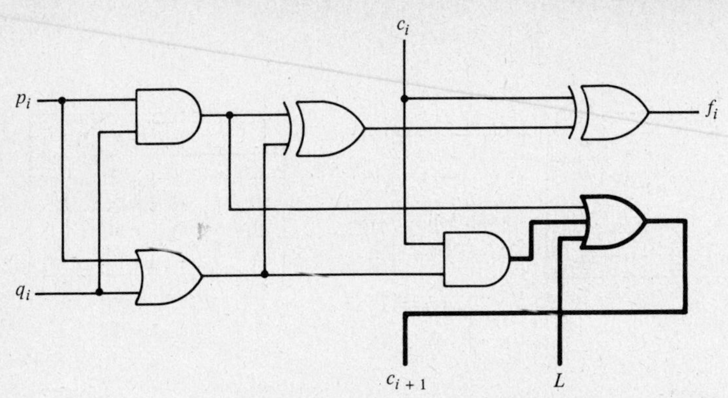

Figure 6.18 Additional modification of adder cell. Signal $L = 1$ forces $c_{i+1} = 1$.

in Figure 6.20(a) without changing the function of the network. Next, since

$$q_i^* = p_i + q_i = a_i + (S_3 b_i + S_2 \bar{b}_i) = a_i + S_3 b_i + S_2 \bar{b}_i$$
$$p_i^* = p_i q_i = a_i(S_3 b_i + S_2 \bar{b}_i) = S_3 a_i b_i + S_2 a_i \bar{b}_i \qquad (6.11)$$

the network of Figure 6.20(a) is equivalent to that of Figure 6.20(b). Suppose now that the "control signals" S_3 to AND gate 1 and S_2 to AND gate 2 are replaced, respectively, by two *new* control signals S_1 and S_0, respectively, as shown in Figure 6.20(c); obviously if we select $S_3 = S_1$ and $S_2 = S_0$ we have the same behavior as that of the network in Figure 6.20(b), that is, the original adder cell behavior when $L = 0$.

When, however, we let $c_0 = L = 1$ and let S_0, S_1, S_2, and S_3 be arbitrarily selectable, we obtain the following expression for the output f_i of the bit unit [refer to Figure 6.20(c) to obtain the expressions of p_i' and q_i']:

$$\begin{aligned}
f_i &= p_i' q_i' + \overline{p_i'} \cdot \overline{q_i'} \\
&= (S_3 a_i b_i + S_2 a_i \bar{b}_i)(S_1 b_i + S_0 \bar{b}_i + a_i) \\
&\quad + \overline{(S_3 a_i b_i + S_2 a_i \bar{b}_i)} \cdot \overline{(S_1 b_i + S_0 \bar{b}_i + a_i)} \\
&= S_3 a_i b_i + S_2 a_i \bar{b}_i + (\bar{a}_i + \overline{S_3 b_i} \cdot \overline{S_2 \bar{b}_i}) \cdot \overline{S_1 b_i} \cdot \overline{S_0 \bar{b}_i} \cdot \bar{a}_i \\
&= S_3 a_i b_i + S_2 a_i \bar{b}_i + \bar{a}_i(\bar{S}_1 + \bar{b}_i)(\bar{S}_0 + b_i) \\
&= S_3 a_i b_i + S_2 a_i \bar{b}_i + \bar{S}_1 \bar{a}_i b_i + \bar{S}_0 \bar{a}_i \bar{b}_i
\end{aligned}$$

Figure 6.20(c) gives the final layout of the bit unit. It is clear from the last

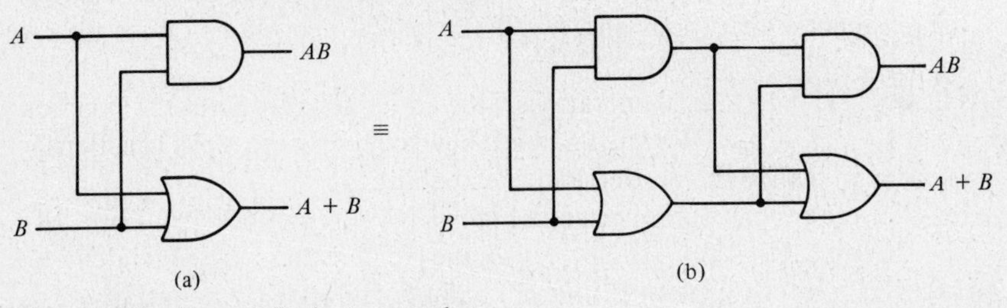

Figure 6.19 Two equivalent networks.

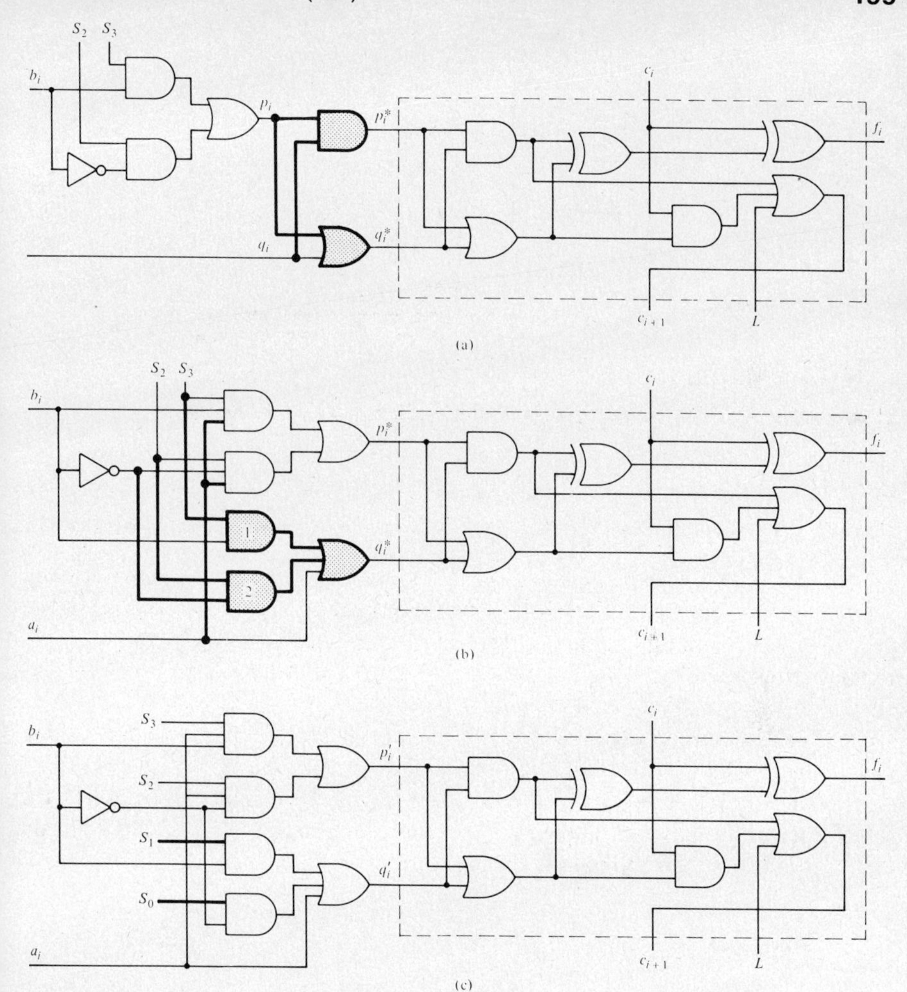

Figure 6.20 Successive modifications of the adder cell to obtain the ALU bit unit, shown in figure (c).

expression that S_3, S_2, $\overline{S}_1$, $\overline{S}_0$ can be chosen as the truth table of any function of two variables. In Table 6.1 we show the control signal selections corresponding to some typical arithmetic and logical operations.

Thus, a flexible ALU has been designed with control signals S_3, S_2, S_1, S_0, c_0, and L. To complete the unit we shall introduce the overflow detector (Figure 6.21) and an output EQ which is 1 exactly when the two operands are identical. Since $A = B$ if and only if $a_i = b_i$ for each index i, or, equivalently, $a_i - b_i = 0$, the condition $A = B$ is tested by first performing the subtraction $(A - B)$ and then verifying the condition $f_0 = f_1 = \cdots = f_{n-1} = 0$. The latter task is implemented by a NOR gate. Notice that if we choose the operand B as $00 \ldots 0$ (this can be done by setting $LS_3S_2S_1S_0c_0 = 000000$) then signal EQ is 1 when A is itself $00 \ldots 0$, and so it can be used to test the condition $A = 0$.

TABLE 6.1

Operation	Control signals					
	L	S_3	S_2	S_1	S_0	c_0
A PLUS B	0	1	0	1	0	0
A MINUS B	0	0	1	0	1	1
A PLUS 1	0	0	0	0	0	1
$A \cdot B$	1	1	0	1	1	1
$A + B$	1	1	1	0	1	1
$\overline{A}$	1	0	0	0	0	1

The layout of a typical, although rather simple, ALU is given in Figure 6.21. The organization of the individual bit unit is a simplified version of the one implemented in the SN74181 integrated circuit chip.

Finally, it is convenient to augment the unit we have just designed by feeding the output lines $f_0, f_1, \ldots, f_{n-1}$ to a *shifting network*. This network has $(n + 1)$ inputs $b_{IN}, f_0, f_1, \ldots, f_{n-1}$ and $(n + 1)$ outputs $b_{OUT}, g_0, g_1, \ldots, g_{n-1}$ and consists of a collection of multiplexers. The interesting shifting operations are of two types: *logical* and *arithmetic*.

Logical shifts are simply data moves:

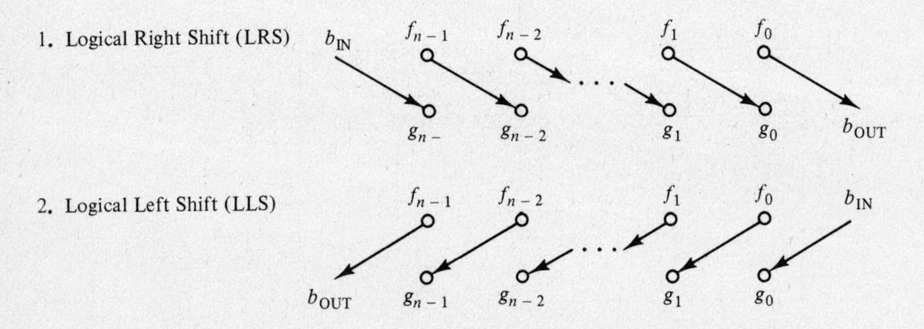

Arithmetic shifts, instead, are full-fledged arithmetic operations, that is, multiplication by 2 and division by 2 (with possible truncations or roundoffs), which are implementable as shifts (left and right, respectively). Specifically we have:

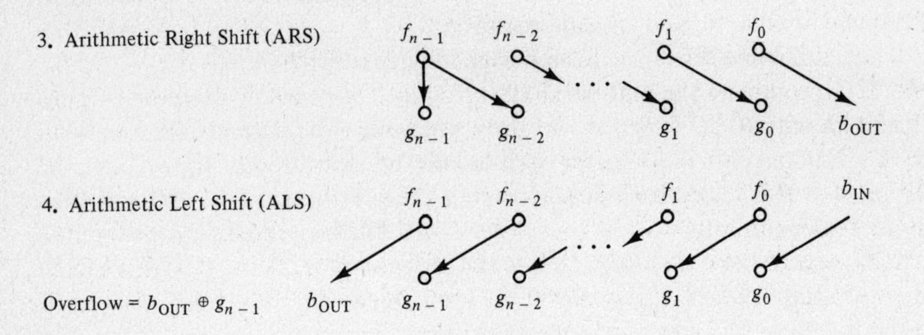

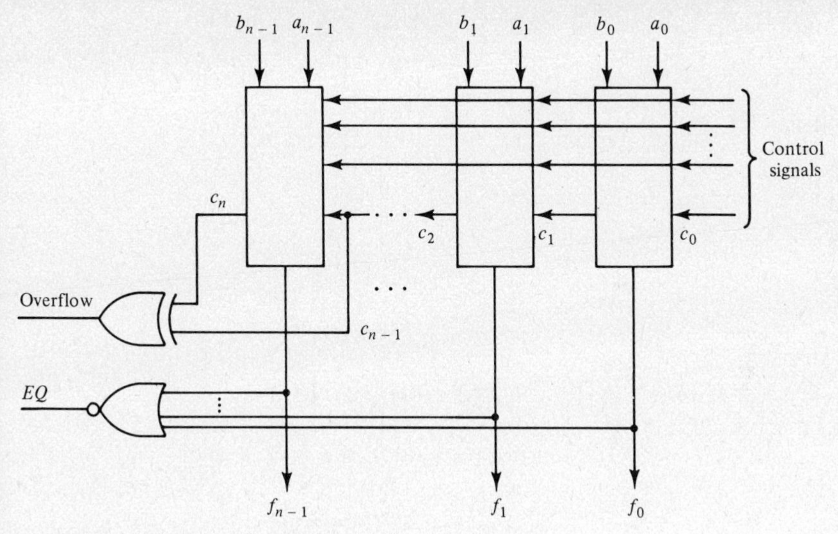

Figure 6.21 Scheme of a simple ALU.

Two comments are now in order with regard to arithmetic shifts.

1. In ARS (a division by 2 with truncation), we have

$$G = \left\lfloor \frac{F}{2} \right\rfloor$$

where the notation "$\lfloor \ \rfloor$" denotes "integer part of." Using Eq. (6.2) we rewrite $G = \lfloor F/2 \rfloor$ as

$$-g_{n-1}2^{n-1} + \sum_{j=0}^{n-2} g_j 2^j = \left\lfloor \left(-f_{n-1}2^{n-1} + \sum_{j=0}^{n-2} f_j 2^j \right) \Big/ 2 \right\rfloor$$
$$= -f_{n-1}2^{n-2} + \sum_{j=0}^{n-3} f_{j+1} 2^j$$

Since $-f_{n-1}2^{n-2} = -f_{n-1}2^{n-1} + f_{n-1}2^{n-2}$, we see that both g_{n-1} and g_{n-2} are equal to f_{n-1}, as indicated in the diagram above.

2. The connection pattern for ALS is identical to that of LLS; however, ALS, which is an arithmetic operation (multiplication by 2) may result in overflow any time $f_{n-1} \neq f_{n-2}$, that is, when $b_{\text{OUT}} \neq g_{n-1}$. Thus, this condition must be tested any time an ALS is executed. The completion of the ALU, with the modified overflow circuit, is in Figure 6.22. Note that b_{IN} is always set to 0 in this realization. Also shown is the narrow-sense multiplexer for output g_i, for $i = 1, 2, \ldots, n-2$, with the appropriate control signals, including NS (No Shift). The construction of the multiplexers for g_0, g_{n-1}, and b_{OUT} is left as a problem for the reader (see Problem 6.14). In later references, the input lines will be called A-lines and B-lines (where the B-lines can be complemented in arithmetic operations), and the output lines will be called G-lines.

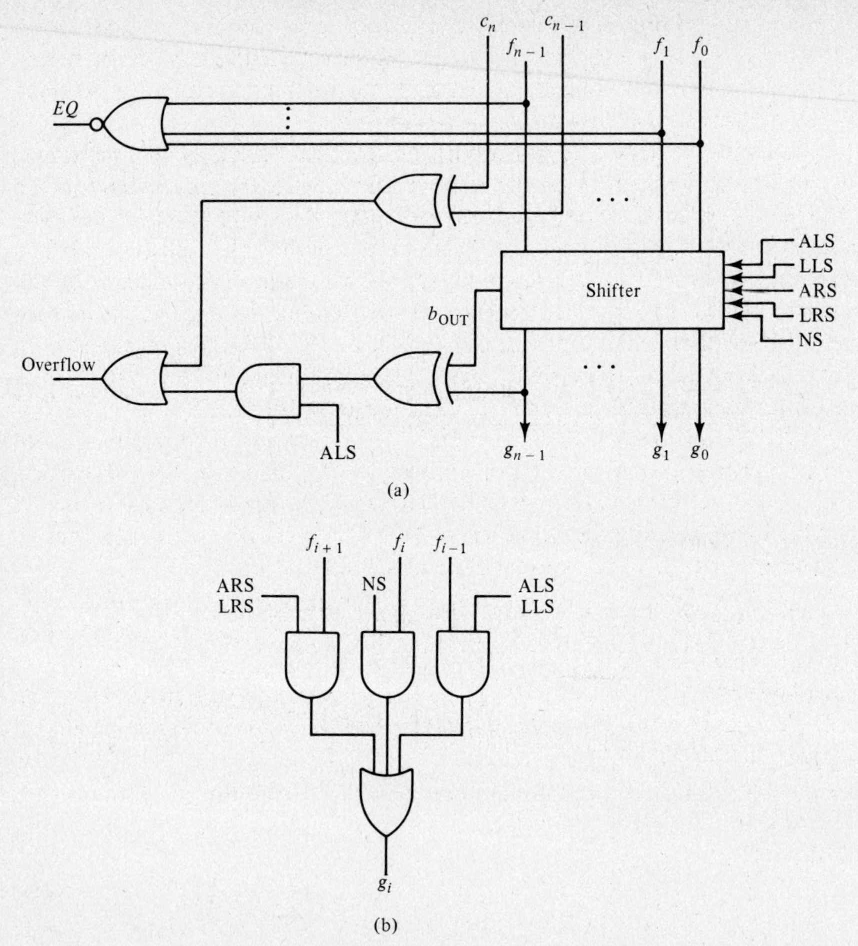

Figure 6.22 (a) Addition of the shifter to the previously designed ALU. (b) Logic circuit of g_i.

NOTES AND REFERENCES

The complement notation for the representation of signed integers is, in some sense, natural for use in calculating machines. Indeed, negative integers were represented as complements in the early mechanical calculator of Pascal in the eighteenth century.

As regards arithmetic operations, in this elementary text we have limited ourselves to addition–subtraction and the corresponding hardware. Although addition of signed integers is adequate for the realization of the other arithmetic operations—multiplication and division—(just think of the so-called "schoolboy" methods, where only additions and subtractions are used), it is possible to construct specific networks for the execution of each of the more complex operations. The topic of computer arithmetic has been of central importance in the development of digital computers and is today a mature body of knowledge. Besides the classical text by Flores (1963), and the excellent discussion by Knuth (1969), the reader may consult the more recent work by Hwang (1979) for a wealth of details on the more advanced topics.

Moreover, we have only considered the so-called "fixed-point" operations, in particular integer operations (the reader should realize that the techniques are valid irrespective of the fixed position of the binary point). However, it has been noted that, especially in scientific calculations, it is preferable to adopt a notation (the so-called "scientific notation") which allocates the digit positions of the representation to the significant portion of the number, so that (in decimal) a number like 0.00002793 is represented as 0.2793×10^{-4}. This different notation, called "floating point," represents each number by means of two variables: a normalized fraction, called "mantissa" (in our example, 0.2793) and a signed integer, the "exponent" (in our example, -4). Of course the execution of an arithmetic operation in floating point is more complex than that of its fixed-point counterpart, since separate mantissa and exponent operations have to be carried out. The cited texts, and many others, such as Mano (1976) and Booth (1971), and so on, give detailed descriptions of this technique.

Finally we mention that with the advent of very large scale integration there is a renewed interest in the study of complex arithmetic circuits to be realized as single chips, or as portions of chips. A bibliography on this topic is beyond our scope; the reader may consult the professional journals, such as the *IEEE Transactions on Computers,* for the most recent developments in this direction.

PROBLEMS

6.1. Write the 7-bit two's complement representation (sign bit + 6-bit numerical part) of the following integers: 45, -38, 61, -11, -49, -22, 51, -57.

6.2. (1) Negate each of the following numbers represented in the two's complement notation (leftmost bit is sign bit).

(a) 1101001010, 01101, 0110100, 1100, 11111

(b) 1010101, 0111000, 0000001, 10000, 0000

(2) Repeat the exercise, but assume that numbers are represented in the one's complement notation.

6.3. Perform the following additions (the integers below are given in base 10 modulus-and-sign form). Represent the numbers in the two's complement notation with 8 bits (sign bit included) and say whether or not overflow occurs.

(a) $-73 + (-91)$, $-73 + 91$, $73 + (-91)$, $73 + 91$

(b) $55 + 75$, $-75 + (-37)$, $-82 + 75$, $-48 + (-82)$

6.4. The following numbers are all represented in a 6-bit two's complement notation. Perform the operations indicated below (results must be in the same notation as the operands). Indicate for which cases overflow occurs.

001110 + 110010   010101 − 000111
000011 + 010101   001010 − 111001

010101 + 001100   111001 − 001010
111001 + 001010   010110 − 100010
101001 + 111000   101011 − 100110
101011 + 100110   010010 − 100101

6.5. "Overflow exists if any only if the sign of the result is different from the common sign of the addends." Is this statement correct? Justify fully your answer.

6.6. Consider the "longhand" binary multiplication method for integers discussed in Problem 1.15, Chapter 1. Anytime we multiply two n-bit factors we obtain a $2n$-bit product. Suppose now to perform such multiplication with numbers in the two's complement notation. Clearly, the result is correct when both operands are nonnegative.

(**a**) Is the result correct when one or both of the operands are negative? Explain.

(**b**) If in case (a) the result is incorrect, which correction must be effected if one operand is negative?

(**c**) Same as (b), when both operands are negative.

6.7. You are to design a 24-stage parallel adder, employing carry-look-ahead units. Assuming that all inputs are available at time 0, you must evaluate after how many gate delays output s_{21} (i.e., the sum output of the 22nd stage) becomes stable in the following cases

(**a**) the carry-look-ahead unit spans three stages.

(**b**) the carry-look-ahead unit spans four stages.

(**c**) the carry-look-ahead unit spans six stages.

6.8. Consider a carry-look-ahead unit which spans four adder stages. The expressions for the outputs c_0, c_1, c_2, c_3 are given on p. 189. We now modify the unit by eliminating output c_4 and creating two new outputs G and P with the following expressions:

$$G = g_3 + p_3 g_2 + p_3 p_2 g_1 + p_3 p_2 p_1 g_0$$

$$P = p_3 p_2 p_1 p_0$$

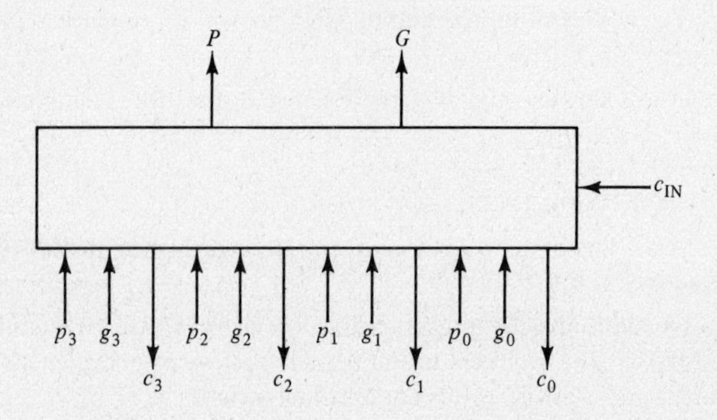

Notice then that c_4, the carry-in into the next look-ahead unit, is given by $c_4 = G + Pc_{IN}$, that is, G and P are respectively the "generate" and "propagate" functions for the look-ahead unit. Assuming that all inputs to the adder system are simultaneously available at time 0, after how many gate delays are the functions P and G stable?

6.9. With reference to the modified carry-look-ahead unit in Problem 6.8, evaluate after how many gate delays the output s_{15} of the adder given on page 205 becomes stable.

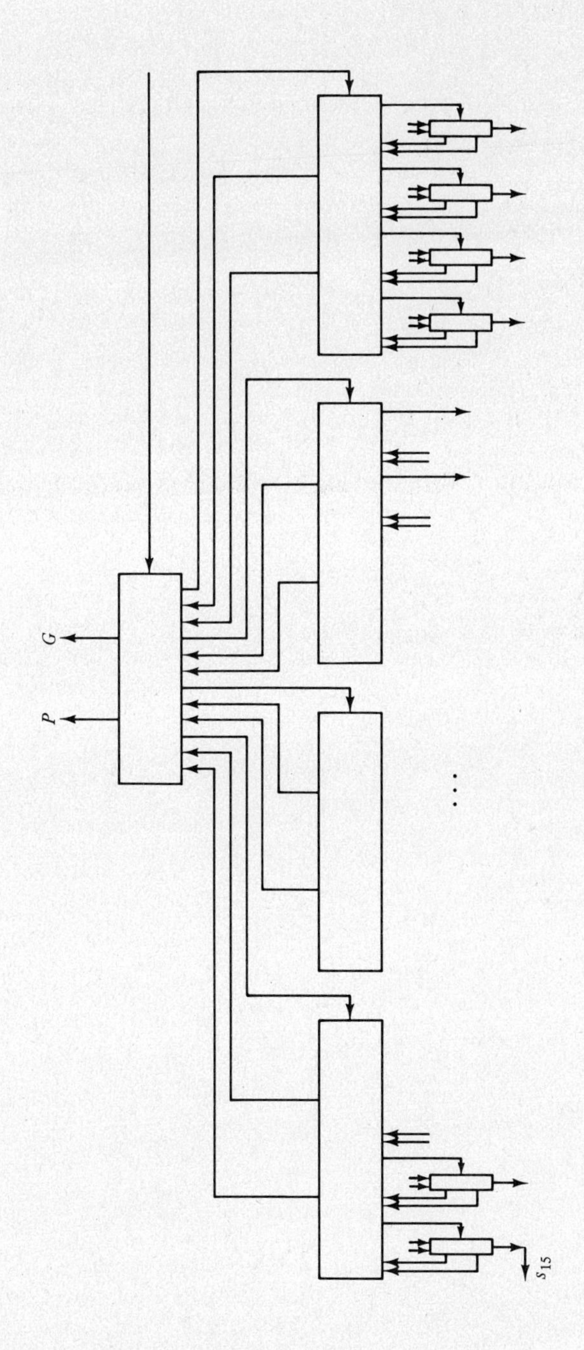
G
P
s_{15}

6.10. With reference to the modified carry-look-ahead unit in Problem 6.8, and extending further the approach presented in Problem 6.9, design a 64-stage adder system which achieves the *least* delay and evaluate the delay for s_{63}.

6.11. With reference to the ALU described in Sec. 6.6 [bit unit is given in Figure 6.20(c)] describe what functions are realized by the following control signal configurations:

	L	S_3	S_2	S_1	S_0	C_0
(a)	0	0	1	0	1	0
(b)	0	1	1	1	1	1
(c)	0	1	0	1	0	1
(d)	1	1	1	1	1	0
(e)	1	1	0	1	1	1

6.12. With reference to the ALU described in Sec. 6.6 [Figure 6.20(c)], design the control signal configuration for each of the following operations:

(a) A MINUS 1

(b) A PLUS B PLUS 1

6.13. With reference to the ALU described in Sec. 6.6 [Figure 6.20(c)] describe which functions are realized by the following control signal configurations:

	L	S_3	S_2	S_1	S_0	C_0
(a)	0	0	0	0	1	0
(b)	0	1	0	1	1	1
(c)	1	0	1	1	0	1
(d)	0	1	1	0	0	1
(e)	0	1	1	0	0	0
(f)	1	0	0	0	0	1

6.14. Sketch circuit diagrams for the multiplexers in the shifting network of the ALU for the following outputs (assuming that the functions LRS, LLS, ARS, ALS, NS are implemented).

(a) b_{OUT}

(b) g_{n-1}

(c) g_0

Chapter 7

Computer Organization: CPU and Memory

7.1 REGISTER TRANSFER LANGUAGE

As we mentioned in Chapter 2, information processing can be viewed as a sequence of transfers of information from registers to registers, either via direct interconnections or through combinational modules (such as the ALU). In the subsequent chapters we have studied the design of the basic subsystems—such as registers and adders—and we are now ready to undertake the design of a computing system.

Boolean algebra and state diagrams provided the formalisms for the description of combinational and sequential networks, respectively; the higher level of design we are going to deal with now also has an appropriate descriptive formalism, called Register Transfer Language (RTL). As the phrase itself indicates, this formalism is a programming language whose "statements" are transfers of information between registers. Although not all readers may be accustomed to the approach we are about to describe, it is very convenient to introduce this language by means of its *grammar*. This will enable us to use the language in a correct, or "grammatical," way.

It is well beyond our present purposes to define "formal" grammars in their generality. It suffices to say that a grammar deals with objects which we shall call *terms* (for example, in the English grammar—a "natural" grammar—terms are *parts of speech* and actual *words*) and consists of a set of rules, called *productions,* which specify admissible substitutions of strings of terms for strings of terms. For example, if A, B, and C are terms, the production A: = BC means that BC may be substituted for A in any string of symbols in which A occurs. Moreover, terms are conveniently classified into *abstract* (normally called *nonterminal*) and *concrete* (normally called *terminal*); for example, in English the term "verb" (a part of speech) is abstract and the term "to go" (a specific verb) is concrete. For convenience of notation each abstract term will be enclosed within angular brackets (e.g., <verb>). Finally, if there is more than one production with identical left side, such as A: = BC, A: = D, A: = CD, we combine them into the notation

$$A{:}= BC \mid D \mid CD$$

This means that any of the strings separated in the right side by a vertical bar "|" can be substituted for A. (This notation is called BNF, or Backus-Naur Form.)

Essentially, a grammar is a machinery to generate all possible sentences of a language. To start this process, that is, the generation of a sentence, we begin from a distinguished abstract term, usually called <sentence>, and by successive substitutions (i.e., applications of productions) we obtain a grammatically correct sentence. It must be emphasized that a generative grammar does not by itself generate a meaningful sentence, but only enables us to express correctly what we want to say; conversely, it enables us to verify the grammatical correctness of sentences which are offered to our understanding. We are now ready to describe the grammar of RTL.

In the grammar of RTL the distinguished term is called <microsequence>, since we wish to generate sequences of actions called microsteps (as in English we wish to generate sequences of words). Other abstract terms are: <concurrent step>, <microstep>, <assignment>, <conditional assignment>, <go to>, <register>, <condition>, <op>, <function>, <label>, <constant>.

The latter six abstract terms, that is <register>, <condition>, <op>, <function>, <label>, <constant>, are the only ones that can be replaced by concrete terms (the "words" of RTL). We shall first illustrate these substitutions.

A "concrete" register, as it appears in the design of an actual computer, is normally denoted by one or more capital letters, such as PC, R, MAR, and so on. In some cases, <register> may actually be a subregister, that is, a portion of a register. For example, if the bits of register R are indexed from 0 to s, R [j:i] denotes the bit positions with indices $i, i + 1, \ldots, j$. (If $i = j$, then the subregister is simply denoted by R[i] rather than by R[i:i].) In other cases, <register> may be one cell of a multiregister storage facility, such as a cell of a computer memory: typically, if M denotes the main memory, M(j) is the memory cell whose address is j. Combining these two cases, M(j)[k:h] denotes bit positions with indices $h, h + 1, \ldots, k$ of the memory cell whose address is j. It must be emphasized that when we speak of "a register," we always mean "the current content of that register."

<Condition> is normally an event whose occurrence can be tested. We prefer to stay with this somewhat vague but perfectly reasonable statement, rather than to embark on a detailed specification of the format of possible "conditions." For example, we may have $(R \geq 0)$, "the content of register R is nonnegative"; $(A = B)$, "the contents of registers A and B are identical"; $(A[12] = 1)$, "the bit with index 12 of register A is equal to 1"; and so on.

<op> is an abbreviation for "operation," specifically, for *binary* operation,

that is, an operation involving *two* operands. Typically, <op> can be replaced by one of the following symbols: +, −, ×, AND, OR, EXCLUSIVE OR, and so on. Similarly, <function> denotes an operation on a *single* operand (also called *unary* operation) and is typically of the form "−" (unary MINUS, or sign change), COMPL (boolean complement). Finally, <label> is an integer, and <constant> is self-explanatory.

We can now present and discuss the productions involving the abstract terms of RTL. These productions will generate strings consisting of the six abstract terms <register>, <condition>, <op>, <function>, <label>, and <constant>, and of special *language symbols* (· , ←) and *language terms* (IF, THEN, END, GO TO).

Production 1. <microsequence>: = <concurrent step> | <microsequence> <concurrent step>

This production says that a <microsequence> is an arbitrarily long nonempty string of <concurrent step>'s.

Production 2. <concurrent step>: = <label>. <microstep> | <concurrent step>, <microstep>

A <concurrent step> is of the form <label>. <microstep>, <microstep>, . . . , <microstep> and the meaning is that all these microsteps can be executed concurrently, that is, at the same time; <label>. is like a marker denoting the beginning of a concurrent step. Notice the language symbols *period* ".", which uniquely identified the label marker, and comma ",", which separates two consecutive microsteps.

Production 3. <microstep>: = <conditional assignment> | <assignment> | <go to>

Microsteps are of the three types listed in the production.

Production 4. <conditional assignment>: = IF <condition> THEN <microstep> END | IF <condition> THEN <microsequence> END

If the condition holds, then the entire microsequence following THEN and preceding END is executed, otherwise it is not executed. Notice the language terms IF, THEN, and END, which uniquely identify this microstep.

Production 5. <assignment>: = <register> ← <constant> |
<register> ← <register> |
<register> ← <function><register> |
<register> ← <register><op><register> |
<register> ← <register><op><constant>.

This is the fundamental production of RTL, and the one from which the language derives its name. In fact, consider, for example, <register> ← <register>. It is almost natural to regard the arrow "←" as denoting a transfer from a source register to a destination register. Upon closer analysis, we realize that we are really transferring the "content of the source register" to "the destination register." So the same term, <register>, would be used to denote in one case a physical storage module, in the other, its content. This ambiguity disappears if instead we interpret "←" as an *assignment,* that is, the action of setting the content of the destination register to whatever specified to the right of the arrow. Thus, we may have the following concrete instances: the value of a register can be set to a constant value (R ← 0); the content of a register can be copied into another register (R ← Q), possibly with some modification (R ← −Q); the contents of two registers can be combined by some operation and the result placed into a register (R ← R + Q, T ← R − Q, etc.), and so on.

Production 6. <go to>: = GO TO <label>

This production, uniquely characterized by the language term GO TO, is a self-explanatory transfer of control (see Sec. 2.3).

As indicated earlier, a microsequence will be a string of concrete terms and language terms and symbols, such as:

1. MAR ← PC
2. MBR ← M(MAR), PC ← PC + 1
3. IR ← MBR

Given any such string we shall momentarily ignore its meaning and concern ourselves with its "grammatical correctness," that is, if it is indeed a microsequence. To test it, we check whether we can successfully parse the string, that is, whether by successive applications of productions in the backward direction (that is, from the right side to the left side) we can reduce the string to the single term <microsequence>. For example, the string 1. MAR ← PC is parsed as follows. We begin by replacing each concrete term by its corresponding abstract term, that is, MAR and PC are both replaced by <register> (abbreviated here as <reg>), while 1. is replaced by <label>.

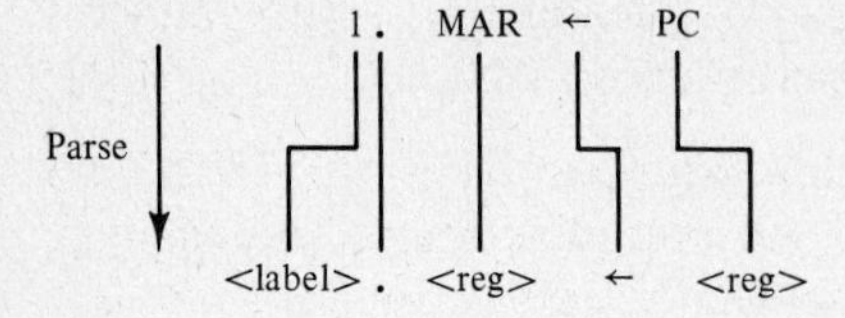

Next we recognize that <reg> ← <reg> is an instance of <assignment>, so that we have

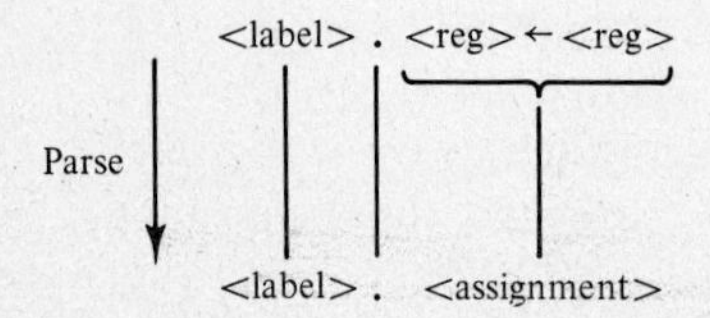

Again, <assignment> is an instance of <microstep>. So we have

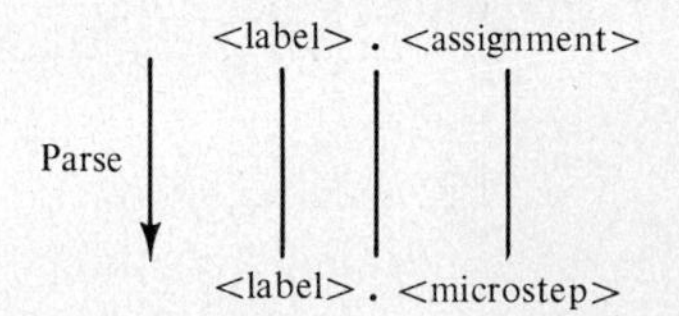

and so on. The activity of parsing is more fully illustrated on the entire string given earlier, which is indeed found to be a correct microsequence:

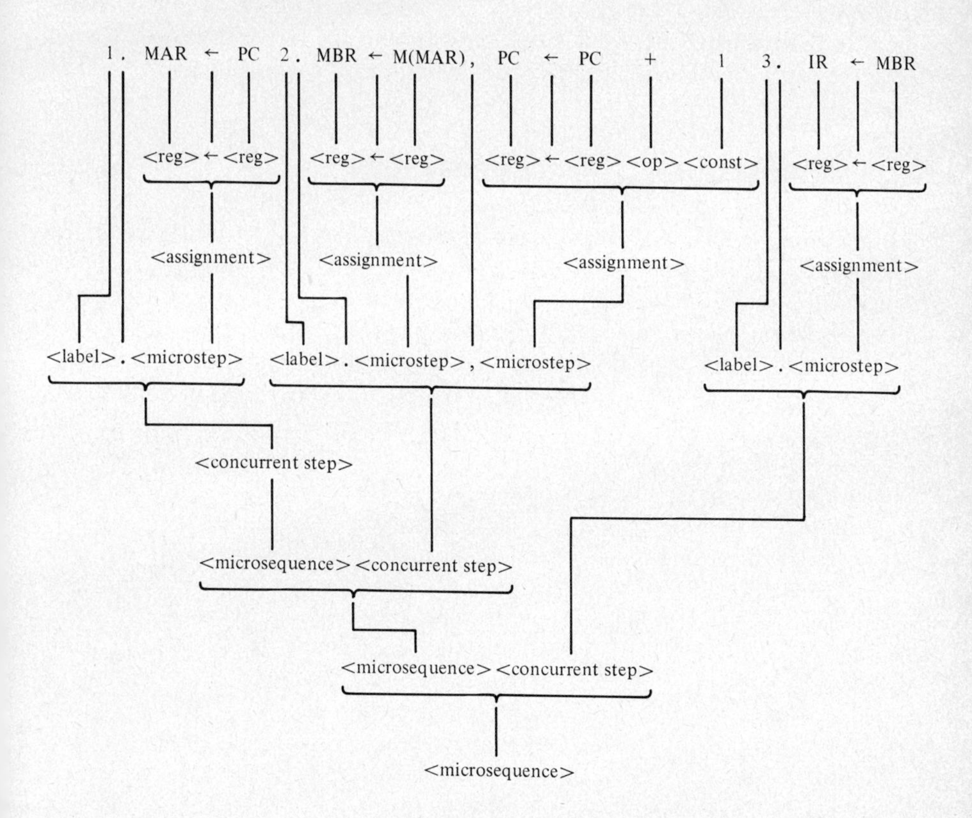

Although this example suggests that parsing is best done on a horizontal string, it is, however, customary to display a microsequence by writing one concurrent step per line; this gives an intuitive perception of the temporal sequencing of events. Indeed, concurrent steps are actions that occur one after the other (*sequentially*) in time, while the microsteps of a concurrent step are *simultaneous* actions. Thus, the preceding microsequence is conveniently displayed as follows:

```
1. MAR ← PC
2. MBR ← M(MAR), PC ← PC + 1
3. IR ← MBR
```

Another typical example of microsequence is the following:

5. MAR ← IR[11:0]
6. MBR ← M(MAR)
7. A ← MBR + A
8. GO TO 1

For the time being, we are not concerned with the "meaning" of these microsequences; we shall dwell on that point in the following sections. Suffice it to emphasize here, however, that concurrent steps are executed one after the other (top-to-bottom in the preceding display). Labels of concurrent steps, each of which simply marks *the beginning of a concurrent step* and *uniquely identifies it,* can be selected in a totally arbitrary way. However, it is customary to select them so that of two concurrent steps, the one with the larger label is executed after the one with the smaller label (both labels are, of course, integers).

7.2. MICROSEQUENCE IMPLEMENTATION

At this point we have gained sufficient motivation for the fact that RTL is the formalism to describe how a computer executes its own repertoire of instructions. Specifically, the behavior of the computer will be expressed as a collection of microsequences. Strictly speaking—and we shall realize this fully in Chapter 9—the whole behavior of a computer is just one single "giant" microsequence. However, for ease both of analysis and of design, it is convenient to break down this giant microsequence into several much simpler microsequences, each of which corresponds to a clearly identifiable activity (for example, the microsequence corresponding to the execution of the instruction ADD in the SEC).

Before undertaking the task of designing these microsequences one by one, it is necessary to outline some general, important features of the system.

A fundamental feature is the *sequential character* of a microsequence, that is, the fact that the concurrent steps of a microsequence must be executed one after the other. To *reliably* achieve the desired sequencing, the digital computer is realized as a (very large) *synchronous sequential network,* and it is therefore provided with a periodic clock which establishes the instants of the change of states of the system's flip-flops. As a rule of thumb (and only as an intuitive guideline), we may think that the execution of a concurrent step takes one clock period; we shall examine the significant cases where this simplistic rule does not hold in Chapter 9, when we shall study the detailed *timing* of the operation. In this chapter we shall be exclusively concerned with the correct *sequencing* of the concurrent steps.

Next, we shall consider the general features of the *assignment microstep.* Suppose we must implement the microstep

B ← A

that is, the transfer of the content of register A into register B. Obviously, our computing system must have a data path from register A to register B, and,

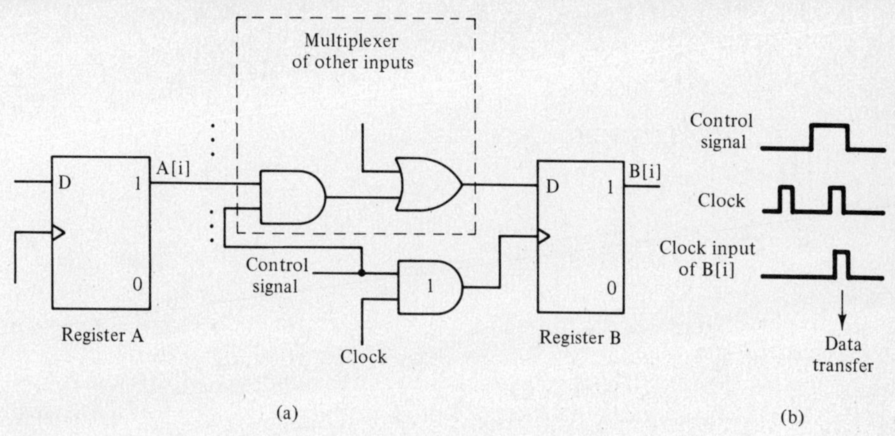

Figure 7.1 Illustration of the transfer B ← A. (a) Hardware to effect B[i] ← A[i]. (b) Timing.

assuming from now on that all transfers occur in parallel, this data path has as many lines as there are flip-flops in the registers. Assuming that all flip-flops are D F/Fs, the content of flip-flop A[i] must be transferred into flip-flop B[i] (see Figure 7.1). This transfer is effected under the control of an appropriate control signal, which gates one clock pulse to the clock input of B[i] (via AND gate 1). The timing is shown in Figure 7.1(b). Gate 1 is required by the necessity to affect the content of B[i] only when information is to be transferred to it, leaving it unaltered at all other times. Note that if more than one register can be connected to register B, then multiplexers must be placed on the D and clock inputs of B[i].

This leads us to the next important feature. In general, the output of a single functional block (register or combinational module) may have to be selectively connected to the inputs of *several* similar modules; and conversely, the

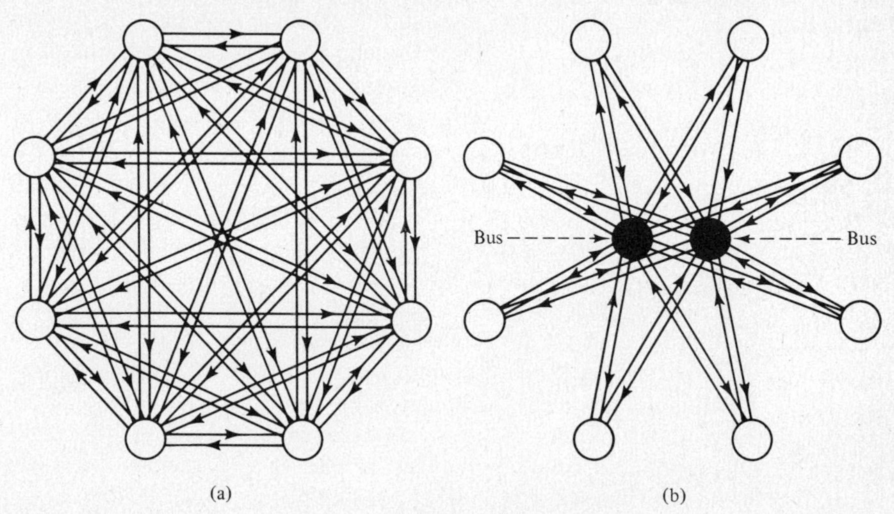

Figure 7.2 Interconnection of registers, direct (a) and through two buses (b) (shown as solid circles).

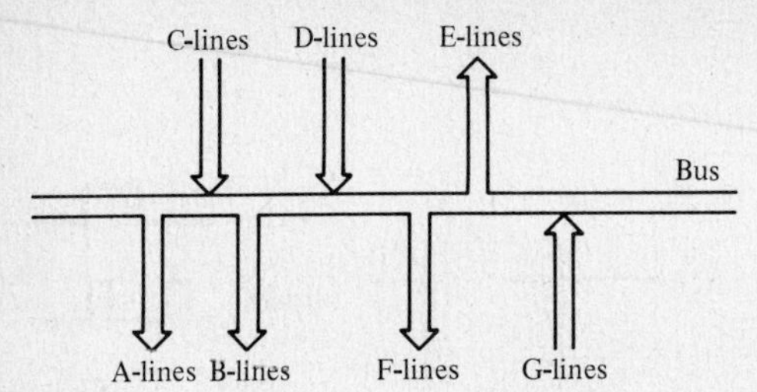

Figure 7.3 A bus with input and output links.

inputs of one such module may have to be selectively connected to the outputs of several other modules. To make a concrete example, suppose there are s different registers and it is required that the content of any one of them be transferable to any other (see Figure 7.2 where "nodes" denote registers). The obvious but expensive way to achieve this is by interconnecting them in all possible ways, that is, with $s(s - 1)$ unidirectional links [Figure 7.2(a)]. Note that this interconnection allows the establishment of $s/2$ simultaneous pairwise connections. If on the other hand we never have to establish more than, say, *two* simultaneous pairwise connections, then it is more economical, in terms of links, to introduce *two buses* and to realize the interconnection shown in Figure 7.2(b), which has $2s$ links only. In terms of sheer economics the saving is dramatic, since $2s$ instead of $(s^2 - s)$ links are needed, and the difference could be enormous when s is large. This situation is also typical of a telephone interconnection, where the number of users is much larger than the number of conversations likely to occur at any one time.

A *bus* is said to *collect its input links* and to *distribute its output links.* Functionally, we have the situation illustrated in Figure 7.3, where the C-, D- and G-lines are input links, while the A-, B-, E-, and F-lines are output links. (Links are represented in Figure 7.3 by wide strips rather than by single lines, to emphasize that each is a bundle of parallel lines.) Since an input link can be selectively fed to the bus and, in turn, the bus can be fed to an output link, it is clear that the functions of collecting and distributing are respectively those of a narrow-sense multiplexer and of a narrow-sense demultiplexer, as described in Sec. 4.8. Thus when two registers are connected via a bus, the scheme of Figure 7.1 must be modified as shown in Figure 7.4. (Note that the "other inputs" are now collected by the bus multiplexer rather than by a flip-flop multiplexer as in Figure 7.1.) *Consistently with Figure 7.4, a bus line is assumed to be at logical* 0 *unless a logical* 1 *is multiplexed to it.*

**Remark:* The reader should not conclude that the multiplexers and demultiplexers alluded to above are actual MSI modules as described in Sec. 4.8. Indeed, they are part of the general circuitry of the computer system and not separately identifiable modules. Although they are *logically* narrow-sense multiplexers and demultiplexers, in the *physical* realization the OR gate may not appear as an identifiable component. This is due to two special forms of realization of the AND gates, either as (1) *open-collector* output devices; or as (2) *tristate* output devices.

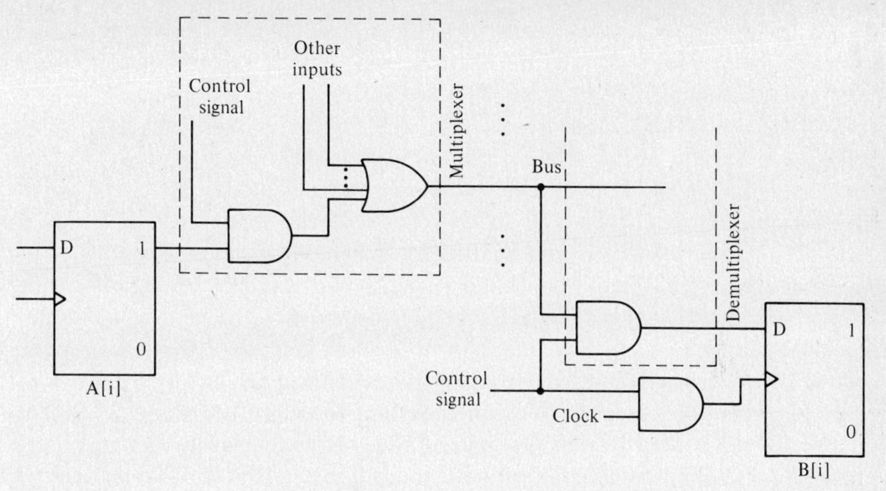

Figure 7.4 Narrow-sense multiplexer and a narrow-sense demultiplexer must be used to control several input and output links to a bus.

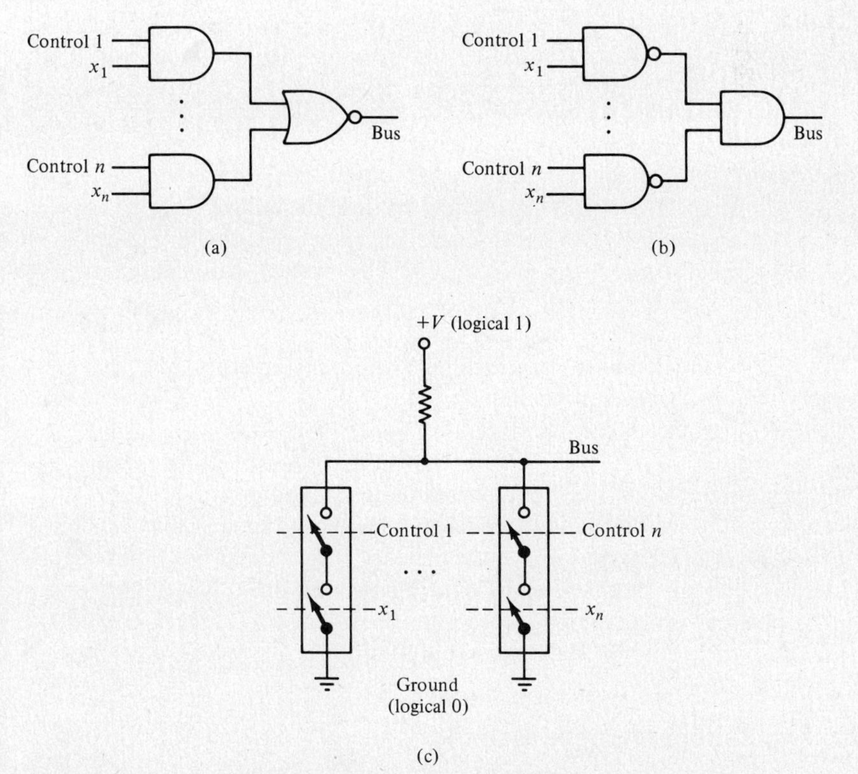

Figure 7.5 (a) A variant of the narrow-sense multiplexer. (b) An equivalent network. (c) The equivalent contact network.

Although a discussion of these two realizations in electronic terms is well beyond our scope (as indicated in Sec. 2.6), we attempt an explanation of one such realization in terms of mechanical switches.

In the "open-collector" realization, the logical network actually being realized is a variant of the original multiplexer, as shown in Figure 7.5(a) (the output is complemented). It is a trivial exercise to show the equivalence of the networks in Figures 7.5(a) and 7.5(b). Each of the input NAND gates is electronically realized as if it consisted of two switches in series, one controlled by x_i and the other by Control i (so that a switch is *closed* when the corresponding variable has the logical value 1). Referring to Figure 7.5(c) and assuming as usual positive logic, when both inputs of a NAND gate are 1 (say, $x_1 = 1$ and Control $1 = 1$) then the corresponding switches connect the bus to ground, that is, the bus is set to logical 0. We conclude that the circuit of Figure 7.5(c) is functionally equivalent to that of Figure 7.5(a). Note, however, that in Figure 7.5(c) there is no identifiable output gate; indeed, this gate is realized by directly *wiring* the gate outputs to the bus. For this reason, this is referred to as *wired logic,* and, according to one's preference, either as wired OR [referring to Figure 7.5(a)] or as wired AND [referring to Figure 7.5(b)].

It must be pointed out that the conceptual issue (realization of a multiplexer) should not be obscured by the details of its implementation nor by the whims of jargon.

7.3 BASIC ORGANIZATION OF THE SEC PROCESSOR

We now undertake the register level design of the SEC processor, which we introduced in Chapter 2. Before proceeding further, the reader is urged to refer back to Chapter 2 in order to refresh his or her knowledge of the (elementary) instruction repertoire we have introduced. The system will consist of the following basic subsystems:

1. The RAM (random access memory) or, simply, memory M.
2. The CPU (central processing unit).
3. Input/output, or peripheral units.

We defer the discussion of the input/output units to Chapter 8.

The layout of the CPU and the memory is shown in Figure 7.6; not shown is a set of *control lines* from the CPU to the memory. The address bus (abbreviated as addr. bus) has 12 lines, while the data bus has 21 lines. Not shown is the system clock which is distributed throughout the system (see Figures 7.1 and 7.4 with regard to the control of each clocked flip-flop).

7.3.1 The Random Access Memory

The *memory* consists of $2^{12} = 4096$ 21-bit registers (memory cells). It is displayed below as an array of 1-bit memory elements shown as "boxes" in

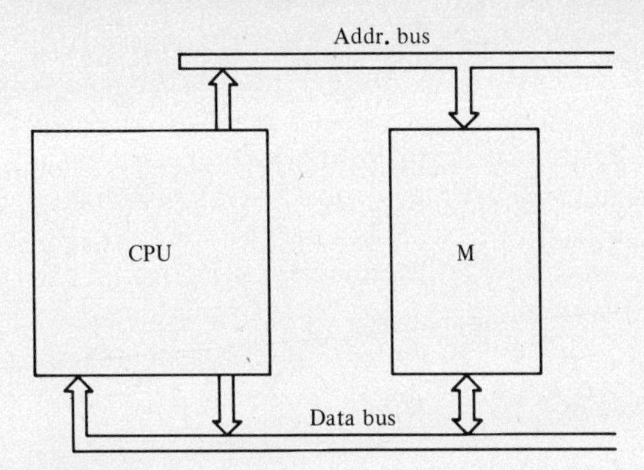

Figure 7.6 Basic layout of CPU and memory.

Figure 7.7(a). This array consists of 4096 rows (each corresponding to a computer word) and 21 columns. The structure of each memory element is shown in Figure 7.7(b). It contains an *unclocked latch,* whose input is gated by a control signal WRITE, and whose output is gated by a control signal READ. Both WRITE and READ consist of pulses synchronized with the main clock and generated whenever a memory operation (write or read) is to be effected. Each memory element is multiplexed to the 21-bit Memory Buffer Register (MBR) and the latter is demultiplexed to any memory cell. The selection of the memory cell is done by standard techniques: the content of the 12-bit Memory Address Register (MAR) which specifies the memory address to be accessed is supplied to a 12-to-2^{12} decoder (see Chapter 4, Sec. 4.8.2); each output line of the decoder is further demultiplexed, under control of the WRITE and READ signals, onto the two lines WORD SELECT WRITE and WORD SELECT READ. Note that by the very function of the decoder and by the mutually exclusive nature of the WRITE and READ signals, only one of the 2×2^{12} WORD SELECT lines can be active at any time. Two operations may occur:

1. Write (the content of MBR is transferred to the selected memory word). The WORD SELECT WRITE signal, applied to the latch input gates, and the WRITE signal, applied to the MBR output gates, establish a path from the MBR to the selected word.
2. Read (the content of the selected memory word is transferred to MBR). The WORD SELECT READ signal, applied to the latch output gate, and the READ signal, applied to the MBR input gate, establish a path from the selected word to the MBR.

Note that the BIT WRITE lines demultiplex the MBR to the memory, while the BIT READ lines multiplex the memory to MBR (in actuality, each BIT READ line is an OR gate). Also, since the flip-flops in the memory elements are unclocked latches, only the flip-flops of the selected word are affected by the combined action of WORD SELECT WRITE and WRITE, jointly acting as a clock.

What we have shown in Figure 7.7 is just one possible organization and realization of a random access memory; indeed we have assumed semiconductor storage devices and word selection (also called "linear selection"). Other technologies were quite popular in the past (typically magnetic-core technology), as well as different selection strategies (typically, coincidence selection), but we shall not discuss these alternatives here.

In what follows, we shall always refer to the memory as shown in the compact layout of Figure 7.8. In addition, the two fundamental memory operations are summarized in RTL as follows:

READ: MBR ← M(MAR) (transfer to MBR the content of the memory cell specified by MAR)

WRITE: M(MAR) ← MBR (transfer to M(MAR) the content of MBR)

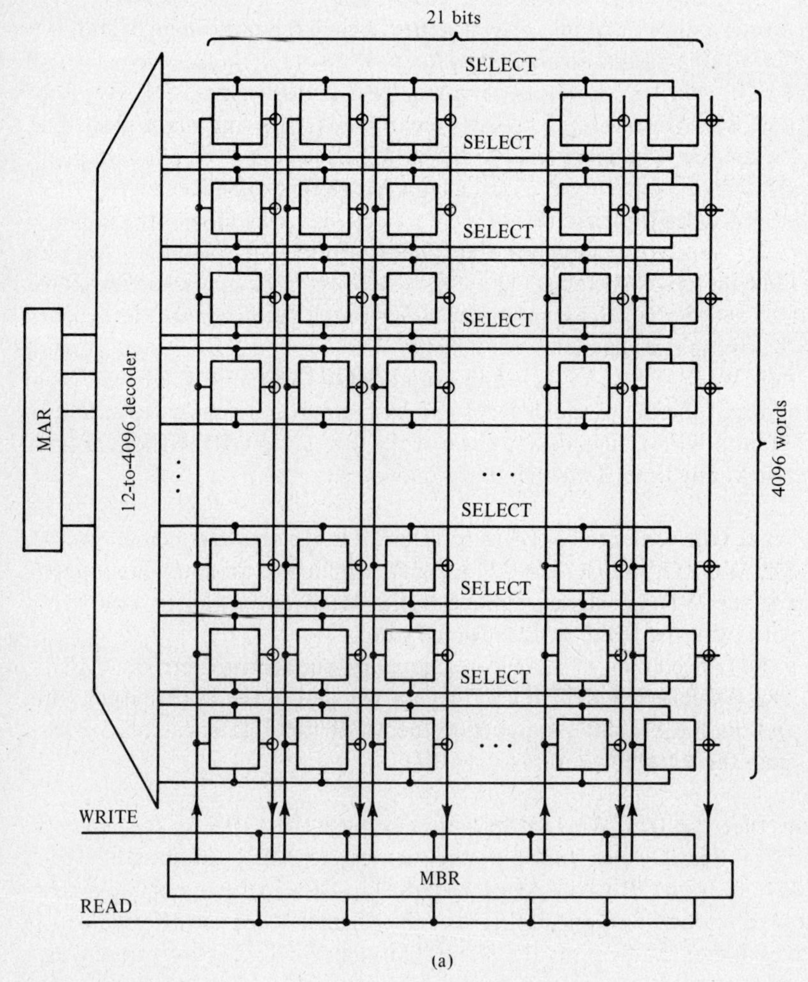

Figure 7.7 General structure of the random access memory.

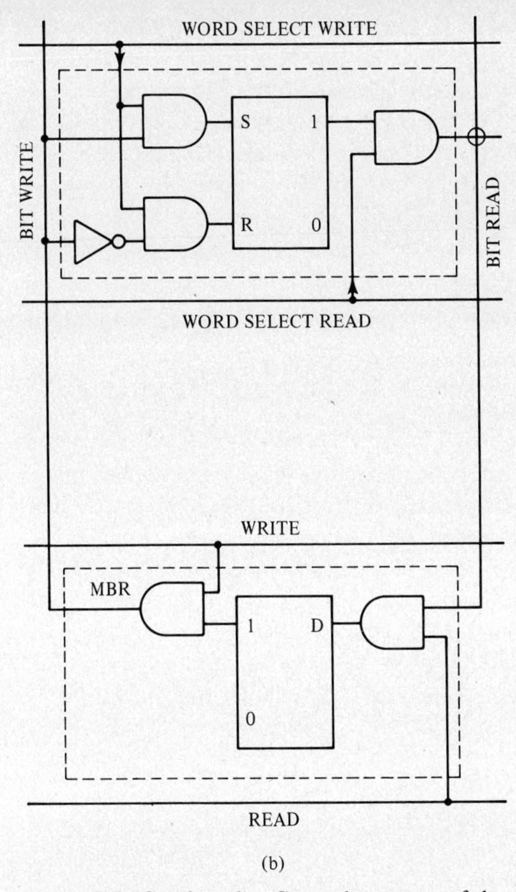

Figure 7.7 ***(Continued)*** General structure of the random access memory.

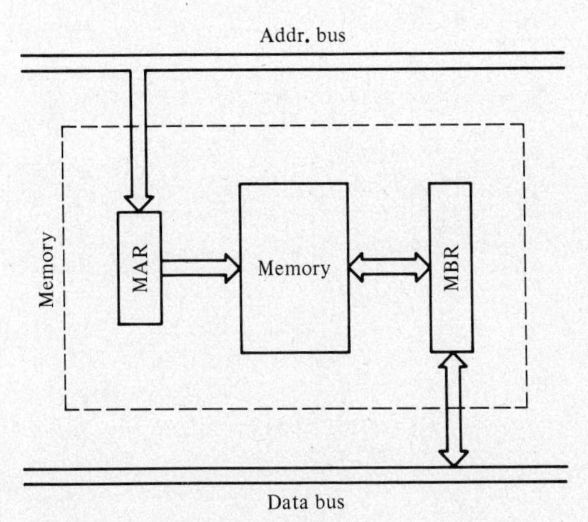

Figure 7.8 Layout of the memory subsystem.

One last comment is in order. If the memory array is realized with semiconductor devices, we know well that the operation MBR ← M(MAR) leaves the content of M(MAR) unaltered. It should be pointed out, however, that in the traditional magnetic core technology, the readout of M(MAR) is destructive, that is, the content of M(MAR) becomes 0 after the read operation; in such systems, therefore, a WRITE STEP:M(MAR) ← MBR occurs immediately after READ, to restore the original content of M(MAR). Since such a write step is automatically provided by a local control subunit, we shall always assume that *memory readouts appear as nondestructive,* independent of the adopted technology.

7.3.2 The Central Processing Unit

Our approach consists of building up the unit step by step, by adding new pieces of hardware as we develop the motivation for them. In this manner we shall obtain a family of CPUs of increasing sophistication. While they are all "functional" (i.e., capable of performing automatic computation), increased complexity produces increased flexibility, that is, ease in programming. The simplest "stripped-down" version—called the SEC-0 computer—will be described in this and the next sections.

The CPU contains the following hardware items (shown in Figure 7.9):

ALU	With two sets of input lines, conventionally called A-lines and B-lines (see Sec. 6.6) and one set of output lines called the G-lines. The ALU is capable of performing at least the following operations: 1. Arithmetic: A PLUS B, A MINUS B, A PLUS 1, A MINUS 1, A PLUS B PLUS 1, and arithmetic shifts on 21-bit numbers represented in the two's complement notation. 2. Logic: all logic operations on pairs of 21-bit strings, and logical shifts.
OF	Overflow flip-flop.
Register A (Accumulator):	A 21-bit register.
Register PC (Program Counter):	A 12-bit register designed to store the address of the instruction to be fetched.
Register IR (Instruction Register):	A 21-bit register designed to store the instruction. The outputs of IR feed the control unit, which generates the control signals in their appropriate sequence (Chapter 9).

The internal interconnection of the CPU is provided by the distributor bus and the collector bus. The latter feeds the ALU A-lines, while the B-lines are fed by the (external) data bus. Notice that, since the ALU is a purely combinational module the distributor bus can feed the data bus only when the B-lines inputs are blocked.

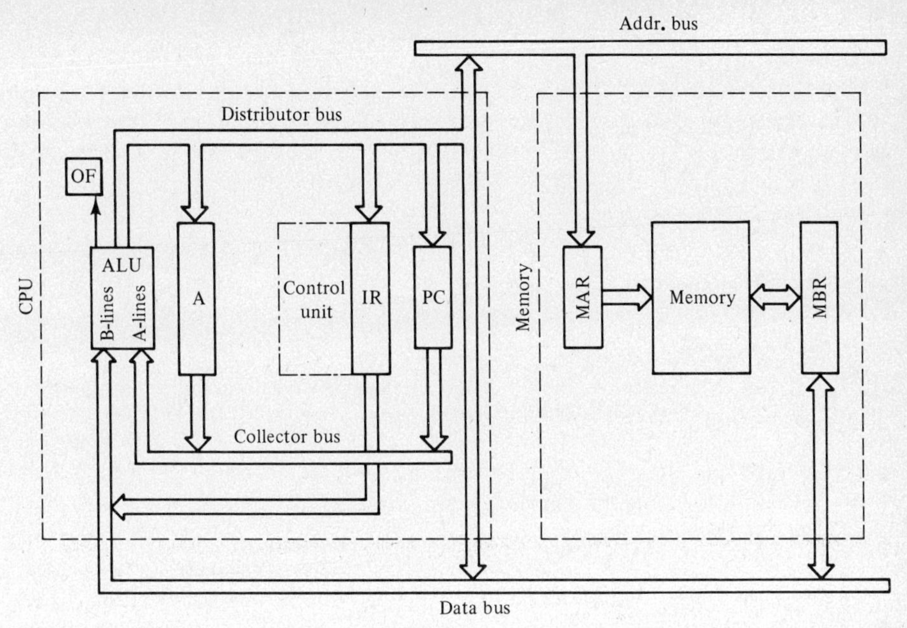

Figure 7.9 Layout of the SEC-0 computer.

7.4 FETCH AND EXECUTE MICROSEQUENCES

The execution of an instruction is carried out in two consecutive phases (sometimes called cycles): FETCH and EXECUTE. With reference to the layout of Figure 7.9, the FETCH microsequence is as follows (*common to all instructions*):

	Comments
1. MAR ← PC	The address of the instruction is sent to MAR.
2. MBR ← M(MAR), PC ← PC + 1	The memory is read; concurrently the content of PC is incremented and contains the address of the next instruction.
3. IR ← MBR	The instruction is stored in IR.

Once the instruction is stored in IR, its operation code completely determines the EXECUTE microsequence, as we shall see below.

We shall now describe the EXECUTE microsequences for some of the instructions introduced in Sec. 2.3; specifically, we shall discuss LDA, STA, ADD, ENT, JMP, and JZA, since for all the others the corresponding microsequence is a slight variant of the microsequence of one of these six instructions. It is worth pointing out that the RTL can be used not only to describe microsteps,

TABLE 7.1 EXECUTE MICROSEQUENCES OF LDA, STA, ADD, ENT, JMP, JZA

Code	Description	Microsequence[a]	Comments
LDA	$A \leftarrow M(Y)$	5. $MAR \leftarrow IR[11:0]$ 6. $MBR \leftarrow M(MAR)$ 7. $A \leftarrow MBR$ 8. GO TO 1	The operand address is transferred to MAR, the specified memory location is read, and the readout data are transferred to the accumulator.
STA	$M(Y) \leftarrow A$	5. $MAR \leftarrow IR[11:0]$ 6. $MBR \leftarrow A$ 7. $M(MAR) \leftarrow MBR$ 8. GO TO 1	The specified memory address is loaded into MAR and the operand is transferred to MBR; next WRITE takes place.
ADD	$A \leftarrow A + M(Y)$	5. $MAR \leftarrow IR[11:0]$ 6. $MBR \leftarrow M(MAR)$ 7. $A \leftarrow A + MBR$, $OF \leftarrow c_{21} \oplus c_{20}$[b] 8. GO TO 1	This microsequence is quite similar to that of LDA, except that now MBR + A, rather than MBR alone, is transferred to A.
ENT	$A \leftarrow Y$	5. $A[11:0] \leftarrow IR[11:0]$ $A[20:12] \leftarrow 0$ 6. GO TO 1	Contrary to the three preceding instructions, this microsequence involves no memory reference.
JMP	$PC \leftarrow Y$	5. $PC \leftarrow IR[11:0]$ 6. GO TO 1	All that is needed to effect this command is to replace the content of PC (which is set in the fetch cycle according the "normal" transfer of control) with the "jump" address contained in Y.
JZA	IF $(A = 0)$ THEN $PC \leftarrow Y$	5. IF $(A = 0)$ THEN $PC \leftarrow IR[11:0]$END 6. GO TO 1	In this case the microstep $PC \leftarrow IR[11:0]$ occurs under condition that the accumulator contains 0.

[a]The reader may be curious as to why we number the EXECUTE concurrent steps starting from 5 (rather than 4). This will be apparent in Sec. 7.5. Notice, however, that we are complying with the general guideline expressed at the end of Sec. 7.1.

[b]This microstep records a possible "overflow" into the OF flip-flop. Recall that c_{21} and c_{20} are, respectively, the carry-out and the carry-in for the sign bit unit of the ALU. The same microsteps occur in the SUB instruction. We also need a "test" instruction JOF, Jump on Overflow, analogous to JPA and JZA, described by IF $(OF = 1)$ THEN $PC \leftarrow Y$.

but also to describe the action performed by a machine language instruction. In what follows

OP, 0, Y

is the format of the typical SEC-0 instruction, where OP is a 6-bit operation code and Y denotes the 12 least-significant bits of the instruction word. Thus the action of ADD, 0, Y is described as $A \leftarrow A + M(Y)$. Notice that the EXECUTE microsequence immediately follows the FETCH microsequence, and that at the end of FETCH, Y is stored in the subregister IR[11:0] of IR.

It is appropriate to introduce here five new instructions which involve exclusively the content of the accumulator; hence, only the function field need be specified. They are:

TABLE 7.2 OPERAND-FREE INSTRUCTIONS

Instruction	Operation	Mnemonic
50, –, –	Logical right shift	LRS
51, –, –	Logical left shift	LLS
52, –, –	Arithmetic right shift	ARS
53, –, –	Arithmetic left shift	ALS
54, –, –	Change the sign of the content of accumulator (i.e., apply the operation of "negation" to the content of A)	NEG

Since the Y-field of the instruction word is not used in LRS, LLS, ARS, ALS, and NEG, such instructions will be referred to as *operand free.* The development of the microsequences of the instructions just introduced is left as an exercise for the reader. (See Chapter 6, Sec. 6.6, with regard to the use of bit b_{OUT} generated in the ALU shifting network.)

7.5 INDEX REGISTERS

We now return to considering our SEC-0 computer from the programmer's standpoint, and specifically we revisit the loop program we described in Chapter 2, Sec. 2.4, repeated for convenience in Figure 7.11(a). As we pointed out earlier, essential in a loop program are the operations of updating counters and pointers; since loop programs are very frequently used, it is desirable to simplify these operations from the user's standpoint. A first step in this direction would be to store the counter in a readily accessible register, without resorting to the main memory: since the number of counters simultaneously used in a program is normally quite small, this provision should entail an acceptable amount of extra hardware. Next, the counter could be used to automatically modify a pointer, appearing as the address in an instruction. These are the ideas that lead to the introduction of *index registers.* In our computer we introduce a set of seven 12-bit index registers: R(1), R(2), . . . , R(7), laid out as in Figure 7.10(a): we shall call the resulting machine SEC-XR. The content of an index register is always a nonnegative number; the selection of an index register is done by means of the field of 3 bits—which we call the *b-field*—comprised between the operation code (OP code) field and the operand field. Register R(i) is selected by setting b to represent i in binary; thus, b = $(000)_2$ means either that *no* register is being selected or, alternatively, we may imagine that there is a register R(0) whose content is always $(0000)_8$.

The introduction of index registers adds an extra concurrent step, called ADDRESS MODIFICATION, between the FETCH and the EXECUTE microsequences (a concurrent step labeled 4). This step consists of adding the contents of the selected index register and the IR[11:0] subregister of IR, and of placing the result in IR[11:0], that is,

4. IR[11:0] ← IR[11:0] + R(b)

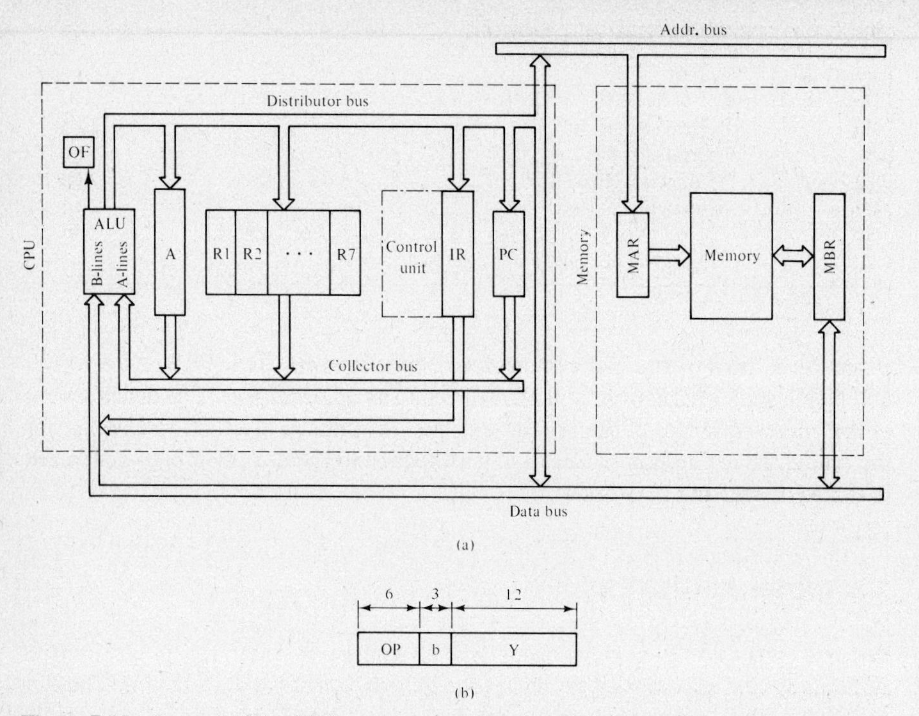

Figure 7.10 Layout of the SEC computer equipped with index registers (SEC-XR).

Alternatively, we could have the conditional microstep "IF (b $\neq$ 0) THEN IR[11:0] $\leftarrow$ IR[11:0] + R(b) END," which would then be skipped when b = 0; however, the previous formulation simplifies the structure of the control unit, as we shall see in Chapter 9. For simplicity of notation, we define

$$N = Y + R(b)$$

in other words, Y is a sequence of 12 bits specified in the instruction and is contained in IR[11:0] *before* the ADDRESS MODIFICATION step; N is a sequence of 12 bits, contained in IR[11:0] *after* the ADDRESS MODIFICATION step and used in subsequent processing. *The ADDRESS MODIFICATION step is executed for all instructions, unless otherwise explicitly noted.*

Suppose now that we want to tally the contents of locations 0100_8 through 0137_8 in the SEC-XR machine. The corresponding program is given in Figure 7.11(b). In this program, let us concentrate on the instruction ADD, 1, 0077; the new instructions ENR, DER, and JPR will be explained below. This instruction performs the following action:

$$A \leftarrow A + M(0077_8 + R(1))$$

Thus if R(1) contains the integer j, the content of $M(0077_8 + j)$ is added to the content of the accumulator. Since our numbers to be tallied are contained in locations 0100_8–0137_8, we can let R(1) run as a counter between 0000_8 and 0040_8

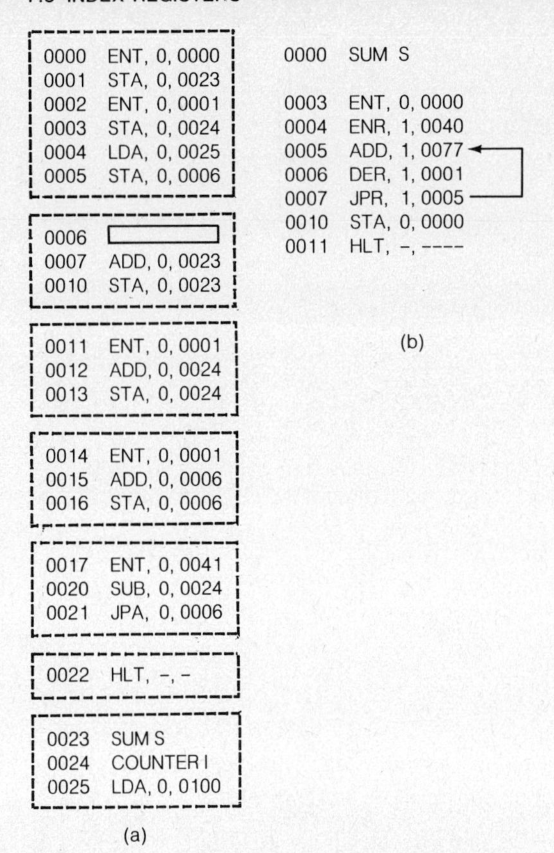

0000 ENT, 0, 0000
0001 STA, 0, 0023
0002 ENT, 0, 0001
0003 STA, 0, 0024
0004 LDA, 0, 0025
0005 STA, 0, 0006
0006
0007 ADD, 0, 0023
0010 STA, 0, 0023
0011 ENT, 0, 0001
0012 ADD, 0, 0024
0013 STA, 0, 0024
0014 ENT, 0, 0001
0015 ADD, 0, 0006
0016 STA, 0, 0006
0017 ENT, 0, 0041
0020 SUB, 0, 0024
0021 JPA, 0, 0006
0022 HLT, –, –
0023 SUM S
0024 COUNTER I
0025 LDA, 0, 0100
(a)
0000 SUM S
0003 ENT, 0, 0000
0004 ENR, 1, 0040
0005 ADD, 1, 0077
0006 DER, 1, 0001
0007 JPR, 1, 0005
0010 STA, 0, 0000
0011 HLT, –, ----
(b)

Figure 7.11 A loop program, in the SEC-0 computer and in the SEC-XR computer.

and use the counter itself to automatically update the *pointer* address in ADD, 1, 0077. What we need, therefore, is some simple machinery to operate on the contents of index registers. This can be done by means of the six instructions described in Table 7.3: *note that for these instructions the ADDRESS MODIFICATION step must be omitted.*

In Table 7.4 we give the RTL description of the EXECUTE microsequence for each of the above instructions.

Returning to the loop program we considered earlier (Figure 7.11) only one additional program variable, the *counter,* is needed and is stored in R(1). It is convenient to scan the array of numbers from the last to the first, rather than the other way around. Thus the initialization of the counter is done by the instruction

```
0004   ENR, 1,0040
```

which replaces those in addresses 0002, 0003 in the original program. Since the accumulator is not involved in the manipulations of the counter as it was in the corresponding SEC-0 program, it will be used to "accumulate" the tally of the

TABLE 7.3 INDEX REGISTER INSTRUCTIONS

Mnemonic code	Instruction	Description	Operation
LDR	41, b, Y	R(b) ← M(Y)[11:0]	Load register R(b) with the 12 least-significant bits of location Y
STR	42, b, Y	M(Y)[11:0] ← R(b), M(Y)[20:12] ← 0	Store in memory location Y the 21-bit string 000 R(b)
INR	43, b, Y	R(b) ← R(b) + Y	Increment the content of register R(b) by the number Y
DER	44, b, Y	R(b) ← R(b) − Y	Decrement the content of register R(b) by the number Y
ENR	45, b, Y	R(b) ← Y	Enter into register R(b) the number Y
JPR	46, b, Y	If (R(b) ≠ 0) THEN PC ← Y	Jump to Y if the content of register R(b) is positive, that is, if R(b) ≠ 0

numbers: therefore it is initially cleared by the instruction 0003 ENT, 0,0000 which initializes the SUM to 0. Next, the instructions in 0006, 0007, 0010 are replaced by the single instruction

```
0005   ADD, 1,0077
```

which scans the array of numbers and accumulates them as the counter is updated. The latter operation is done by means of the instruction

```
0006   DER, 1,0001
```

TABLE 7.4 EXECUTE MICROSEQUENCES OF INDEX REGISTER INSTRUCTIONS

LDR	4. MAR ← IR[11:0] 5. MBR ← M(MAR) 6. R(b) ← MBR[11:0] 7. GO TO 1
STR	4. MAR ← IR[11:0] 5. MBR[11:0] ← R(b), MBR[20:12] ← 0 6. M(MAR) ← MBR 7. GO TO 1
INR	4. R(b) ← R(b) + IR[11:0] 5. GO TO 1
DER	4. R(b) ← R(b) − IR[11:0] 5. GO TO 1
ENR	4. R(b) ← IR[0:11] 5. GO TO 1
JPR	4. IF (R(b) ≠ 0) THEN PC ← IR[0:11] END 5. GO TO 1

Finally, the termination test is carried out by the instruction

```
0007   JPR, 1,0005
```

since the counter becomes 0 once the 32 numbers have been tallied. Thus the original program is replaced by the much more compact one shown in Figure 7.11(b).

What we have developed is a more sophisticated machine than SEC-0, which we shall call SEC-XR. Notice, however, that for any SEC-XR program, generally only with the expenditure of extra storage, we can develop a SEC-0 program which performs the same action. In other words, some functions implemented by *software* in SEC-0 are implemented by *hardware* in SEC-XR; these functions, such as loop management, occur so frequently in the user's practice that the increased equipment complexity pays off. What we have seen is an instance of a general engineering design principle of computing systems: the dividing line between functions implemented by equipment (hard-wired) or by program is mainly due to their relative frequency in actual operating practice.

Note: It is appropriate at this point to carefully examine the flow of operands from and into a 12-bit register [such as IR, PC, and R(1), . . . , R(7)] through the ALU. The latter, as we know, consists of 21 parallel bit units and so has 21 inputs and 21 output lines. For concreteness, let us refer to the microstep R(1) ← R(1) + IR[11:0] (Figure 7.12). The 12 output lines of R(1) are gated on the 12 least-significant (LS) lines of the collector bus, while nothing (i.e., 000000000_2) is gated onto the 9 most-significant (MS) lines of the same bus; the same happens for the content of IR[11:0] fed to the data bus. Thus the incoming operands to the ALU are 000_8R(1) and 000_8IR[11:0]; their sum is

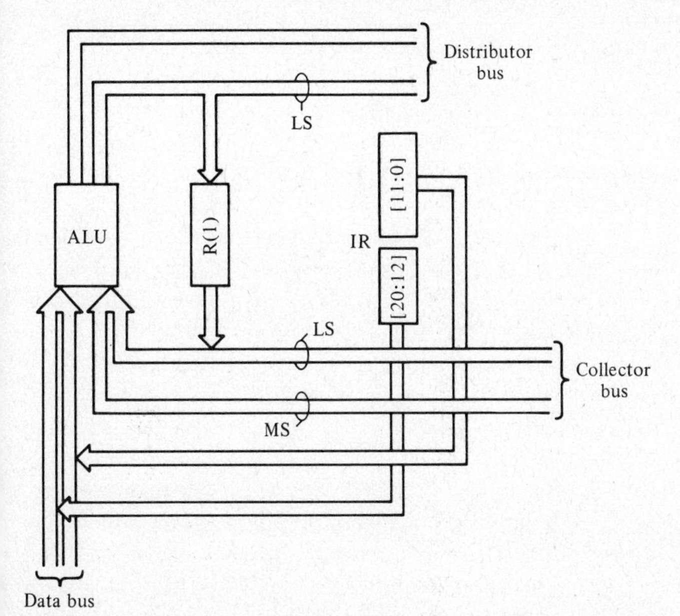

Figure 7.12 Data flow in the microstep R(1) ← R(1) + IR[0:11].

formed by the ALU as a 21-bit number, which emerges from the ALU on the 21-line distributor bus. Of these 21 lines only the 12 LS ones are gated back onto the input lines of R(1), so that only the 12 LS bits of the results are really saved. For example,

	Contents (octal)	
R(1) (before):	0103 → Collector bus	000 0103
IR[11:0]	1701 → Data bus	000 1701
R(1) (after)	2004 ← Distributor bus	000 2004

7.6 SUBROUTINES

In this section we shall discuss another interesting instance of the general principle of hardware software trade-offs mentioned in the last paragraph of the preceding section.

When writing a program for a specific application, we may need to execute over and over again a rather complex operation which is implemented by several machine instructions. Typical of such operations could be the multiplication of two integers (for which SEC has no machine instruction), code conversions, arithmetic operations with multiple-precision operands (that is, where an operand occupies two or more memory words, rather than just one), and so on. For any such operation, it appears natural to prepare a small program once and for all and to "adapt" it to all occurrences in the body of the main application program. Quite naturally, such small program is called a subroutine, which is defined as follows:

> A *subroutine* is a program that computes m functions (results) $f_1(x_1, \ldots, x_n), f_2(x_1, \ldots, x_n), \ldots, f_m(x_1, \ldots, x_n)$ of n variables $x_1, \ldots, x_n$ (input data).

We shall now consider an example of a subroutine.

Problem. A memory location u contains seven octal digits $d_6d_5d_4 \ldots d_1d_0$; it is requested to break up this string and to store digit d_i in the three least-significant bit positions of memory location v + i, where v is a given address and i = 0, 1, . . . , 6 [so that the content of M(v + i) becomes $000000d_i)_8$]. (This operation is typical in the preparation to print out the content of cell u.)

To implement this task, we write the small program shown in Figure 7.13, where several addresses—u, v, and p—are left in "symbolic" form (we leave the interpretation of the program as a problem for the reader). The reason for leaving those addresses unspecified is that their actual specification depends not only upon the given values of u and v, but also upon where (address p) the subroutine is going to be inserted in the main program. Therefore, for each use of the given subroutine the programmer will have not only to rewrite the sequence of instructions, but also to correctly select the values of the unspecified addresses. Such mode of operation is referred to as *open subroutine* [see Figure 7.14(a)]

```
p       ENR,1, 7771
p+1     LDA,0, u
p+2     AND,0, p+12
p+3     STA,1, v+7
p+4     INR,1, 0001
p+5     LDA,0, u
p+6     LRS,-, -
p+7     LRS,-, -
p+8     LRS,-, -
p+9     STA,0, u
p+10    JPR,1, p+2
p+11    JMP,0, p+13
p+12    0000007
p+13
```

Figure 7.13 Example of subroutine.

and has two obvious disadvantages: first, several copies of the subroutine must be incorporated in the main program, with ensuing memory waste; second, the selection of the unspecified addresses for each use of the subroutine is both burdensome and error-prone.

These two drawbacks are avoided in another mode, called *closed subroutine,* where the subroutine is stored just once in a fixed portion of memory. The main program interacts with the subroutine in two major aspects [Figure 7.14(b)]:

1. Transfers of control (*linkage*).
2. Parameter exchanges.

A *transfer of control* occurs initially from the program to the subroutine: this is commonly referred to as *subroutine call.* Once the subroutine has been executed, control is returned by the subroutine to the main program (*return of control*).

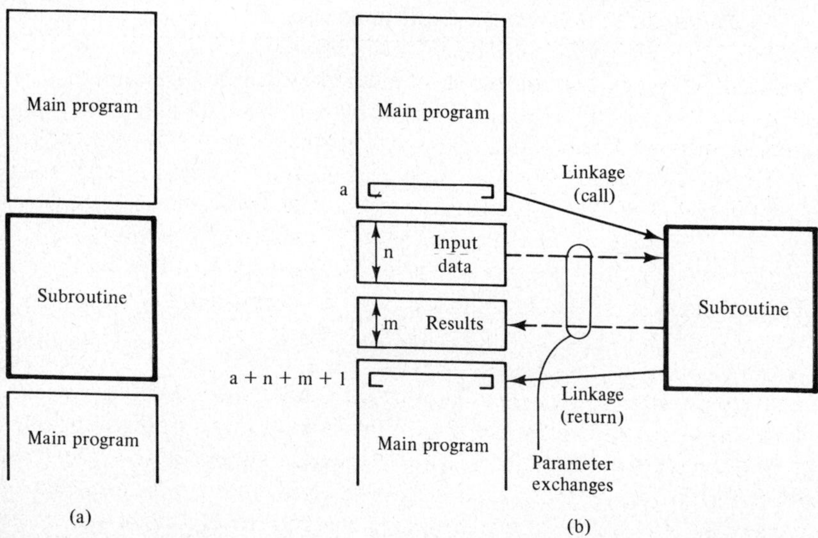

Figure 7.14 Schemes of open and closed subroutines.

The *parameter exchanges* involve the acquisition of input data and the delivery of results by the subroutine.

Transfers of control and parameter exchanges are implemented by establishing a *protocol* (i.e., a standard format) of communication between main program and subroutine. We shall now illustrate one such protocol, by making at first reference to our subroutine shown in Figure 7.13. In our example we have one input word and seven output words. Let a be the address of the main program instruction which transfers control (from the main program) to the subroutine; we choose to reserve cell a + 1 to store the input and cells a + 2, a + 3, . . . , a + 8 to the outputs; return of control will occur at address a + 9.

In general, if the subroutine computes m functions of n variables and a is as above, then: cells a + 1, a + 2, . . . , a + n are reserved to store the input data; cells a + n + 1, a + n + 2, . . . , a + n + m are reserved to store the results; return of control will occur to address a + n + m + 1. The integers n and m are characteristic of the subroutine; thus once the address a is known to the subroutine (or, for that matter, any other address a + j, for fixed j), all of the other important items are also known, such as: (1) where the data are, (2) where the results are to be delivered, and (3) where to return control.

Thus to achieve correct operation, the following actions must take place:

1. The main program must store address a in *a fixed location known to the subroutine.*
2. Return of control to the main program occurs to an address at *a fixed displacement* from address a.

We can now describe the implementation of these actions. The first technique makes no use of index registers and is therefore suited for both the SEC-XR and the SEC-0 machines; the second makes use of index registers, and is not suited for the SEC-0 machine.

Assume that the subroutine is stored in consecutive locations of memory starting at p (where now p is *a fixed address,* not a parameter). We stipulate that location p is used to store the address (a + 1) and that (p + 1) contains the first instruction of the subroutine. Then the *subroutine call* must jump to (p + 1) and it can be implemented as follows:

```
a - 2     ENT,0,a+1
a - 1     STA,0,p
a         JMP,0,p+1
```

The *return of control* requires "constructing" and then executing an instruction JMP, 0, a + 1 + n + m. This is accomplished by the following subroutine segment starting at some location c:

```
c         LDA,0, p
c+1       ADD,0, c+4
c+2       STA,0, c+3
c+3
c+4       JMP,0, n+m
```

where location c + 3 is arranged to contain the "Jump" instruction constructed by the sequence LDA, ADD, STA.

The next implementation makes use of index registers. Here the protocol is that an index register R(b) is used to store address (a + 1) and the first instruction of the subroutine is stored at p. Thus the subroutine call is realized by

```
a - 1        ENR,b,a+1
a            JMP,0,p
```

and the return of control is effected by the single instruction

```
c        JMP,b,n + m
```

The use of subroutines is very frequent in programming practice. Thus it may be desirable to enrich the instruction repertoire of the computer by slightly complicating the hardware with an instruction that explicitly performs the entire operation of subroutine call.

The first instruction we shall introduce to achieve the desired behavior is named Jump to Subroutine (mnemonic code, JS). This instruction uses the field b in the normal "address modification" mode, that is, the starting address of the subroutine is Y + R(b) + 1; notice, however, that the use of an index register is not essential, that is, JS, 0, Y is an instruction also of the SEC-0 machine. It must be pointed out that after fetching the JS instruction from address a, the PC register contains the number (a + 1).

Instruction	Operation	Mnemonic
23, b, Y	Jump to Subroutine in location N + 1[1] and store PC in location N (1. M(N) ← PC, 2. PC ← N + 1)	JS

The EXECUTE microsequence (beginning after the ADDRESS MODIFICATION) is as follows:

5. MAR ← IR[11:0]
6. MBR[11:0] ← PC, MBR[20:12] ← 0
7. M(MAR) ← MBR, PC ← IR[11:0] + 1
8. GO TO 1

When index registers are available (i.e., only in the SEC-XR machine) we can use a new Jump to Subroutine instruction (mnemonic code: JSR) which makes essential use of index register. This instruction is in the same class as those of Table 7.3, that is, *address modification does not take place for this instruction.*

[1]Recall that N = R(b) + Y.

Instruction	Operation	Mnemonic
47, b, Y	Jump to Subroutine in location Y and store PC in index register R(b) (1. R(b) ← PC 2. PC ← Y)	JSR

with the following EXECUTE microsequence:

4. R(b) ← PC
5. PC ← IR[11:0]
6. GO TO 1

Returning to our running example, we show in Figure 7.15 the program corresponding to the closed subroutine format; the jump to subroutine is performed by JSR. Notice again that in both cases p is a fixed address.

Before leaving this brief discussion, it is appropriate to mention an important feature of subroutines, which is standard practice. A given subroutine may use some CPU registers—such as A and a subset of the index registers—which may contain data being used by the main program. In general the main program expects that the contents of all registers be unaffected by the subroutine, that is, it expects to find upon reentry from the subroutine the same register contents preexisting the subroutine call [possibly with the exception of the

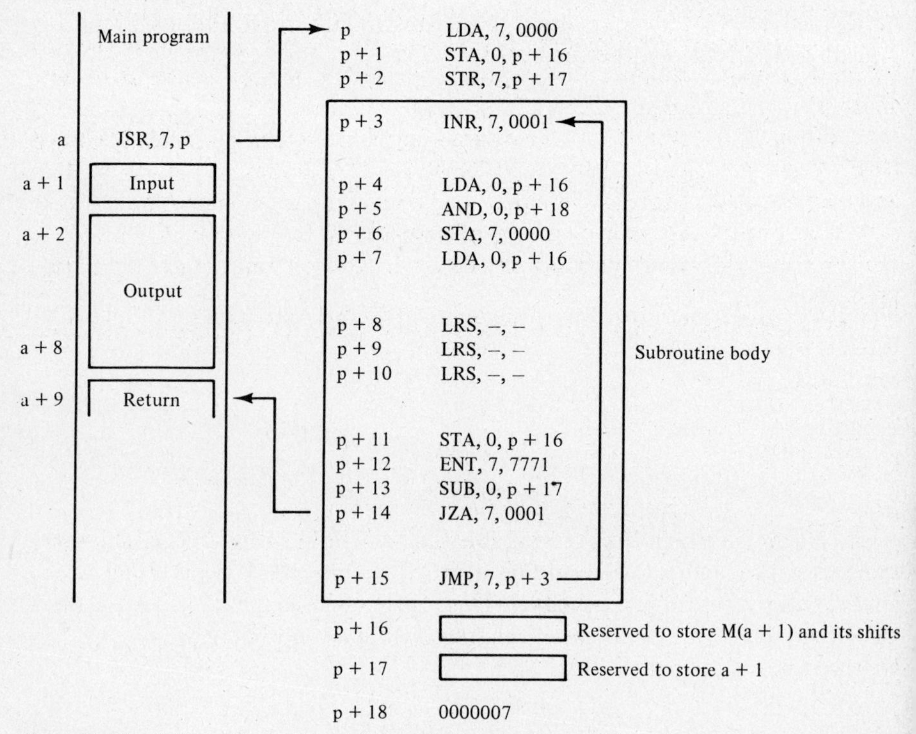

Figure 7.15 Subroutine in closed form (for running example).

content of R(b) used by JSR]. This objective can be achieved if the subroutine contains two program segments, one just after the beginning and one just before the end, to respectively "save" the CPU register contents in designated memory locations and to "restore" these register contents before control return to the main program is effected.

7.7 NESTED SUBROUTINES—STACKS

In the preceding section we have examined the situation of a program "calling" another program, specifically, the main program calling a subroutine. This is the simplest situation. In common practice, however, subroutines can be themselves rather long and complex programs, which may in turn call other subroutines, and so on and so forth. We thus have the situation of "nested subroutines," for which Figure 7.16 is the analog of Figure 7.14(b), regarding the simple case of a single subroutine. The sequence of actions is illustrated in Figure 7.17, where the indentations give evidence of the nesting structure. It is clear that the policy, which correctly handles a chain of nested subroutines, is a *last-in–first-out* (LIFO, in jargon) policy, whereby the last called subroutine (in our example, subroutine k) is the first one to return control (to subroutine k − 1).

For each call, some *reference* address must be available to the *called* subroutine, both for exchanging the parameters and for returning control to the *calling* program (whether the main program or another subroutine). The handling of these reference addresses presents some problems in each of the two schemes (with or without index registers) described earlier, as we shall now analyze.

In the SEC-0 (subroutine handling) scheme, the address is stored in a fixed location (conventionally, a subroutine location). This creates no problem as long as we do not allow a subroutine either to call itself, or, more generally, to occur more than once in the chain of nested calls illustrated in Figure 7.16. Therefore, the only way to avoid incorrect operation in this scheme is to forbid this practice.

The above drawback is not present in the SEC-XR subroutine handling

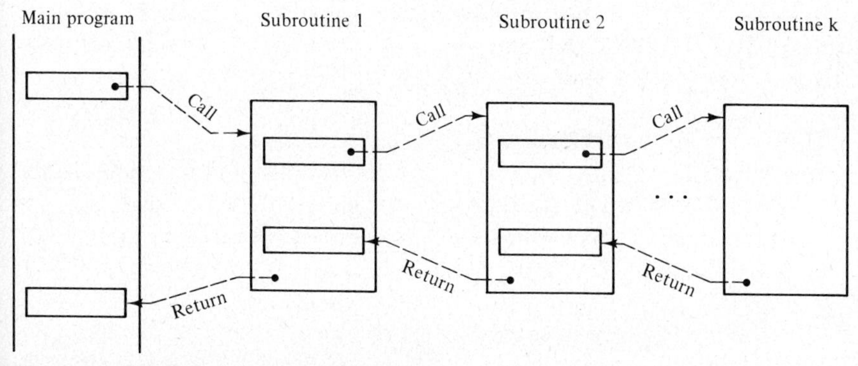

Figure 7.16 Scheme of nested subroutines.

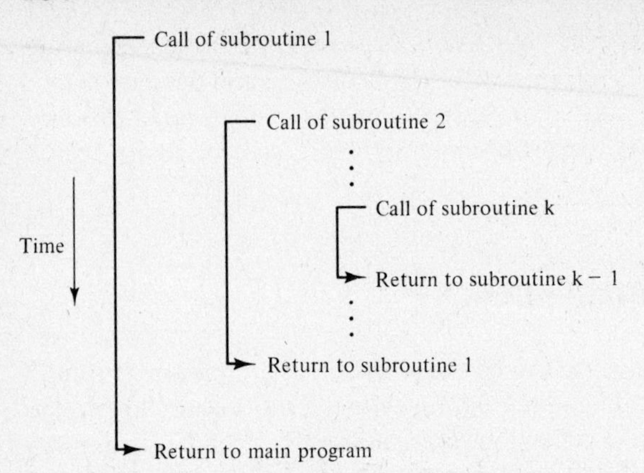

Figure 7.17 Sequence of actions in the execution of a chain of nested subroutines.

scheme, described earlier, since, for each call, the calling program selects an index register designed to contain the reference address. However, the emerging drawback is that, since the supply of index registers is rather small, we may substantially limit our capabilities both in the depth of nesting and in general indexing facilities.

But there is another effective way to handle nested subroutines, which makes use of a structure naturally attuned to the last-in–first-out policy. This structure—which can be realized either in software or in hardware—is called a *stack*. The intuitive notion behind it is that of a stack of trays, which is accessed only at the top both when adding a tray or when taking one. Keeping with this metaphor (and with the additional imaginary feature of springloading) the operation of adding an item is figuratively called PUSH (down) and the reverse operation of removing an item is called POP (up). In our context, a stack is a set of registers, typically a set of consecutive memory locations s, s + 1, s + 2, . . . , s + h. Location s is identified with the stack bottom. Provided that the memory area assigned to the stack has adequate size, all that is needed to handle the stack in our application is to be able to point to the stack top. In this manner, a PUSH corresponds to incrementing by 1 the pointer to the TOP, while a POP corresponds to decrementing it by 1 (see Figure 7.18).

In describing a possible *software* implementation of a stack, we shall restrict ourselves to the SEC-XR machine. We assume that 16 memory locations are adequate for the stack and:

1. Memory locations 7700–7717 are reserved to the stack.
2. The pointer to STACK TOP is conventionally stored in a fixed index register [say, R(7)]. [Rather than the pointer itself, we shall store in R(7) a counter, from which the pointer can be obtained by adding the base address 7677.]

The stack can be used in subroutine handling in the manner illustrated in Figure 7.19. The calling program will PUSH the reference address (usually, the

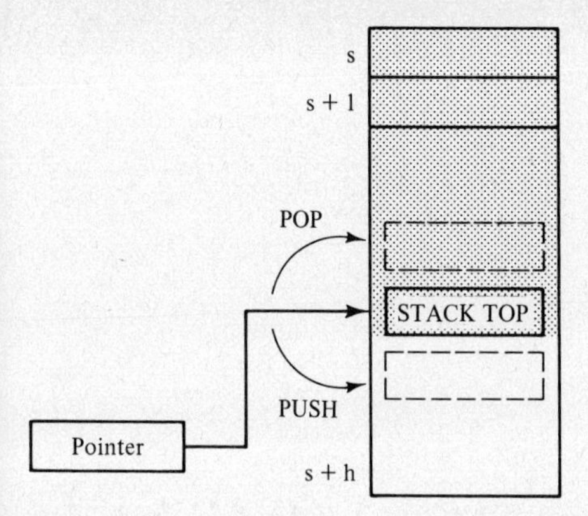

Figure 7.18 Illustration of a stack. The shaded area is the active stack.

current content of PC) onto the STACK, and subsequently jump into the subroutine. The subroutine, in turn, will POP the stack before returning control; the subroutine may itself perform another PUSH STACK operation in calling a subroutine. Notice, however, that while the subroutine is being executed, STACK TOP always contains the reference address for that subroutine. The subroutine will access the reference address anytime it is needed for the parameter exchange, but it will POP it only prior to the transfer of control back to the calling program.

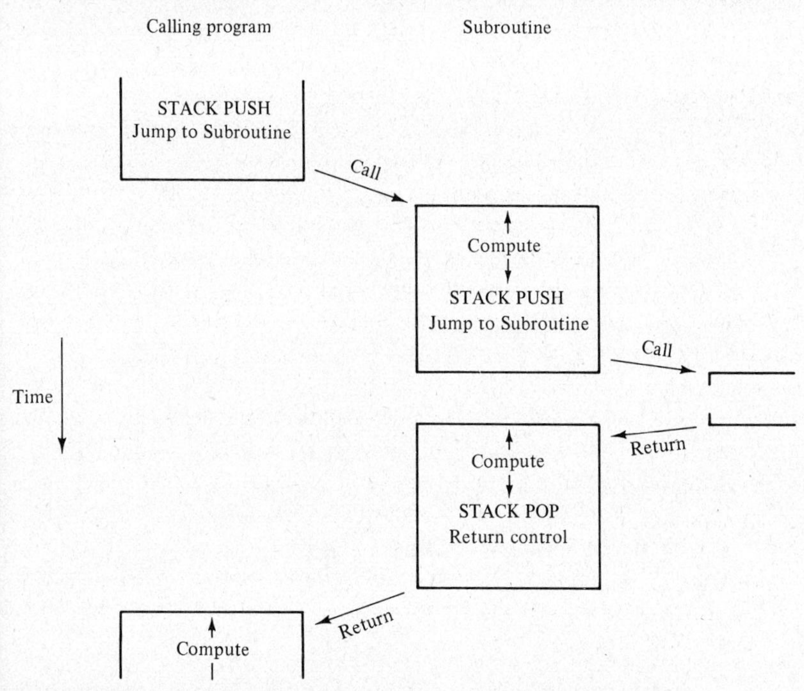

Figure 7.19 Sequence of actions in using a stack to control the linkage between the calling program and the called subroutine.

Specifically, the subroutine call can be effected by the following standard instructions:

```
a − 3   INR, 7, 0001  } STACK     increment stack counter
a − 2   ENT, 0, a + 1 } PUSH      reference address (a + 1) in accumulator
a − 1   STA, 7, 7677  }           place reference address in stack
a       JMP, 0, p                 jump to subroutine in p
```

Analogously, the return of control can be implemented as follows:

```
c       LDA, 7, 7677  } PREPARE     reference address in accumulator
c + 1   ADD, 0, c + 5 } RETURN      add template to reference address
c + 2   STA, 0, c + 4 }             set up return
c +     DER, 7, 0001    STACK POP   decrement stack counter
c + 4   [            ]              return control
c + 5   JMP, 0, n + m               template
```

The above implementation is a (somewhat awkward) way to realize stack handling of subroutines, with no addition to our instruction repertoire. Of course, we could develop specific instructions to handle this new mode of operation such as PUSH, b, – and POP, b, –, or we could even introduce new hardware in the form of a set of stack registers, exclusively dedicated to this purpose. We shall not, however, belabor this point, except to note that the increase in speed obtained at the expense of hardware complications is yet another example of the hardware–software trade-off.

*7.8 PAGED MEMORY AND INDIRECT ADDRESSING

In this section we study the problem of using a memory whose size largely exceeds the number of different addresses which can be specified by the operand field of the instruction. (For example, suppose that for our SEC computer we have a memory of 2^{15} words, rather than 2^{12} as in our original layout.)

There seems to be no way around this difficulty. However, if we reflect on the facts that instructions to be executed in sequence are placed in consecutive locations of memory, and that it is natural for the user to store operands in locations not too remote from those of the instructions that use these operands, then there is a *practice of locality* which can be conveniently exploited.

Specifically, suppose that we subdivide our larger memory into blocks of 2^9 consecutive locations: these blocks are normally called *pages* and the memory is referred to as "paged." In our example, we shall have $2^6 \cdot 2^9 = 2^{15}$ words; the organization of the memory is given in Figure 7.20.

According to the previously mentioned practice of locality, normally both the operand and the next instruction are to be found in the same page to which

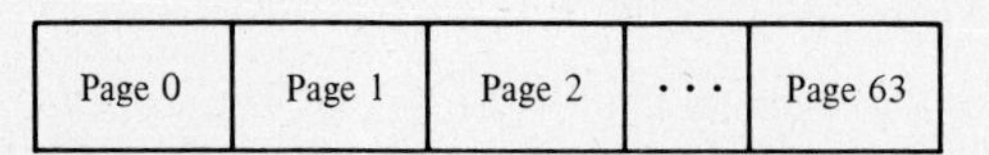

Figure 7.20 Paged memory organization.

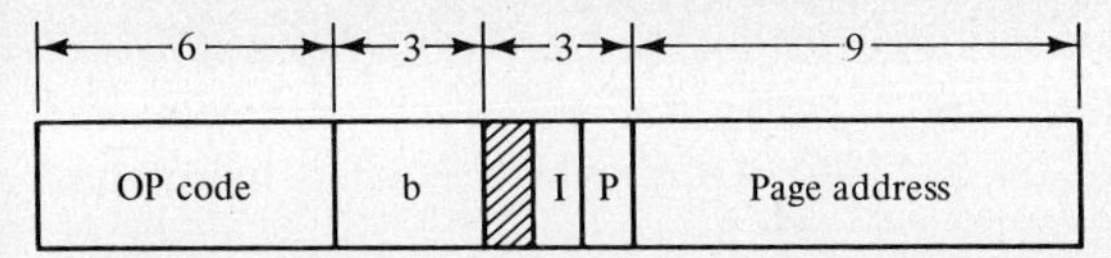

Figure 7.21 Instruction format for paged-memory SEC computer.

the current instruction belongs; otherwise, a *page jump* (or change) is required. In the first case, the nine least-significant bits of the operand field are perfectly adequate to specify the address within the current page. In the second case, the remaining 3 bits will be used to manage the page jump.

The instruction format for paged memory is shown in Figure 7.21. By Y[11:0] we denote, as usual, the 12 least-significant bits of the instruction word; bit Y[9] is called P (page bit) and bit Y[10] is called I (indirect addressing bit); Y[11] is not used. The <memory address> is formed as follows:

<memory address>: = <page> <page address>

where <page> is a 6-bit string given by

IF (P = 1) <page> : = <present page>
IF (P = 0) <page> : = 000000_2

Therefore as long as the program remains in the same page, P = 1 and the page address bits are entirely adequate. When the need arises to change page, then we must use "indirect addressing." *Indirect addressing* (specified by I = 1) means that the memory location specified in the address portion of the instruc-

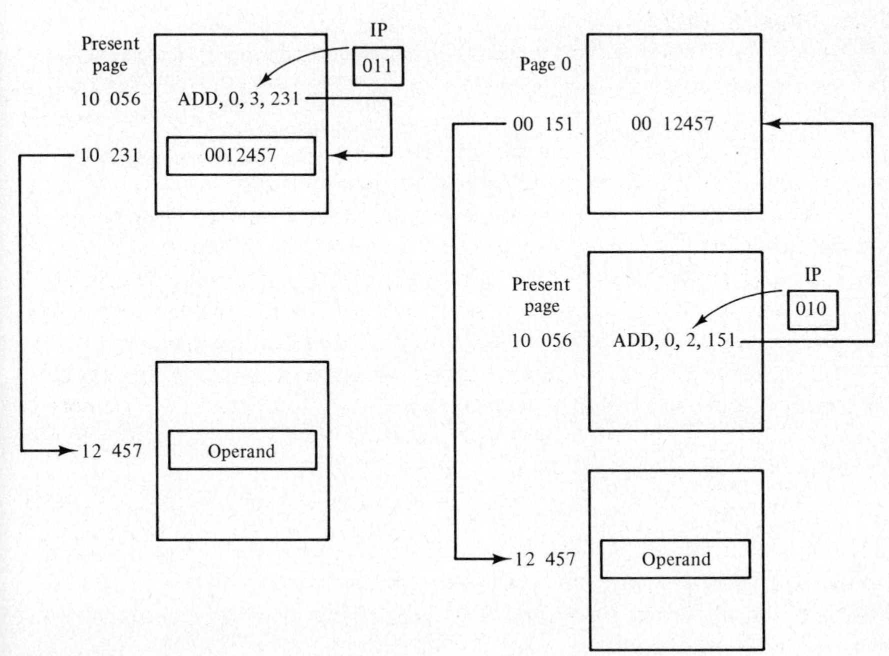

Figure 7.22 Two modes for "page jumps."

tion word does not contain the operand itself, but rather *it contains the address of the operand.* It should be clear how this feature solves our problem: the instruction specifies the address of a 21-bit word, which is then more than adequate to contain a 15-bit address for any location in the paged memory. Clearly, page jumps can be effected by indirect addressing, either in the same page (P = 1) or in page 0 (P = 0). These two examples of page jumps are shown schematically in Figure 7.22. The same operation occurs when page jump is used in connection with transfer of control.

We now examine how this new feature can be managed at the microstep level. In essence, what is involved is an *address calculation* on the basis of the I- and P-bits and the Y-field of the instruction word. This address calculation will take place *between the FETCH cycle and the ADDRESS MODIFICATION effected by the index register.* Notice that in the paged-memory SEC computer MAR is a 15-bit register, of which MAR [8:0] contains the address in the page and MAR[14:9] contains the page number; also, the size of PC is increased to 15 bits. On the other hand, the size of the index registers will be reduced to 9 bits, since the index registers are now needed to modify only the 9-bit page address. With these modifications, the FETCH cycle for the SEC computer with index registers (XR) and paged memory (the SEC-XR-PM computer) will be (note that P = IR[9] and I = IR[10]):

Microsteps	Comments
1. MAR[8:0] ← PC	
2. MBR ← M(MAR), PC ← PC + 1	
3. IR ← MBR	
4. IF (IR[9] = 0) THEN MAR[14:9] ← 0 END	This microstep sets to 0 the page number; obviously, this step is omitted when P = 1.
5. IF (IR[10] = 1) THEN	Indirect addressing.
6. MAR[8:0] ← IR[8:0]	MAR contains the address of the address.
7. MBR ← M(MAR)	The address is readout of memory.
8. IR[8:0] ← MBR[8:0], MAR[14:9] ← MBR[14:9] END	The page number is stored in MAR[14:9], while the address in the page is brought to IR for possible modification (see microstep 9).
9. IR[8:0] ← IR[8:0] + R(b) END	Address modification step.

In the paged-memory SEC, the EXECUTE cycles of the various instructions are basically as described in Sec. 7.4, the only difference being that for any instruction requiring a memory reference (LDA, STA, AND, SUB, AND, OR), the microstep MAR ← IR[11:0] is replaced by MAR[8:0] ← IR[8:0]. Notice that the content of MAR[14:9] is left unchanged until the completion of the EXECUTE cycle. For any transfer of control instruction (JMP, JPA, JZA, JSR, JPR, JS) the microstep PC ← IR[11:0] (appearing in the microsequences of SEC-XR) is replaced by the concurrent step PC[8:0] ← IR[8:0], PC[14:9] ← MBR[14:9].

Remark: In the scheme outlined above, the address modification step occurs *after* the indirect addressing sequence; this scheme is naturally called *postindexing.* We must mention that it is also possible to have a *preindexing* scheme, whereby the address modification is performed on the address of the address of the operand, rather than on the address of the operand.

7.9 A REVIEW OF THE ADDRESSING MODES OF THE SEC

At several places in this chapter, as well as in Sec. 2.3 of Chapter 2, we have made references to the ways in which different instructions obtain their operands. These different ways are technically called *addressing modes* and represent a very important aspect of the organization of a computer. It is appropriate at this point—after our detailed study of the CPU—to give a synopsis of the addressing modes used by the SEC.

We consider first the data-manipulating instructions (arithmetic, logic, and transfer). Each of these instructions has one result (destination) and one or two input operands (sources). Thus a two-operand instruction must specify, either explicitly or implicitly, three locations (two sources and one destination); similarly, a one-operand instruction must specify two locations (one source and one destination). Frequently, as we shall see, the destination is either implicit or identified with one of the two sources. Recall the definitions of Y and N = Y + R(b); an addressing mode using N is further qualified with the phrase "with index." Thus the SEC-XR has the following addressing modes:

1. *Immediate.* The parameters of Y and N are used as *operands* themselves. We have the following variants:
   a. *Accumulator immediate with index:* <source> = N, <destination> = A. Instruction ENT uses this mode. (We shall see later that JPA, JZA, JOF, JS, and HLT use N in an analogous way, but the destination is PC.)
   b. *Register immediate:* Either <source> = Y, <destination> = R(b), or <source 1> = Y, <source 2> = <destination> = R(b). Instructions ENR, INR, and DER fit one of these two modes. [Note that JSR and JPR use Y in an analogous way, although R(b) is not a destination, but PC is.]
2. *Direct.* The parameters Y or N are used as *address of operands,* in forming, for example, M(N). We have the following variants:
   a. *Accumulator direct with index:* Either <source 1> = M(N), <source 2> = <destination> = A, or <source> = M(N), <destination> = A, or <source> = A, <destination> = M(N). Instructions fitting one of these modes are ADD, SUB, AND, OR, LDA, STA. [Similar to these are ACT, STP, RD, and WRT (Chapter 8) with unit(N) replacing M(N).]

   b. *Register direct.* Either <source> = M(Y), <destination> = R(b), or <source> = R(b), <destination> = M(Y). Instructions fitting one of these modes are LDR and STR.

3. *Implicit.* There is no explicit operand. The implied source and destination coincide with the accumulator. These instructions are NEG, ALS, ARS, LLS, and RLS.

The SEC-XR-PM machine (described in the optional Sec. 7.8) has an important additional addressing mode, called *indirect addressing,* where the Y parameter does not determine the address of an operand, but rather *the address of the address of an operand.* Specifically, denoting by Y* the subregister IR[8:0], the operand address is given by the following complicated RTL expression: M(M(<page>Y*)[14:9] (M(<page>Y*)[8:0] + R(b))); that is, the operand address is the concatenation of the bit strings M(<page>Y*)[14:9] and M(<page>Y*)[8:0] + R(b).

The control instructions must specify one destination, the address to which control is transferred. All of these instructions (except HLT) use the direct mode of addressing either with index (JMP, JZA, JPA, JOF, JS) or without (JPR, JSR). The HLT instruction, in its usage seen so far, makes reference to no address; however, we shall see later (in Sec. 9.4) how the b and Y fields of HLT can be used to form the restarting address after a halt. In this sense, HLT may be viewed as employing the direct addressing mode with index.

NOTES AND REFERENCES

The notion of stored-program computers was clearly formulated in a classical paper by Burks, Goldstine, and von Neumann in 1946 (see References), which marks, in some sense, the beginning of the modern computer era.

Since that time, the pace of development has defied any imagination. (One must also make the sobering comment, however, that some overoptimistic predictions of that day have failed to materialize.) The early workers in the field formed a rather diverse crowd that consisted mainly of electronic engineers, mathematicians, physicists, and even philosophers. Perhaps their diverse professional origins contributed to the polarization of the field in the two aspects of hardware and software, which only today have coalesced in the discipline of computer science and engineering.

The early machines (the EDVAC, the IAS of the Institute for Advanced Study in Princeton, the EDSAC of Manchester University, all of which appeared in the late forties or very early fifties) were not basically different from our rudimentary SEC. In some broad sense, the SEC retraces the historical evolution of the stored-program computer. Indeed, the first computers did not have index registers, which appeared for the first time in the Manchester machines in the early fifties.

The technology of the logic networks evolved from the vacuum tube to the transistor in the fifties, and to the integrated circuit in the late sixties. The available technology has been for a long time a major limitation for the achievable memory sizes. Since a large vacuum tube memory was impractical, the chosen technologies were—in chronological order—memory delay lines, cathode-ray tubes, magnetic drums, and magnetic cores. Of these, only the cathode-ray tubes and the magnetic cores exhibited random-access capabilities. Magnetic cores made possible the realization of large-size random-access memories and have been the memory technology of the sixties, to be supplanted in the seventies by the more compact and economical semiconductor memories.

In the domain of the von Neumann stored-program machines there is an extremely rich variety of system solutions, regarding the selection of the instruction set, the instruction format, the addressing scheme, indexing, subroutine handling, and so on. This subject is known as *computer architecture* and is a thriving area of research and development. The reader may consult the excellent texts of Baer (1980) and of Hayes (1978), for a detailed treatment of the topic. The book of Bell and Newell (1971) reflects a systematic approach to the symbolic description of computer organizations, and contains reprints of many fundamental papers and a historical overview of the most significant architectural innovations.

Finally, a word of caution. We mentioned earlier the rapid pace of development of the computer field. This pace is perhaps responsible for the fast growth of a "lingo," which sometimes borders the misnomer. Indeed, the terminology is contributed to by an extremely numerous population of workers, and gets established before undergoing a serious scrutiny by the community. Although glossaries can cope with this problem, we shall point to some dubious usages of terms in the following chapters.

PROBLEMS

7.1. The following microsequences are "grammatically" incorrect. Identify the errors.
- **(a)** 1. IF (A < B) THEN 2. A ← B D ← C
- **(b)** A + (B) → M(D)
- **(c)** 1. A ← A + B 2. C + D END
- **(d)** IF (A ← C + D) THEN A + B

7.2. Parse the following RTL strings in the format used in Sec. 7.1.
- **(a)** 5. MAR ← Y, 6. MBR ← A, 7. GO TO 1
- **(b)** 4. A[11:0] ← Y, A[20:12] ← 0, 5 GO TO 1
- **(c)** 4. IF (A = 0) THEN PC ← IR[11:0] END 5. GO TO 1
- **(d)** 5. IF (IR [10] = 0) THEN 6. MAR ← IR 7. MBR ← MAR END

7.3. Parse the following strings to recognize if they are, or are not, "microsequences" in RTL.
- **(a)** 1. A ← B + C 2. IF (A < B) THEN GO TO 1 END
- **(b)** 4. MBR ← A 5. M(MAR) ← MBR, B ← −A
- **(c)** 12. A ← A 3. B ← A
- **(d)** IF (A > B) THEN A ← A − B END
- **(e)** 1. MBR ← A, M(MAR) ← MBR 2. GO TO 3
- **(f)** 4. GO TO 6 5. A ← B, B ← A 6. C ← 3

7.4. For each of the following SEC instructions write its description using the register transfer language: Example. 03 0 1472: A ← A + M(1472_8).

```
20 0 0112   ,   04 0 0100
01 0 4311   ,   06 0 1012
10 0 0001   ,   21 0 0001
22 0 0103   ,   51 0 0010
02 0 0171   ,   03 0 1043
05 0 2311   ,   54 0 1101
```

7.5. Write the EXECUTE microsequence in the SEC-0 system of the following instructions:

(a) AND, SUB, ALS, LRS

(b) OR, JPA, ARS

7.6. The following is the RTL description of the EXECUTE microsequence of instruction XYZ (a "new" instruction):

4. IF (A < 0) THEN PC ← MBR [11:0] END
5. GO TO 1

Describe in words the action performed by this instruction.

7.7. Write the EXECUTE microsequence of the JOF (Jump on Overflow instruction) described on p. 222.

7.8. Consider the following new instructions for the SEC-0 computer, and the meanings shown for each one:

Instruction	Meaning (description of action)
(a) NENT, 0, Y	A ← −Y
(b) DB, 0, –	A ← A + A
(c) ADD*, 0, Y	A ← A + M(M(Y))
(d) ADD2, 0, Y	A ← A + M(Y) + M(Y + 1)

Referring to the machine structure of Figure 7.9, give for each instruction the EXECUTE microsequence.

7.9. Write the EXECUTE microsequence for the instruction NEG, –, –: A ← –A.

7.10. Consider the following SEC-0 program

```
1021    LDA, 0, 0041
1022    NEG, 0, 0000
1023    ADD, 0, 0042
1024    STA, 0, 0043
```

(a) How many memory operations have been performed?

(b) How many memory write operations?

(c) How many times is information sent through the ALU?

All of the following problems refer to the SEC-XR system, unless otherwise noted.

7.11. Using index registers, write a SEC-XR loop program starting in location 1000_8 to count the number of 1s in the bit strings stored in the accumulator. The result is to be stored in 0000_8.

7.12. Using index registers, write a SEC-XR loop program, starting in location 0100_8, which searches locations 0200_8 through 0247_8 to find the smallest number stored in them: this number is to be stored in location 0250_8.

7.13. Same as Problem 7.12, except that the problem must find the largest number.

7.14. Using index registers, write a short SEC-XR loop program which computes the 21-bit product of two nonnegative numbers ≤ 1023 stored in locations 0100_8 and 0101_8.

7.15. Write the microsequence (FETCH and EXECUTE) of a *new* jump to Subroutine instruction which works as follows:

JMS, b, Y: Jump to subroutine in Y + R(b) and store PC in location 0000

7.16. Write a SEC-XR program to copy the contents of cell 0100_8 into all the cells 0200_8 to 0277_8. (Note: use index registers.)

7.17. Write a SEC-XR loop program to count the number of 0s in the bit string stored at memory location 0100_8, and to write such a number in location 0101_8.

7.18. Two positive numbers A, B are stored in locations a + 1, a + 2, respectively. Location a is designed to contain a jump to a subroutine that computes the (integer part of the) average of A and B, stores the result in a + 3, and returns control to a + 4. Write the contents of a, and the subroutine, for:
(**a**) The SEC-0 computer.
(**b**) The SEC-XR computer.

7.19. Consider the following program segment. Operands 1 and 2

a	
a + 1	Operand 1
a + 2	Operand 2
a + 3	Product

are both nonnegative integers $\leq 1023_{10}$; location a is designed to contain a jump to a multiplication subroutine starting in location p, which computes the product of the two operands, delivers it in a + 3, and returns control to a + 4. Write the content of a and the subroutine for the following cases:
(**a**) The SEC-0 system.
(**b**) The SEC-XR system.

7.20. Write the microsequence of each of the following two instructions:
(**a**) PUSH, b, –: Push the content of A onto the stack. The top of the stack is at location 7700_8 + R(b).
(**b**) POP, b, –: Pop the stack, that is, the content of stack top is transferred to A. The top of the stack is at location 7700_8 + R(b).

Chapter 8

Input/Output

8.1 GENERAL CONSIDERATIONS

Up to this point we have assumed that data were present in memory when the execution of a program was started, and—by the same token—that results were left in memory. In this chapter we shall see how a computer will communicate with the outside world.

This communication takes place through the so-called input/output (I/O) devices. These are of various types, depending upon the medium used to hold information; the latter could be paper tape, punched (paper) cards, magnetic tapes, magnetic drums or disks, paper to be printed, signals on conventional communication lines, and so on. We shall not analyze all of these media and the corresponding devices; we shall only consider typical I/O devices, from which the general features of the I/O process can be learned and possibly applied to other cases.

In some sense, the communication between the CPU and the memory could be viewed as an input/output operation (by so doing, we would put the memory in the "outside world"). So, why should we have any concern about input/output operations involving the devices mentioned above? The fundamental reason is that the CPU and the memory are realized with very similar technologies, whereas the other I/O devices involve an extremely wide range of technologies, with features remarkably different from those of the CPU (and the memory). This difference in features is basically a difference in operating speeds and in the forms of representation of information. Matching the speeds of the CPU and of the I/O devices is what makes the I/O problem particularly complex.

An attempt to thoroughly classify the input/output devices (also called "peripherals") is well beyond the scope of this text. We shall content ourselves with an outline of the most important features of the input/output process. An input device converts data from a specific external format to the internal computer representation, while an output device performs the inverse conversion.

Consider, for example, a paper-tape reader. In Figure 8.1(a) we show a fragment of paper tape: it contains several *information tracks* (e.g., 8), where a hole corresponds to a 1 bit and the absence of a hole to a 0 bit. A transversal alignment of bits is a *character,* and is the unit of information being exchanged; there is also a *sprocket track,* with a hole for each character position, for timing purposes. The tape is at rest, but the driving motor is running. To read a character [Figure 8.1(b)], the clutch-controlled feed rolls are energized, and the tape is engaged and advanced so that one character goes past a light-photocell arrangement, at which point the feed rolls are stopped. In this process, for each track bearing a hole, the corresponding photocell generates an appropriate signal.

In what follows, a data item involved in a single (parallel) I/O transfer will be referred to as a *character,* independently of its number of bits.

Many I/O devices are of the "transport" type, that is, an external information support (paper tape, punched cards, magnetic tape, magnetic disk, etc.) has to be transported past a fixed station—in the previous example, the photocell—to effect the information transfer (to or from the CPU).

Necessarily a transport device has a start-stop behavior. Referring for concreteness to an input device, a short time interval after the beginning of the mechanical motion the first character becomes available [see the timing diagram of Figure 8.2(a)], and successive characters become available with a fixed period, until the device stops. Clearly, for a transport device that inputs more than one character between a start and a stop, there is a narrow time window during which each character must be transferred from the input unit to the CPU. Similarly, in Figure 8.2(b) we have the analogous diagram of an output device.

We shall see later that distinctly different modes of operation are in order depending upon the frequency of data transfers (for example, a tape cassette is considerably slower than a standard tape drive). We note at this point that a

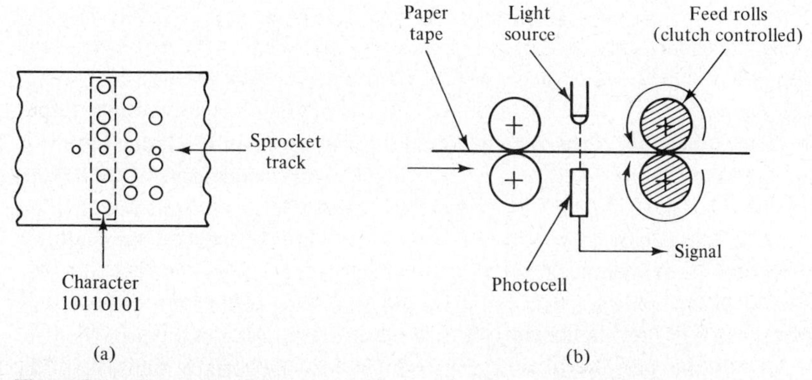

Figure 8.1 Paper tape format and scheme of a paper tape reader.

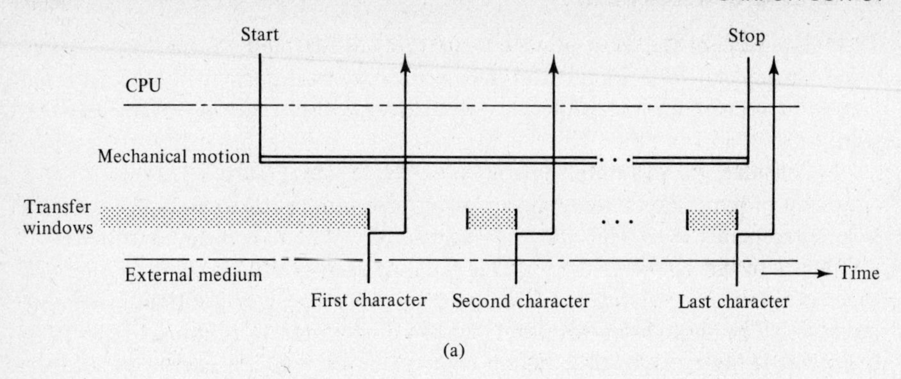

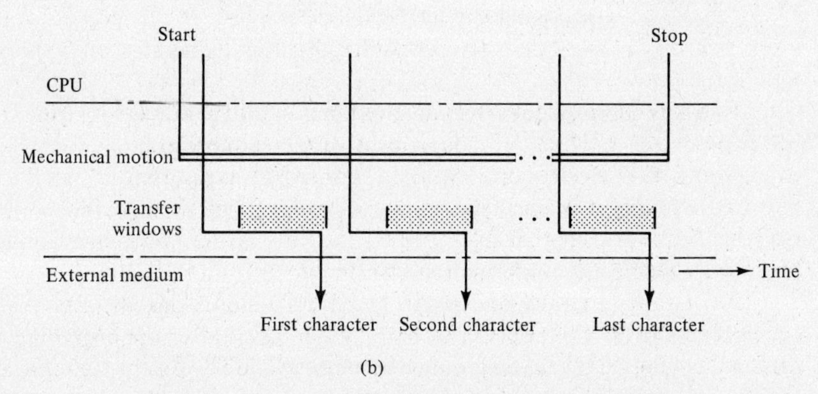

Figure 8.2 Timing diagrams of data transfers for I/O transport devices.

basic feature of transport devices is the necessity to *activate* the motion (under CPU control) so that transfers can take place.

Another class of I/O devices is represented by those for which no activation is necessary (referred to here as "static devices"). In this category are not only devices controlled by the CPU, such as video displays (output) and measuring instruments (input), but also the inputs originating in the outside world (*exogenous*), such as a keyboard or a communication line input.

From the above gross classification of devices (static versus transport) it is clear that in each I/O operation we must distinguish two fundamental functions:

1. *The transfer of data from or to the device* (common to all devices).
2. *The activation of the device* (specific of transport devices).

These two functions can be kept conceptually separate. Moreover, we shall not explicitly consider exogenous inputs.

The simplest and most naive mode of managing the input/output functions is by letting the CPU retain control over the whole operation. This means that the CPU activates the peripheral unit (by issuing an appropriate signal) at the beginning of the EXECUTE cycle of the I/O instruction, and EXECUTE lasts

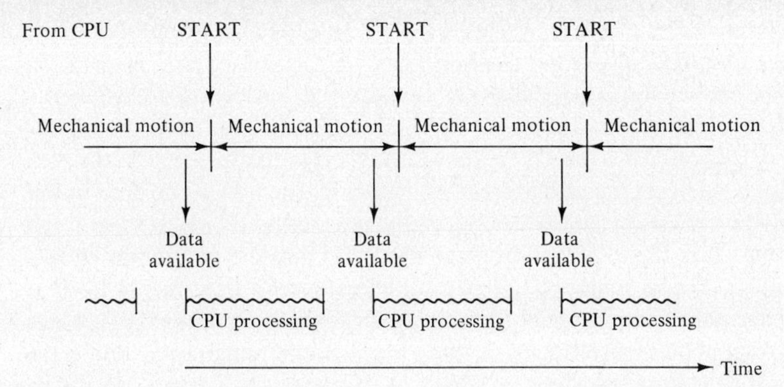

Figure 8.3 Illustration of the overlap of the input operation and the internal CPU processing in a payroll application.

for the entire duration of the mechanical motion; during this time interval information is transferred, and there is no danger of improperly issued I/O commands (while the unit is engaged, so is the CPU). However, if one reflects for a moment on the fact that the duration of the mechanical motion is typically on the order of 20–100 msec, and that for most instructions the FETCH-EXECUTE time is typically of the order of 1 μsec, then, rather than letting the CPU idly wait for the completion of the I/O task, one could profitably put it at work executing something like 20,000–100,000 other instructions. To motivate the desirability of such a mode of operation, consider, for example, a payroll application, where the input operation consists of obtaining all the pertinent data about a given employee (such as his/her salary, overtime service, tax exemptions, voluntary payroll deductions, etc.) and the computation consists of calculating the gross pay, state and federal taxes, and so on, and in preparing the "text" to be printed on the payment voucher. It is clear that it is truly desirable to overlap these two kinds of activities (input/output on one hand and computation on the other) so that the peripheral devices can operate at full speed and the CPU is kept busy while waiting for the next batch of data (*busy wait*). The situation is schematically illustrated by the periodic diagram shown in Figure 8.3. Such a mode of operation entails that the CPU partially relinquish control of the I/O operation, or, more specifically, that the CPU control the I/O process only for a very small fraction of the motion interval and that *separate* specialized I/O control capabilities be established to carry on the task.

Thus, we shall have two interacting information processing systems, the CPU and the I/O unit (i.e., the device and its associated "peripheral" control) which cooperatively carry out the I/O operation. To adapt the widely different operating speeds it is essential to place between the two systems (or "to interface" as the terminology goes) a *buffer register,* which, during a data transfer, is accessed on one side by the CPU, on the other by the peripheral device.[1] This register is called Data Buffer (DB). In addition, the peripheral

[1]A similar register (MBR) was also used to interface the CPU to the memory. While such register can be eliminated in the CPU-memory communication (by connecting the memory directly to a CPU register), it becomes a necessity in I/O communication.

device must have some elementary control unit, possibly consisting of a single flip-flop, called FLAG.

The basic scheme of the peripheral unit (device and control) is shown in Figure 8.4 (for an input unit). Here the *command decoder* interprets the type of action that is requested by the CPU (as we shall explain later), while the *unit selector* uniquely detects the unit identification sent by the CPU. The data buffer register and the control circuitry are normally referred to as an *I/O port,* and physically constitute the interface between the CPU and the peripheral device. The flip-flop FLAG (an unclocked SR latch) indicates with its status (FLAG = 1) when the peripheral unit is ready for a data transfer to or from the CPU. Thus, FLAG gets reset upon completion of the data transfer and remains reset until the unit is ready for the next transfer.

Depending upon the speed of the I/O device, data transfers occur either between DB and the accumulator ("slow" devices) or directly between DB and memory ("fast" devices). In the next two sections we shall discuss two basic modes of operation transfers of the first type, while (the optional) Sec. 8.4 is devoted to the transfers of the second type. In all cases, an input/output character is a bit string whose length is a property of the I/O unit considered; it never has more than 21 bits (the common capacity of the accumulator and of a

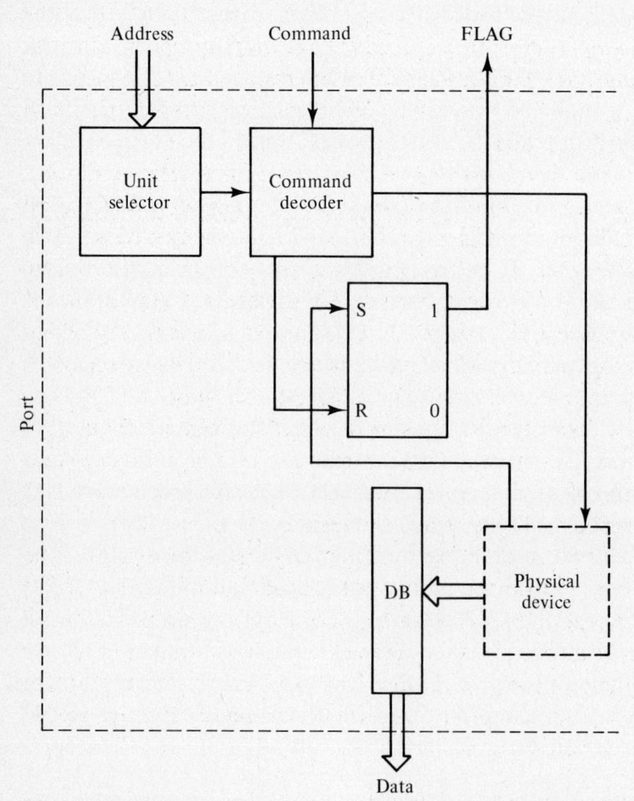

Figure 8.4 Basic organization of an I/O unit (in the figure, an input unit).

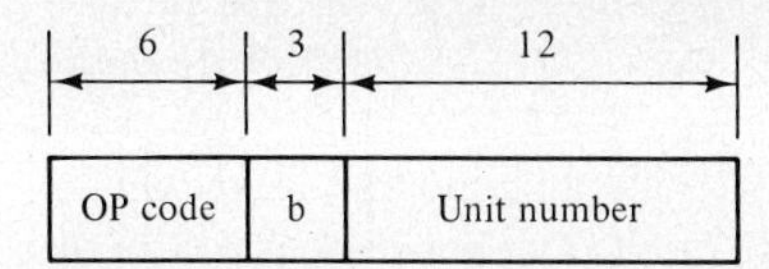

Figure 8.5 Format of the I/O instructions.

memory location) and will *always be stored in the least-significant end of a register* (so a 6-bit I/O character will be stored in A[5:0]).

The format of the I/O instruction is given in Figure 8.5. The operation code specifies, as usual, the type of operation; b and Y have their conventional meanings, and $N = Y + R(b)$ will now be used to denote the number of the I/O unit. Clearly, up to 2^{12} units can be addressed, but obviously a much smaller number will ever be deployed.

Notice that the collection of all unit selectors in the various I/O units could be functionally viewed as a "decoder," in the sense described in Chapter 4 (Sec. 4.8.2).

8.2 PROGRAMMED INPUT/OUTPUT

In this mode the CPU delegates control of the operation of the I/O devices to the peripheral control units but retains complete control of the data transfers. As we mentioned earlier, data transfers occur between the accumulator and the data buffer: $A \leftarrow DB$ for input (read), $DB \leftarrow A$ for output (write).

If the I/O device is of the transport type, both *device activation* and *data transfer* must be implemented. Although numerous variants are possible, we choose a mode of operation that keeps separate the two functions by means of two distinct sets of instructions: ACT (activate) and STP (stop) for the device activation, WRT (write) and RD (read) for the data transfer.[2] The typical layout is shown in Figure 8.6, with reference to an output unit, to which the following discussion pertains. Note that the unit identification (or, unit address) is transmitted over the same bus used for the memory, and, similarly, the data bus is also used for the I/O data transfer. In addition, we need two directional control signal buses, the c-bus and the f-bus, whose structure and functions will be described later in this section.

The output operation is carried out by the joint action of the ACT, WRT, and STP instructions. Referring to Figure 8.7, the ACT instruction sets the selected device into motion. While the device is moving, the WRT instruction periodically transfers a character from the accumulator to the data buffer of the unit; with a fixed delay, the character in the data buffer is transferred to the external medium. At the end of the process (i.e., after the last character transfer to the output medium has been completed), an STP instruction deactivates the device and brings it to a standstill.

[2]See **Remark* at the end of this section.

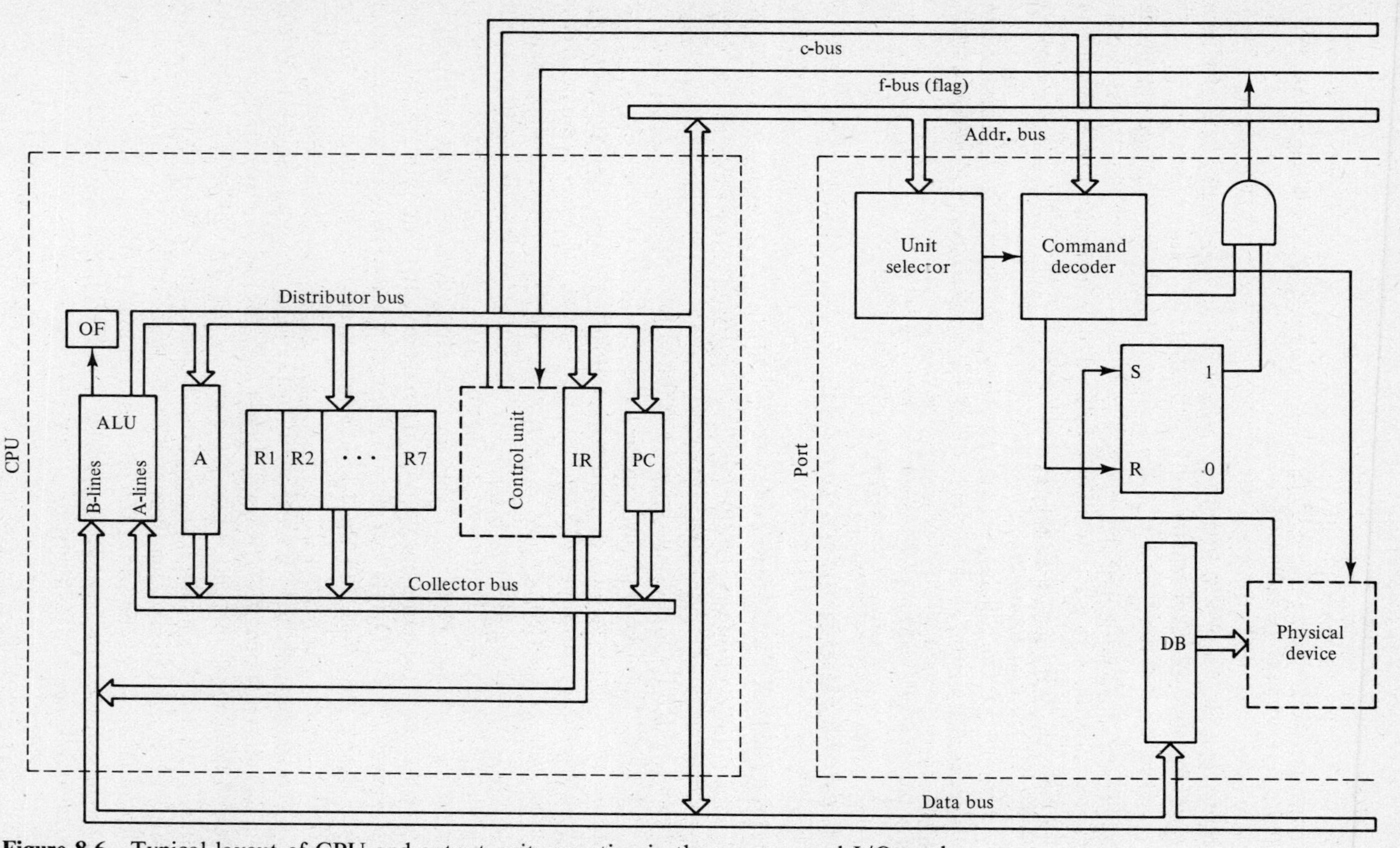

Figure 8.6 Typical layout of CPU and output unit operating in the programmed I/O mode.

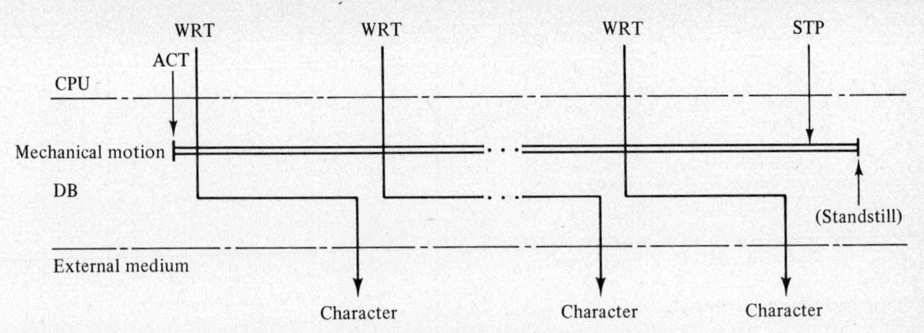

Figure 8.7 General timing of the output process.

In summary, we have introduced the following instructions:

Instruction	Operation	Mnemonic
74, b, Y	Activate unit N (start motion) (unit(N) ← START)	ACT
70, b, Y	Write the character presently in the accumulator via unit N (DB(N) ← A)	WRT
75, b, Y	Stop unit N (unit(N) ← STOP)	STP

We must now devise a convenient way to carry out the process outlined in Figure 8.7. Obviously, the unit cannot initiate a new data transfer if it is already engaged in completing one; on the other hand, an activation command should be ignored by a device already engaged. So, if a WRT instruction is issued while a transfer from DB to the external medium is taking place, *completion of its EXECUTE cycle must be held up until the unit becomes available.* To avoid this undesirable circumstance, which corresponds to wasting valuable CPU time, the CPU must be able to test the status of output unit N (available, not available), in order to issue an output instruction only in the first case. This leads to the following requirements:

1. The output unit N signals its availability by setting signal FLAG to 1.
2. The CPU must be able to test the status of FLAG(N).
3. In programming practice, a WRT instruction will be issued only when the addressed unit is available.

Requirement **1** can be fulfilled by the FLAG F/F in each output unit: FLAG = 1 when and only when the device is available for data transfer. To satisfy **2** we shall introduce the following test instruction (SFH, or Skip on Flag High):

Instruction	Operation	Mnemonic
72, b, Y	Skip the next instruction if the FLAG of unit N is 1 (high) (IF (FLAG(N) = 1) THEN PC ← PC + 1 END)	SFH

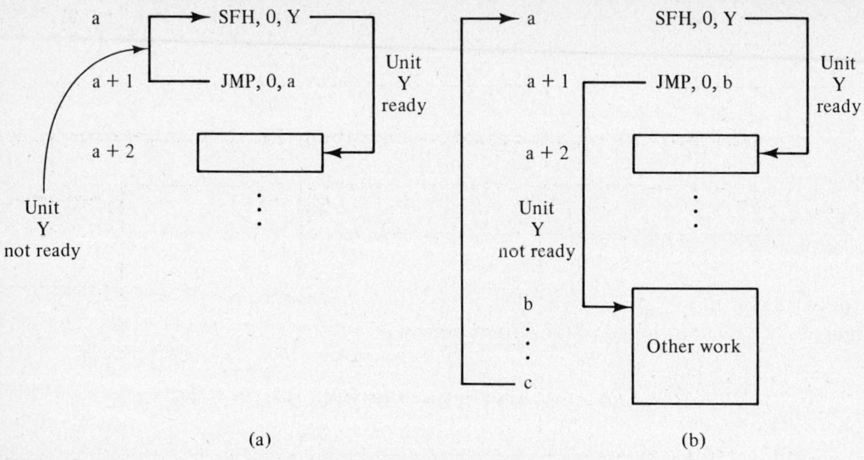

Figure 8.8 General programming patterns in programmed I/O.

Notice, this is a new type of jump instruction: here we cannot use N as the "jump" address, because N is used to specify the peripheral unit; thus, an expedient way to obtain an exceptional transfer of control consists of "skipping" the next instruction, that is, of transferring control two instructions ahead, where the appropriate output instruction (WRT or RD—see below) will be placed. The general programming pattern is shown in Figure 8.8, where the SFH instruction is stored in location a; as long as the unit is not ready, control is transferred back and forth between the SFH and the JMP instructions; as soon as unit N becomes ready, then the instruction in a + 2 is executed. Actually, the "loop" SFH-JMP is just a device to kill time until the unit becomes ready. But SFH becomes very valuable if some productive "other work" can be done while unit Y is not yet ready [Figure 8.8(b)]; in this case control is transferred along the sequence a, a + 1, b, . . . , c; after unit N is *found available,* control is transferred from a to a + 2.

We can now fill in the missing details for the management of the process illustrated in Figure 8.7. The device is set into motion by the ACT instruction. Since the status of FLAG indicates the availability of the output unit to receive data from the CPU, the FLAG F/F must be set by the device and reset by the WRT instruction. Thus, in Figure 8.9 we have an updated version of Figure 8.7, where the status of FLAG(N) is also shown. Normally an SFH test instruction readily detects the condition FLAG(N) = 1, so that the WRT instruction can be issued *timely*. The STP instruction can be issued immediately after WRT, since its action is synchronized with the character cycle of the output unit.

The scheme illustrated in Figures 8.7 and 8.9 is valid also when a *single* character is to be transferred during a start–stop cycle of the device (as with the paper reader discussed in Sec. 8.1).[3]

[3]In this case, however, we could (but do not) modify the port so that a single instruction WRT could accomplish the functions of both ACT and WRT, while the function of STP could be assigned to a timer within the peripheral unit.

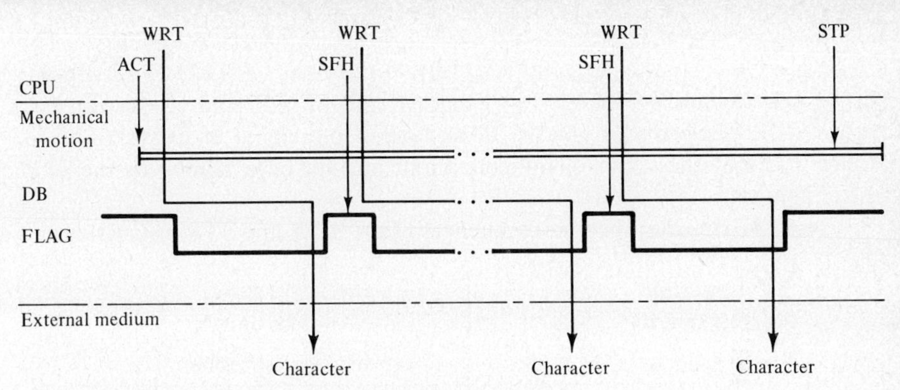

Figure 8.9 The timing diagram of Figure 8.7 revisited.

From the equipment standpoint, we have the following additional items, which were mentioned earlier:

1. A "command bus" (c-bus), acting as a distribution bus for the operation code, which is transmitted in parallel to all peripheral units. Obviously only one unit will be receptive to the OP code, namely the one being "called" (via the address bus) by the CPU. It is necessary to transmit the OP code to the addressed unit to inform it of the requested action (e.g., whether to write or to send its FLAG, etc.)
2. A "flag bus" (f-bus), which consists of a single line and acts as a collector bus for the flag signals (see Figure 8.6).

We shall now discuss the microsequences of the instructions introduced above. It is useful—for convenience of notation—to extend the notion of <register> to include also "bus lines" as possible sources or destinations in the <register> ← <register> microstep; this extension will be very useful also in Chapter 9.

The execution of any I/O instruction requires the coordinated actions of the CPU and of the peripheral control. Thus, these two sections must be synchronized, that is, they must operate under the same clock. Notice, however, that the FLAG F/F of the peripheral unit is accessed by both the CPU (synchronously with the system clock) and by the peripheral device (asynchronously, with respect to the system clock).

Denoting by N the unit number, the EXECUTE microsequence of the SFH instruction is

In CPU	In unit N
5. addr. bus ← IR[11:0], c-bus ← IR[20:15], IF (f-bus = 0) THEN GO TO 7 END	IF (addr. bus = N) THEN f-bus ← FLAG(N) END
6. PC ← PC + 1 7. GO TO 1	

Concurrent step 5, which carries out the polling operation, is a complex one. The information flow is schematically IR[11:0] → data bus → ALU → distributor bus → address bus → decoder → f-bus → control unit, and is illustrated in Figure 8.10. Obviously, if f-bus = 0, no further increment of PC takes place. Notice also that the status of the polled unit has not been altered by the SFH instruction.

Next we consider the microsequences of the ACT and STP instructions:

ACT Instruction

In CPU	In unit N
5. addr. bus ← IR[11:0], c-bus ← IR[20:15]	(activate device if inactive)
6. GO TO 1	

STP Instruction

In CPU	In unit N
5. addr. bus ← IR[11:0], c-bus ← IR[20:15]	(deactivate device if active)
6. GO TO 1	

The microsequence of WRT involves also a data transfer:

In CPU	In unit N
5. addr. bus ← IR[11:0], c-bus ← IR[20:15], IF (f-bus = 0) THEN GO TO 5 END	IF (addr. bus = N) THEN f-bus ← FLAG(N) END
6. data bus ← A	DB(N) ← data bus, FLAG(N) ← 0
7. GO TO 1	

In the microsequence of WRT concurrent step 5 indicates that the CPU will be held up (IF (f-bus = 0) THEN GO TO 5) until the flag of the polled unit becomes "1," that is, the unit is ready; at this point the appropriate action takes place. Note that FLAG(N) is reset upon completion of a character transfer and recall that FLAG(N) will be set again to 1 by a signal generated by the physical device at some point during the device cycle.

Comparing steps 5 of the EXECUTE microsequences of SFH, ACT, STP, and WRT we have an example of why it is necessary to send the OP code IR[20:15] to the peripheral unit via the c-bus. In fact, for SFH, the peripheral unit simply sends its flag to the f-bus, for WRT some additional specific action is performed (step 6), and no action on FLAG occurs for ACT and STP. These cases are distinguished by the OP code.

We now consider the input instruction (read), concisely defined below:

Instruction	Operation	Mnemonic
71, b, Y	Read a character from unit N and transfer it to the accumulator (A ← DB(N))	RD

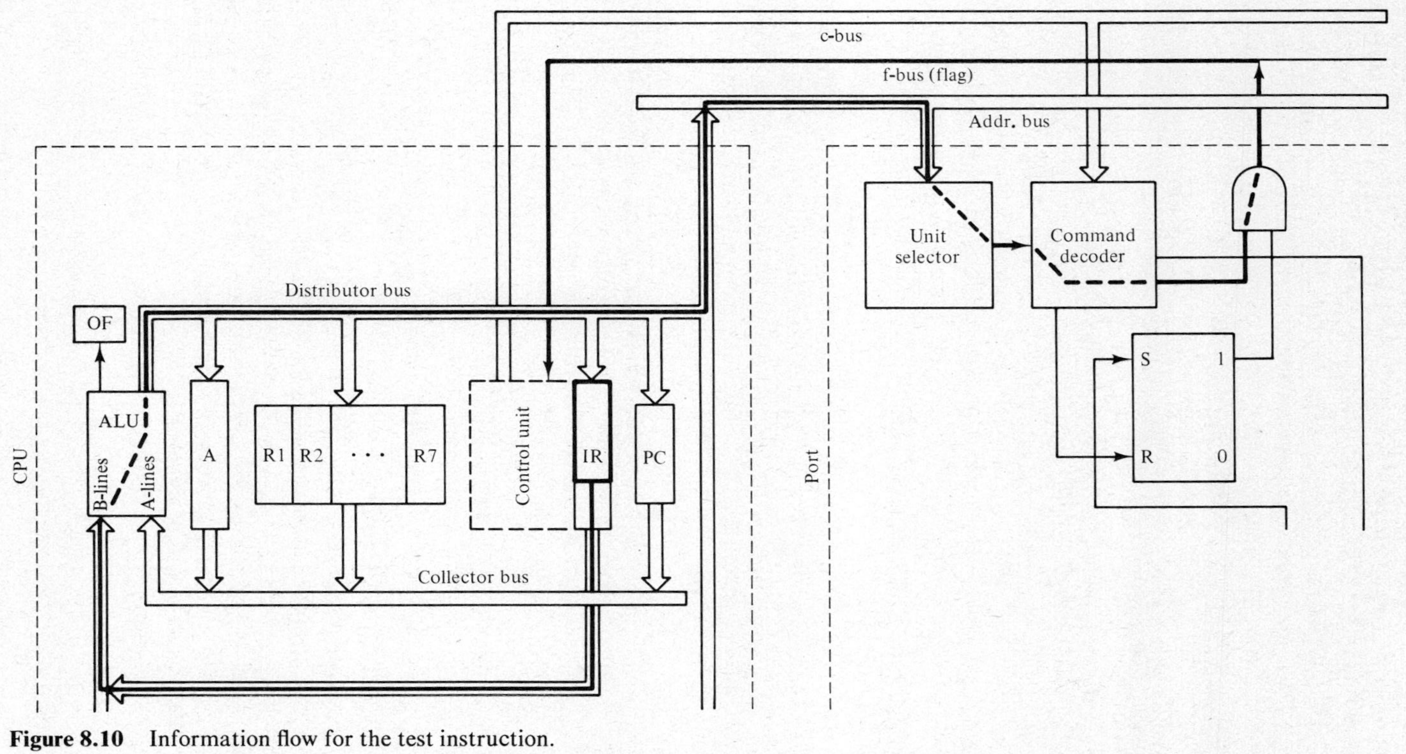

Figure 8.10 Information flow for the test instruction.

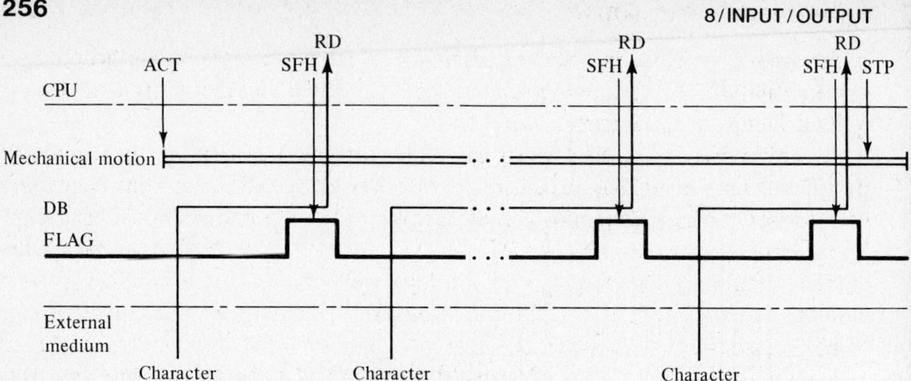

Figure 8.11 Timing diagram of the input process.

The EXECUTE microsequence of this instruction is analogous to that of its counterpart, the WRT instruction. The only difference is the replacement of concurrent step 6 in the EXECUTE microsequence of WRT with the following:

6. A ← data bus ¦ data bus ← DB(N), FLAG(N) ← 0

However, the reversal of the information flow (input versus output) causes some significant modifications in the organization of the process. Indeed, referring to Figure 8.11, after the device activation caused by the ACT instruction, some time must elapse before the first input character is available at the output of the DB(N) register. The availability is indicated by the condition FLAG(N) = 1, which can be detected by the SFH test instruction, and immediately followed by the RD instruction [which also resets the FLAG(N) F/F]. After the last character input (last SFH instruction), the device is to be immediately deactivated by the STP instruction, preceded or followed by the character transfer (RD).

In conclusion, we note that the phrase "programmed I/O" reflects the fact that availability of a unit is tested by *software* means. This mode of operation is conveniently illustrated by the following "telephone" analogy: the peripheral units have messages for the CPU but cannot call it; only the CPU, at its discretion, can call the peripherals to find out whether they have messages.

**Remark:* There could be an alternative way to handle the device activation and the information transfer functions. This alternative way makes use of a single pair of instructions (RD and WRT), but assigns two distinct ports (unit addresses) to the data buffer and to the mechanical control of a given unit. Denoting by N_1 the data port and by N_2 the control port of a given unit, WRT, O, N_2 would activate the device while WRT, O, N_1 would transfer data. The RD instruction could be similarly employed. However, this method is not used in the SEC.

8.3 INTERRUPT INPUT/OUTPUT

In the programmed I/O mode, just described, the CPU must poll the peripheral units for readiness (or, availability). This involves a substantial demand upon the programmer, because in order to efficiently place SFH tests in the program, a careful analysis of the running time of the latter must be made; this is in general quite difficult for the average user.

This difficulty is circumvented in the *interrupt I/O mode,* in which the peripheral unit is no longer waiting *passively* for the CPU poll, but it *actively* calls for service from the CPU as soon as it is ready to exchange data. This is done by *interrupting, at the hardware level* (i.e., without program's awareness) the current computation. In the usual "telephone" analogy, the peripheral units cannot originate a call to the CPU, but can leave a message. In the simplest form, to be discussed next, the message is that "someone" called; the CPU will then poll in sequence its directory of peripheral units to find out who left the message. This general philosophy leads to the following requirements:

1. The peripheral unit signals its availability by setting its FLAG to 1 and transmitting this information—without being polled—to the CPU.
2. *If the current computation is interruptible,* the CPU completes execution of the current instruction, and then handles the "interrupt" by jumping to the first instruction of the I/O SERVICE ROUTINE. This entails saving the current value of PC, so that resumption of the interrupted computation is possible once the I/O service routine has been executed.
3. If *only one* value of PC can be saved, processing *is not interruptible* during the execution of the service routine.

To implement the just described approach, some additional equipment is necessary. Notice, however, that the two modes—programmed I/O and interrupt I/O—may coexist in the same system, since some units may be wired in the first mode and other units in the second one. The additional equipment for the interrupt I/O mode is shown in Figure 8.12, where the illustrated peripheral unit is wired to operate in the interrupt I/O mode:

An "interrupt bus" (i-bus), acting as a collector bus for the FLAG signals

The i-bus is simply an OR of the FLAG F/Fs. Each FLAG is gated to the i-bus by a flip-flop set and reset by the ACT and STP instructions, respectively. Thus the CPU can be immediately alerted to the fact that a peripheral unit has completed its activity and is calling for service. This implements requirement 1.

Obviously, the CPU activity cannot be interrupted instantaneously if computation is to be carried out correctly; a convenient interruption point is at the completion of the current instruction. In this manner the response to the interrupt is practically immediate. To ensure correct operation the following actions must be taken:

1. The content of PC must be saved in a specified location (for example, in 0000_8).
2. Transfer of control must occur *via hardware* to the first address of the

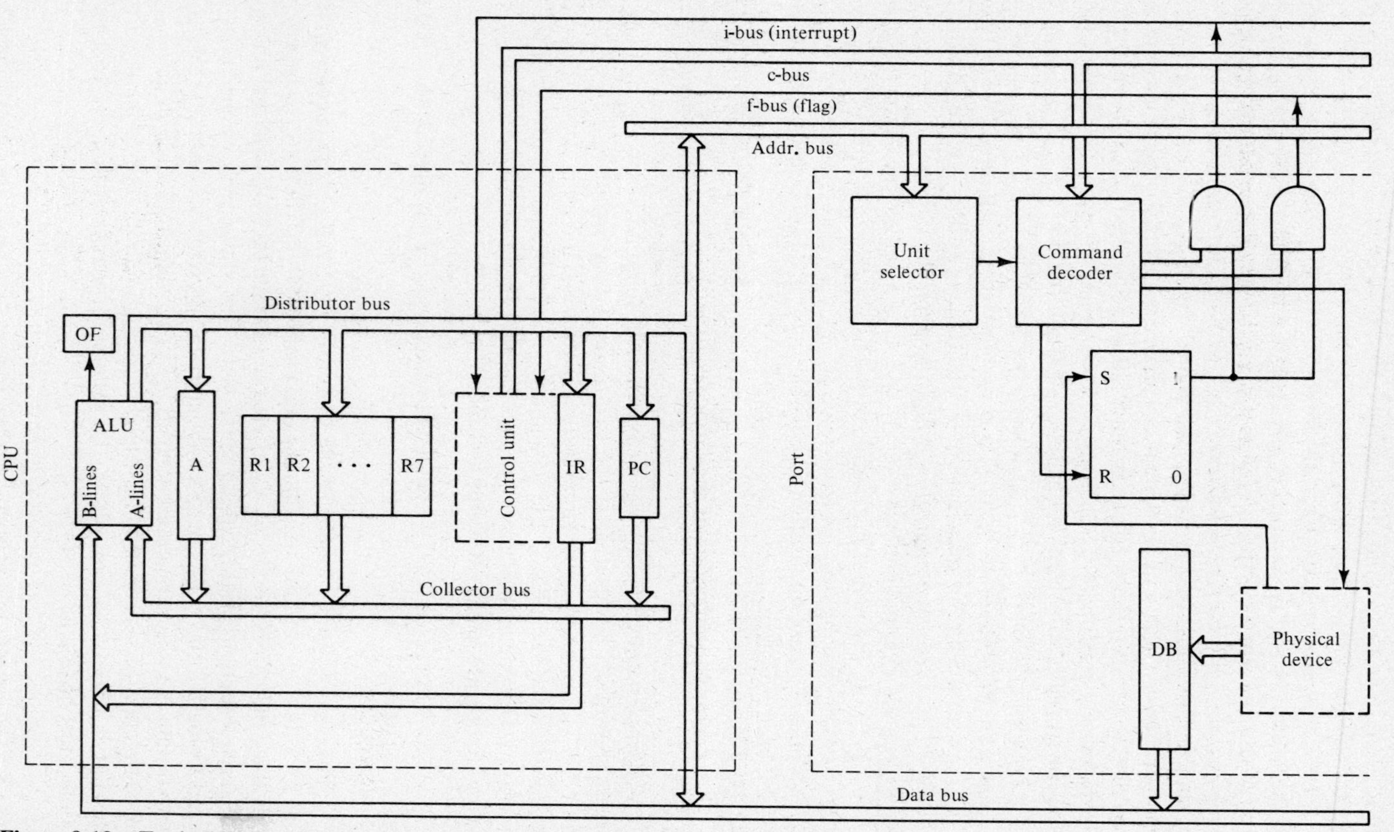

Figure 8.12 Typical layout of CPU and input unit operating in the interrupt I/O mode.

I/O service routine, which is in a specified location (for example, in 0001_8).

3. Assuming that only one value of PC can be saved, during execution of the service routine computation is not interruptible. Therefore, a flip-flop must be provided in the CPU—called IE, Interrupt Enable—which is set during conventional operation and is reset during the execution of the I/O service routine.

We see therefore that the CPU must be constantly on the alert for possible interrupts. Specifically, at the end of *each* EXECUTE cycle (for all instructions, except as noted below), just before the GO TO step which transfers control to the FETCH cycle, the CPU must test if the above three actions are to be executed. This is done by inserting, just prior to GO TO, a new microsequence, called INTERRUPT TEST, described as follows:

```
IF (IE = 1 and i-bus = 1) THEN    8. MBR ← 000 PC, MAR ← 0
                                  9. M(MAR) ← MBR, PC ← 1, IE ← 0 END
```

This microsequence is executed only when there is a request for service (i-bus = 1) which can be accepted (IE = 1); it stores PC in location 0000, resets the IE flip-flop, and transfers control to location 0001.

In 0001 we have the first instruction of the I/O service routine, whose function is the handling—by *software*—of the data transfer. One possible way to organize this activity is by *polling* each of the peripheral units wired on the interrupt mode. As soon as one is found whose flag is set, the data transfer will take place.

At the completion of the I/O service routine we must ensure that computation be again interruptible, that is, we must set flip-flop IE. This is done by the following instruction:

Instruction	Operation	Mnemonic
55, 0, --	Set Interrupt Enable flip-flop	SIE

whose EXECUTE microsequence is:

```
4. IE ← 1
5. GO TO 1
```

Notice the absence of the INTERRUPT TEST microsequence[4]; indeed, if, after setting IE, we had an interrupt request (i-bus = 1) and the interrupt microsequence were executed, the original value of PC—stored in 0000—would be overwritten with the current PC and would therefore be irretrievably lost (with incorrect results). Moreover, note the absence of the ADDRESS MODIFICATION step, since IE is an operand-free instruction.

[4]INTERRUPT TEST is also omitted for the instruction ACT.

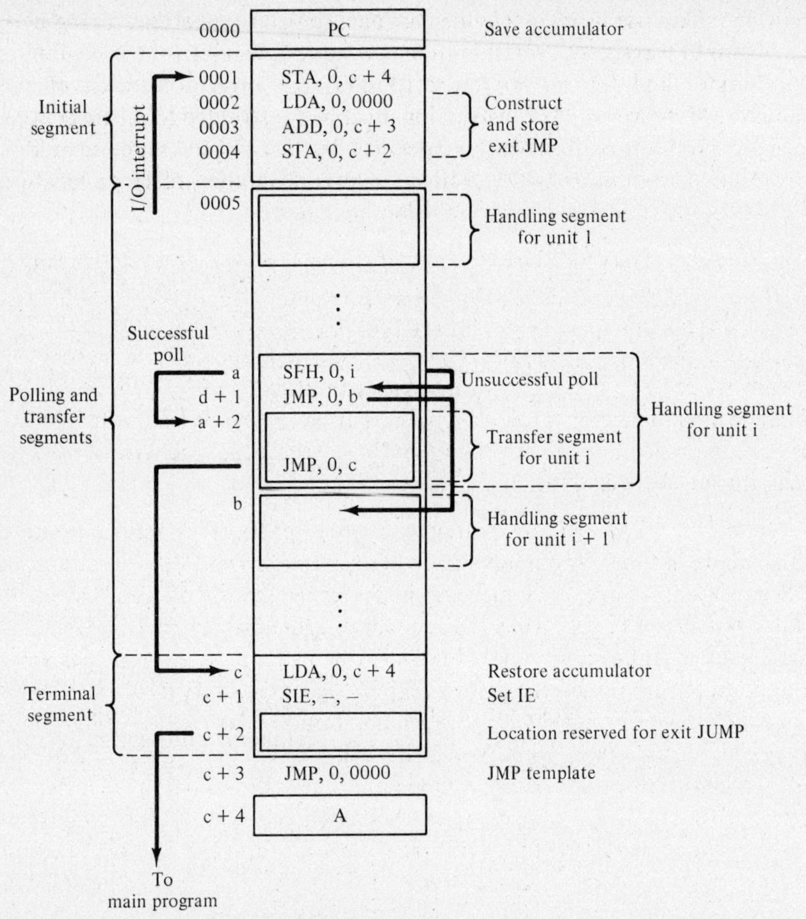

Figure 8.13 General format of the I/O service routine.

We can now describe the operation of the I/O service routine (Figure 8.13). Suppose we have units 1, 2, . . . , n to be polled. For each such unit, say, unit i, the I/O routine contains:

1. A program segment (*polling segment*) which polls the unit and, if it receives a positive response, transfers control to the corresponding transfer segment.
2. A program segment (*transfer segment*) which handles the data exchange for unit i. Typically, if unit i is an input unit, data coming from it are received into the accumulator and then stored into a specified memory location; if unit i is an output unit, data to be output are taken from a specified memory location, brought into the accumulator, and from there transferred to the output unit.

The polling segments are scanned in the order 1, . . . , n until one is found that provides a positive response; next, the data exchange is executed, and from here control is transferred to the terminal segment of the I/O routine. Since this routine makes use of the accumulator (as a temporary storage during both input and output transfers), provisions must be made for saving (at the beginning of the routine) and for restoring (at its end) the original content of the accumulator. Moreover, a JMP instruction must be constructed to return control to the main program at the location specified by the original content of PC, currently stored in location 0000. Notice, at last, the setting of IE in location c + 1. All these actions are illustrated and commented on in Figure 8.13.[5]

Remark 1: The procedure (technically, a *protocol*) whereby a "transmitter" and a "receiver," operating independently and with markedly different speeds, synchronize themselves to exchange information is called *handshake.* As a typical example, an input unit (transmitter) signals to the CPU (receiver) the availability of data by means of an interrupt; the CPU receives the interrupt, acquires the input data (synchronously with its own clock), and acknowledges the successful completion by resetting the FLAG F/F (this acknowledgment embodies the notion of "handshaking").

Remark 2: The polling strategy adopted in the I/O service routine described above has the obvious advantage of an extremely simple f-bus structure (an OR gate), but it appears clumsy and inefficient in the use of time. Indeed, it may work satisfactorily only when the number of I/O units is very small, so that polling is not very time consuming. However, when a large number of I/O units are connected to the system, the polling activity becomes a very costly overhead. In this case it may be preferable to resort to an alternate strategy, called *priority interrupt.* With this policy—described in Problems 8.7 and 8.8 at the end of this chapter—when an interrupt occurs a single unit flag is presented to the CPU for service, thereby avoiding the necessity of polling. Of course, this is accomplished by considerably complicating the hardware of the flag-reporting mechanism with a so-called *priority encoder* (see Problem 8.7). This is another typical instance of the hardware–software trade-off. This mode of operation, in which the interrupting unit supplies the CPU with its own identification, along with the interrupt signal, is commonly referred to in jargon as *vectored interrupt.*

**Remark 3:* The proposed implementation of the interrupt I/O mode contemplates the use of a single location (location 0000) for storing the return address for the interrupted program. This has the observed consequence that the I/O service routine is itself not interruptable. Alternative schemes, which avoid this special handling of the I/O routine, must provide an adequate number of return addresses, indeed, as many such addresses as there are I/O units that must be simultaneously serviced.

One possible realization is the adoption of a *last-in–first-out policy,* that is, the unit currently being serviced, if any, is the one that most recently caused an interrupt. In other words, with this policy, interrupts and completions of service form a "nested" time sequence as illustrated in Figure 8.14.

The last-in–first-out policy and the nesting scheme are familiar patterns, which we discussed earlier in connection with the handling of subroutines (Chapter 7, Sec. 7.7). There, we realized that the natural way to handle the

[5]This mode of operation is similar to that found in the DEC PDP-8 computer.

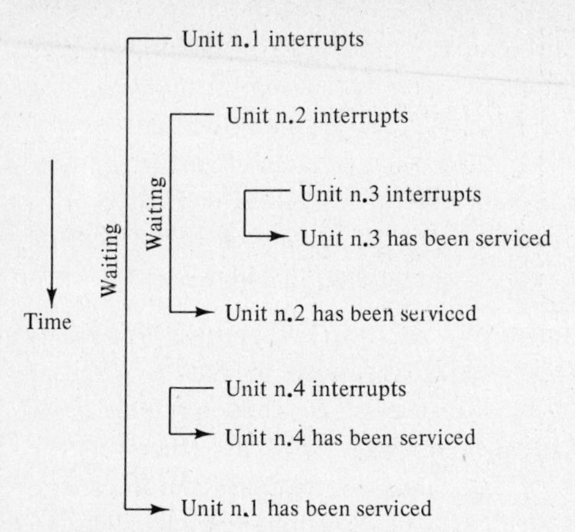

Figure 8.14 Sequence of actions for nested I/O interrupts.

last-in–first-out policy was by means of a "stack," and the reader is referred back to Chapter 7 for a description of this structure. Again, we stipulate to organize the stack as follows:

1. A fixed set of memory locations, for example 7600-7637, are reserved to the stack.
2. The pointer to STACK TOP is conventionally stored in a fixed index register, say R(6). [Rather than the pointer itself, in R(6) we store a counter, from which the pointer is obtained by adding a fixed displacement 7600_8.]

The stack organization is displayed in Figure 8.15. We note that, while a stack operation (i.e., a PUSH or a POP) is being executed, CPU operation cannot be interrupted, otherwise erroneous behavior would arise (such as the pointer to the TOP addressing a memory location different from the actual

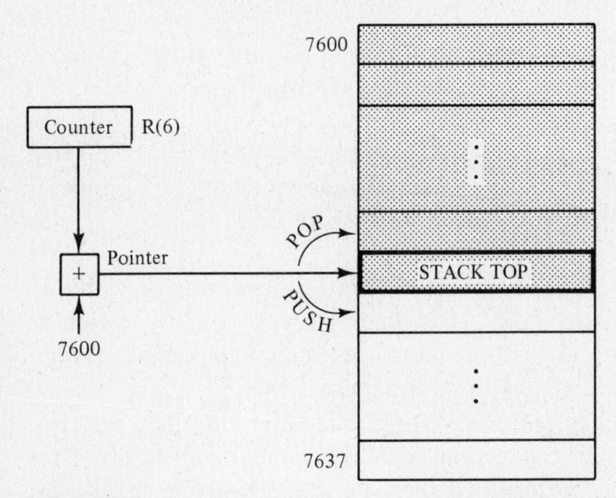

Figure 8.15 Illustration of the stack used to handle I/O interrupts.

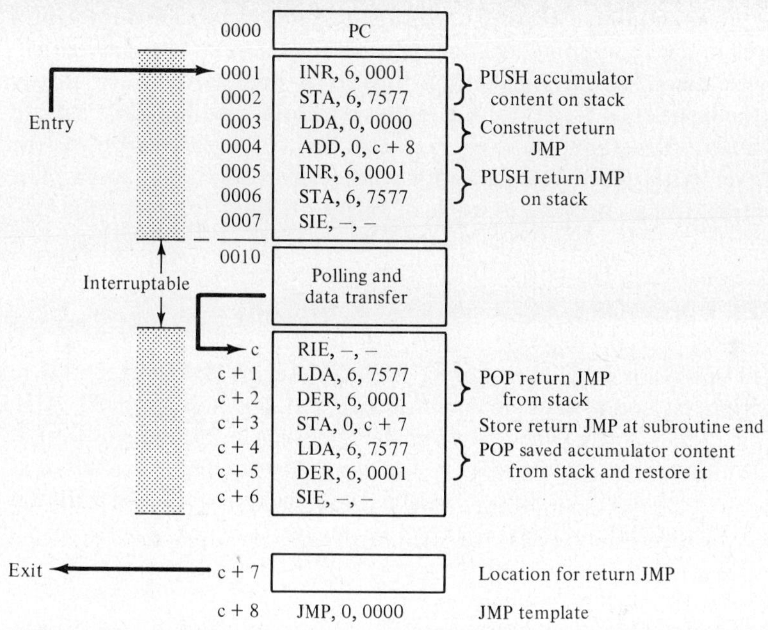

Figure 8.16 Modified I/O routine with stack for accumulator contents and return addresses.

STACK TOP). Therefore, it is necessary that the IE F/F (interrupt enable) be reset for the brief duration of these operations. As concerns PUSH, IE is automatically reset by the interrupt itself and can be set again by the SIE instructions. As concerns POP, on the other hand, reset of IE can only be provided by a new instruction (Reset Interrupt Enable) RIE, 0, –,[6] whose EXECUTE microsequence is (no address modification occurs):

4. IE ← 0
5. GO TO 1

With these conventions, the modifications of the I/O routine described earlier in Figure 8.13 are not substantial. As before, the I/O interrupt must automatically save the contents of both PC (to ensure return to the interrupted program) and the accumulator (which may contain valid data of the interrupted program). This pair of items will be both pushed on the stack in two consecutive locations. As in our general scheme, the interrupt causes PC to be stored in location 0000 and control to be transferred to 0001. This is the initial location of the new I/O service routine, illustrated in Figure 8.16, with comments that should make the diagram self-explanatory. We just note once the content of A and the return JMP have been saved in the stack (instructions 0001–0006) computation becomes again interruptible while the data exchange takes place. At the end of it, the stack operation is again protected against interrupt by the

[6]This instruction, however, is *not* a standard feature of our SEC-XR system.

RIE instruction in location c. Finally, the content of A is restored and the return JMP is stored in the appropriate location c + 7, while the stack is popped twice.

The just described implementation involves a minimal addition to our computer (the instruction RIE). With a more substantial modification we could implement a *hardware* stack, whose management could take place entirely at the hardware level, with added cost but with considerable additional speed. The described alternative is one more example of hardware–software trade-offs.

*8.4 DIRECT MEMORY ACCESS (DATA BREAK)

The mode of operation denoted as Direct Memory Access (DMA) pertains to very fast peripheral units—such as tape drives or disks—which involve a large number of characters or words per data transfer operation. Such large amount of data must be ultimately stored in memory, and it would be quite inefficient—if at all feasible—to temporarily park each word in the accumulator before storing it in memory. Also the protocol involved in the exchange of each word is quite time consuming, as we have seen in the two preceding sections.

Therefore, in the DMA mode data are directly exchanged between the peripheral unit and the main memory without transiting through the CPU. Moreover, the CPU and the PU (DMA peripheral unit) are to be viewed as two independent information processing systems which interact with a third system, the memory M, and *compete for service*. Obviously, in this competition the *PU must have priority over the CPU,* since the latter can be temporarily halted with no detriment, whereas the PU is supplying data at a constant rate and, if service is not readily provided, data may be irretrievably lost. Also, if there is more than one PU on the DMA mode, a service priority scale must be established in order of decreasing data rates; in our discussion, however, we assume that only one such PU exists.

The general layout is shown in Figure 8.17. In a system without DMA the memory (M) exchanges data only with the CPU, whence the *CPU governs* the

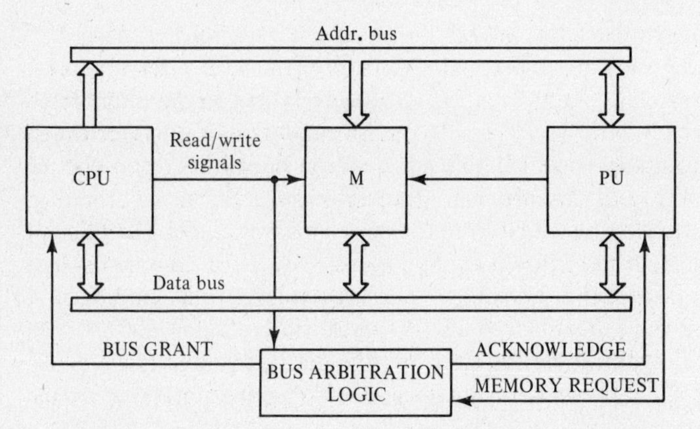

Figure 8.17 Typical layout of CPU, M, and PU in the DMA mode.

data and address buses (and the memory is enslaved to the CPU). At any time, either the CPU or a DMA peripheral unit has control of the buses. Since the CPU is the most frequent user of memory, normally the CPU is the *default* controller, but, at the same time, also the unit with the lowest priority in any competition for the access to memory. Thus, if the PU requests access to memory (because it is ready to transfer data), the most it has to wait is for the completion of the *current* memory use by the CPU (as evidenced by the activity of the CPU memory read/write signals). Once this use is completed, the BUS ARBITRATION LOGIC takes the bus control from the CPU by disactivating the BUS GRANT signal, and transfers it to the PU by acknowledging its request. At this point the data transfer is initiated. At its completion, the PU memory request is withdrawn which causes reactivation of the BUS GRANT signal. Notice that the handshake protocol is implemented entirely at the hardware level.

In what follows, we shall assume that the peripheral unit is a tape drive. Therefore, there are two types of data exchanges, output or "write," input or "read." Each of these exchanges involves several computer words, for example 2^9; such parcel is normally called a *block*. Also we assume that transfer occurs either to or from a *fixed area* in memory for each PU [for example, from location $(1000)_8$ to location $(1777)_8$ for a tape drive unit].

The peripheral unit is a rather complex information processing system (Figure 8.18), equipped with the necessary additional hardware, consisting of a 12-bit register CR (counter register) designed to store the memory address of data, and an INCREMENTER, designed to increment by 1 the content of CR.

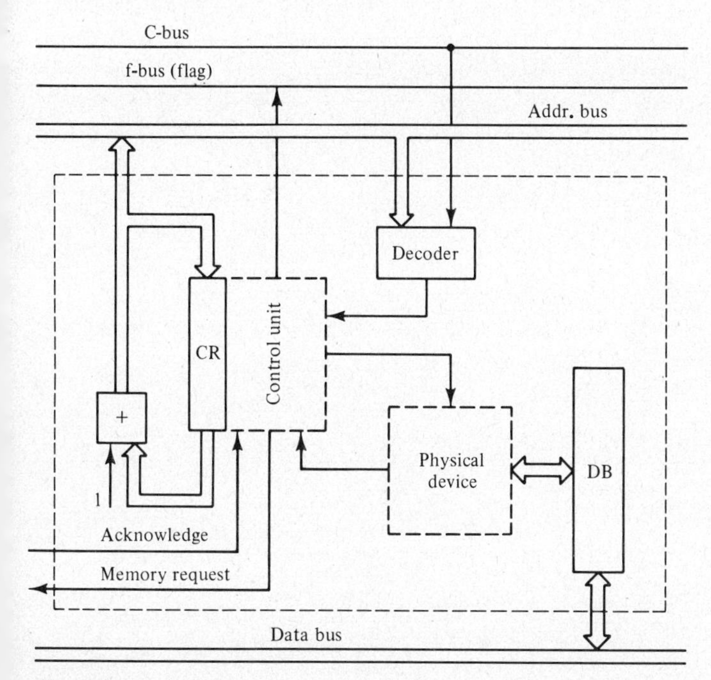

Figure 8.18 Internal structure of the DMA peripheral unit system.

Initially CR is set to contain the initial address $(1000)_8$; operation terminates when the content of CR reaches $(1777)_8$.

Only data transfer instructions will be considered; obviously we must have also a status test instruction (not to be described) to avoid the issuance of transfer instructions when the peripheral unit is in use. The transfer instructions are as follows:

Instruction	Operation	Mnemonic
61, b, Y	Direct Access Read: read a block of data from unit N to its specific memory area	DAR
60, b, Y	Direct Access Write: write a block of data on unit N from its specific memory area	DAW

We shall now describe a DAR instruction, assuming that $(512)_{10}$–$(1023)_{10}$ is the memory area assigned to the unit. By DATA READY we mean a signal generated by the physical device when information (DATA) is available for readout; REQUEST is a signal from PU to M, while ACKNOWLEDGE goes in the reverse direction. The entire operation is summarized by the following microsequences, where, as for the preceding section, the same clock is common to all three interacting subsystems (CPU, M, and PU):

EXECUTE OF DAR

In CPU	In PU
5. addr. bus ← IR[11:0], c-bus ← IR[15:20] IF (f-bus = 0) THEN GO TO 5 END	IF (addr. bus = N) THEN f-bus ← FLAG(N) END
6. GO TO 1	FLAG(N) ← 0, CR ← 1000, (activate device)

This microsequence shows that the CPU hangs up if the unit is not ready; otherwise, in concurrent step 6, the unit is activated, the FLAG is reset, and CR is initialized to $(1000)_8$. At this point the CPU is released, and the competition of CPU and PU for memory service begins.

Transfer from PU to M is described by the following microsequence:

In M	In PU
	1. IF (DATA READY = 0) THEN GO TO 1 END
	2. DB ← DATA, DATA READY ← 0
	3. REQUEST ← 1, IF (ACKNOWLEDGE = 0) THEN GO TO 3 END
4. MAR ← addr. bus, MBR ← data bus	addr. bus ← CR, data bus ← DB
5. M(MAR) ← MBR	IF (CR ≠ 1777) THEN 6. CR ← CR + 1, GO TO 1 END
	7. (deactivate device)

Nothing happens until information is available (DATA READY = 1), so concurrent step 1 simply waits for this event. When data are ready, they are unloaded into DB (concurrent step 2). At this point PU calls for memory service, and it waits if memory is being used (it never waits more than a single memory usage, however); in concurrent step 4, data and address are transferred to memory. Finally in concurrent step 5, input data are stored in memory, CR is incremented, and control is transferred to concurrent step 1, where PU waits for the next input word. However, if the counter CR has reached value $(1777)_8$, the unit is stopped (concurrent step 6). Notice that the unit is directly activated by the DAR instruction, and automatically deactivated at the end of transfer. Instruction DAW is only minutely different and is left as a problem.

*8.5 THE "COLD START" OF THE SYSTEM

The preceding sections discuss the principal input/output modes, but they all assume that the I/O instructions—and possibly the I/O service software—are already stored in memory. Hence a most important question is still avoided: how do we place information into a blank memory? This is what is descriptively referred to as the "cold start" of the computing system.

There are several ways to solve this problem, which all involve some kind of external intervention, such as the use of a keyboard or of other manual pushbuttons: we shall describe the READ-CARD button.

The READ-CARD button is normally placed on the console of the computer system. When pressed, it causes a punched-card reader to read a single card and to transfer its contents to a prescribed area of memory. The card reader is assumed to work in the DMA mode. A (Hollerith-type) punched card contains 80 columns, each of which stores a "character"; in our case, we assume the character is an octal digit. Thus a card carries the equivalent of 11 SEC words, since $11 \times 7 = 77 < 80$. We assume that these words are transferred in order to memory locations $(7765)_8$–$(7777)_8$. Obviously, the READ-CARD button is enabled to operate only when the system is in the "halt" state.

To ensure correct operation, the following sequence of events must automatically take place as a consequence of pressing the READ-CARD button:

1. The card reader is activated and a punched card is read; in the DMA mode the 11 words on the card are transferred to memory locations $(7765)_8$–$(7777)_8$.

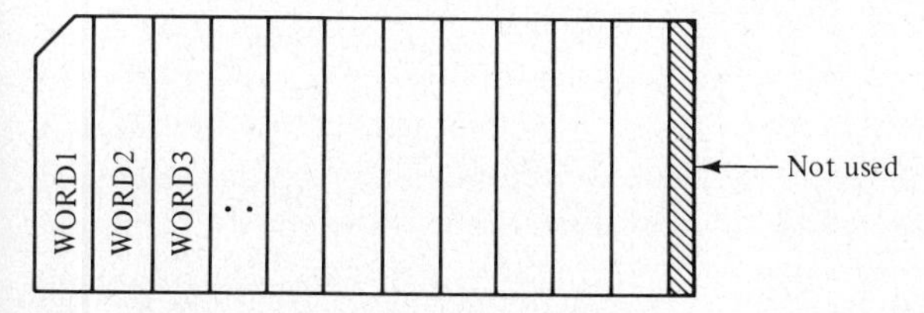

Figure 8.19 Format of a punched card.

2. While the card is being read, the CPU stalls and waits for completion of the input operation.
3. When the data transfer is completed, control is transferred to location $(7765)_8$, whose content is interpreted as an instruction.

These actions are implemented by the following CPU microsequence:

```
10. IF (READ-CARD = 0) THEN GO TO 10 END
11. IF (f-bus = 0) THEN GO TO 11 END
12. PC ← (7765)8
13. GO TO 1
```

Notice that before concurrent step 12 can be executed the card reader is in motion and the data input takes place; once this is completed and the reader becomes available, the FLAG of the reader becomes 1 and so does f-bus; when this happens, PC is loaded with memory address 7765 and a fetch takes place. We see, therefore, that the content of the card just read constitutes a little program of 11 words, which the SEC system will now execute.

The simplest thing we can do with an 11-word program is to load, anywhere in memory, *three* computer words; this is accomplished, for example, by the following 11 words (assume that WORD7, WORD8, and WORD9 on the card are to be stored in locations a, a + 1, a + 2, respectively, and recall that WORDj is read into location $7764_8 + j$):

```
WORD1 (7765): LDA, 0, 7774        WORD5 (7771): LDA, 0, 7776
WORD2 (7766): STA, 0, a           WORD6 (7772): STA, 0, a + 2
WORD3 (7767): LDA, 0, 7775        WORD7 (7773): HLT, -, -
WORD4 (7770): STA, 0, a + 1       WORD8 (7774): DATA
                                  WORD9 (7775): DATA
                                  WORD10 (7776): DATA
```

In this manner a program can be loaded into the system three words at a time, by using the READ-CARD pushbutton each time. However, there are more sophisticated ways of loading programs. The first is to load a short program which can in turn be used to load an arbitrary user's program: such a short program is called a "loader" and represents an important item of the computing system as a whole (hardware and software). Assume that the DAR instruction, when applied to the card reader, transfers the information stored in the card to the same locations 7765–7777 used by the READ-CARD button. Note that since the card reader operates in the DMA mode (see preceding section), the input information is transferred directly to the designated memory location while the read takes place. Assume then that the user's program is punched on N cards, whose contents are to be transferred in their order to locations A through A + 11N − 1 in main memory (note: N cards contain 11N computer words) and that, when the loading is completed, control must be transferred to location A, which contains the first instruction of the user's program. The following simple loader does the job just described (card reader: I/O unit no. 1):

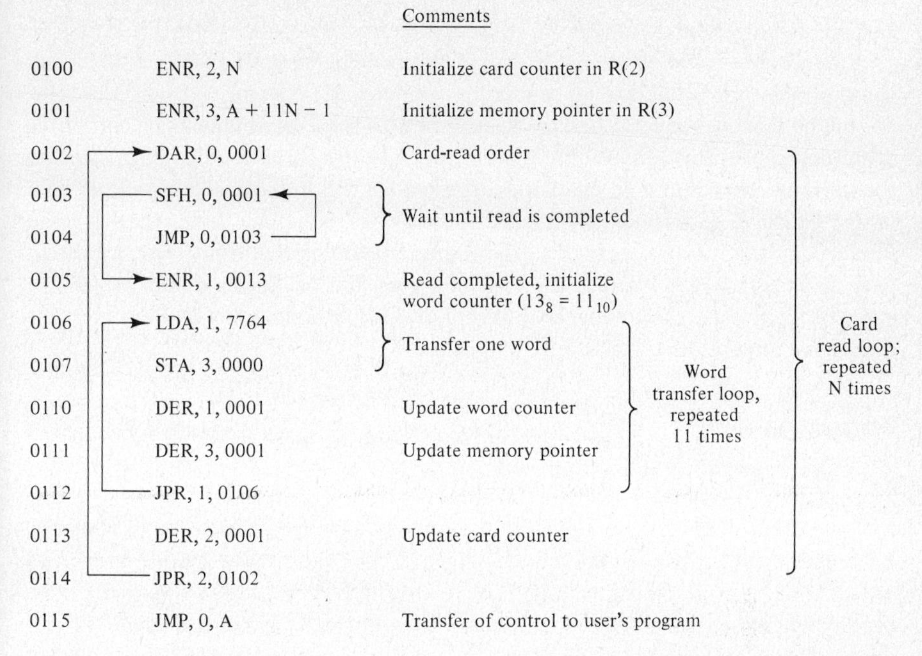

Comments
0100 ENR, 2, N Initialize card counter in R(2)
0101 ENR, 3, A + 11N − 1 Initialize memory pointer in R(3)
0102 DAR, 0, 0001 Card-read order
0103 SFH, 0, 0001
0104 JMP, 0, 0103
Wait until read is completed
0105 ENR, 1, 0013 Read completed, initialize word counter ($13_8 = 11_{10}$)
0106 LDA, 1, 7764
0107 STA, 3, 0000
Transfer one word
0110 DER, 1, 0001 Update word counter
0111 DER, 3, 0001 Update memory pointer
0112 JPR, 1, 0106
Word transfer loop, repeated 11 times
0113 DER, 2, 0001 Update card counter
0114 JPR, 2, 0102
Card read loop; repeated N times
0115 JMP, 0, A Transfer of control to user's program

The 14 instructions of the loader can be loaded by using five times the CARD-READ button, as previously described; at this point the card deck containing the user's program can be placed in the input bin of the card reader and thus be loaded and executed.

But there is an even simpler way, which involves using the CARD-READ button just once. This consists of including in the 11 words contained in the first card a DAR instruction, which will load the second card and so on, in a bootstrap fashion, until the entire loader resides in memory. Thus in the input bin we shall place the user's program preceded by the loader's cards; we press the button once, and the cold start can be effected. As an exercise, show that a bootstrap loader can be stored in five cards.

NOTES AND REFERENCES

The implementation of input/output processes is one of the most complex and studied topics in the organization of digital computers. In this chapter we have attempted to outline some of the basic features of these processes, in the hope that such background may greatly facilitate the detailed understanding of specific realizations.

Although in the very early computers the I/O operations were executed directly under control of the CPU (with the ensuing idle time), already in the early fifties the advantages of introducing peripheral controllers (or "synchronizers") were fully realized. The transfer of information frequently used an intermediate high-speed buffer. In an input operation, for example, blocks of data—such as those coming from a tape, a drum, or a punched card—could be accumulated in a buffer under control of the peripheral unit, and the buffer was subsequently unloaded into main memory under CPU control.

The reader is encouraged to consult some of the available texts on computer architecture, such as Hayes (1979) and Baer (1980), for a detailed discussion of the various approaches. In particular, the reader will find important details on a topic we have not even mentioned in this introductory text, that is, the implementation of an externally driven input (such as the input from a human user at a keyboard). Finally, the text of Leventhal on microprocessors (1978) adeptly contrasts the modern different I/O methods and highlights their respective advantages and disadvantages.

PROBLEMS

8.1. Rather than using instruction SFH, the test of a flag is done by a new type of instruction FTA, b, Y (Flag to Accumulator), which makes the accumulator negative if the flag of I/O unit Y + R(b) is high, and makes it positive otherwise. Write the execute microsequence of the FTA instruction.

8.2. Using the FTA instruction (and assuming that SFH is not available) write a machine program which replaces the following one (FTA is defined in Problem 8.1):

```
0010  SFH, 0, 0002
0011  JMP, 0, 0010
0012  WRT, 0, 0002
```

8.3. In the programmed I/O mode, assume that SFH is not available but that SFL (Skip on Flag Low) is available. Write a short program to test flag and read from input unit no. 6.

8.4. Unit no. 1 is a paper-tape reader operating in the programmed I/O mode and inputs one character per start–stop cycle. The paper-tape characters are BCD numerals (4 bits only are used); when the RD operation is completed the BCD character appears in A[3:0]. Write a short SEC-XR program which reads two consecutive characters from unit no. 1, interprets the first and second characters as units and tens of an integer, respectively, translates the corresponding number into binary, and stores it in location 0100_8.

8.5. The SEC-XR system has exactly two units, input unit no. 1 and output unit no. 2, in the interrupt mode. Write the I/O service routine for these two units. (Recall that, before anything else, the content of any registers used by the routine must be saved; they must then be restored before returning to the main program). Assume that the handling segment for each individual unit simply involves a transfer operation. (Output information is in 0100; input information goes into 0200.)

8.6. SEC-XR has input unit no. 1 and input unit no. 2 operating in the interrupt mode. Data successively read from unit no. 1 (or respectively no. 2) are to be written in consecutive memory locations, starting from the address initially specified in index register R(6) (or respectively R(7)). Write the I/O service routine for units no. 1

and no. 2. (Note: completely specify the instructions for the polling and data-exchange sequences.)

8.7. An s-to-n *priority encoder* is an encoder (see Chapter 4, Sec. 4.8.1) with s input lines $I_s, I_{s-1}, \ldots, I_1$ and n output lines $y_{n-1}, \ldots, y_0$ (normally $n = \lceil \log_2(s + 1) \rceil$). The output lines display the binary equivalent of i, where i is *the largest* of the indices of the input lines which are logically 1. (Note that more than one input line can be at 1 at a given time.) Example:

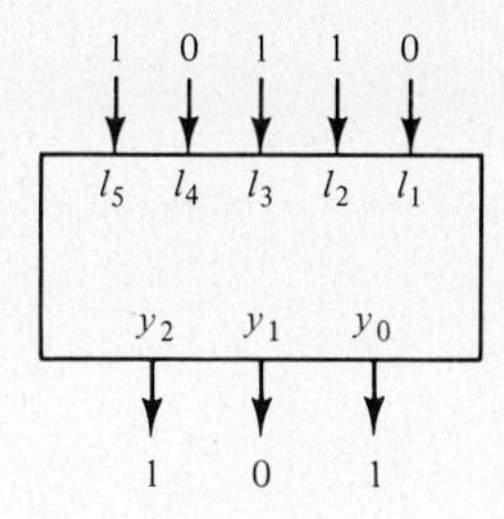

(**a**) Design a 3-to-2 priority encoder.
(**b**) Design a 4-to-3 priority encoder.

8.8. The SEC-XR system is to be modified as follows. It has four I/O units—numbered 1, 2, 3, and 4—operating in the interrupt mode with priority: specifically any one of these units has priority over the lower numbered units. For this purpose the flag lines of the units are brought directly to the CPU. Here, whenever a flag is set to 1, an interrupt occurs, PC is saved in 0000_8, control is transferred to 0001_8, and the interrupt mode is disabled. The system also has an instruction, FTC, –, –, which places in A[2:0] the binary representation of i, where i is the largest of the numbers of the units whose flag is 1. Using a priority encoder (Problem 8.7), do the following
(**a**) Draw the diagram of the INTERRUPT circuit;
(**b**) Write the EXECUTE microsequence of the FTC instruction;
(**c**) Write an I/O service routine which serves the highest priority unit (assume that all units are input units and that input from unit i goes into address 64_{10} + i).

8.9. Modify the INTERRUPT TEST microsequence so that, when the interrupt is due to occur, PC is stored in R(7) and control is transferred to 0000.

8.10. An 80-column (Hollerith) punched card is subdivided into 11 fields of 7 columns each: the first field comprises columns 1–7, the second comprises columns 8–14, and so on. Each field corresponds to a computer word. Columns 78–80 are not used. Each column contains—in some code—a numeral from 0 through 7. The card is read by columns—column 1 first—and the output of the reading device is the binary encoding of the numeral (i.e., an octal digit). The peripheral unit packs all digits of a field into a 21-bit DB register. Data input takes place in the DMA mode, and is transferred to memory locations $7765–7777_8$.
(**a**) Draw the layout of the PU.
(**b**) Design the DMA microsequence corresponding to transfer from PU to M.

8.11. The READ-CARD pushbutton in the SEC system's console activates the card reader, transfers the content of the card (see Sec. 8.5) to locations 7765–7777, and then transfers control to location 7765. Fill the 11 words on a card so that, when read by action of the READ-CARD pushbutton, the contents of the last three words on the card will be transferred to locations 0000–0002 and the computer will halt.

Chapter 9

Control Unit; Microprogramming

9.1 INTRODUCTION

In the two preceding chapters we have illustrated how the various machine language instructions of the SEC computer can be themselves realized as "programs" written in a more elementary language, the RTL. These programs we have called microsequences.

We are now going to study two problems in connection with microsequences. The first is how we can physically implement the operations specified by an individual microstep; the second is how the microsequences, which collectively describe the computer behavior, can be implemented and correctly timed.

For simplicity we shall refer to the SEC-XR machine, and ignore from this point on the more complex SEC-XR-PM option; in connection with input/output, we shall also ignore the DMA mode (Chapter 8, Sec. 8.4).

9.2 MICROSTEP IMPLEMENTATION AND CONTROL SIGNALS

Although a concurrent step is a rather simple operation (especially if it consists of a single microstep), its implementation normally consists of the *simultaneous performance* of several even simpler actions, called *micro-operations*.

Most of the micro-operations can be described in RTL if we extend the notion of "register" to include "buses" (see Chapter 8, Sec. 8.2). Typically, a micro-operation is the connection of a set of lines to a set of lines (for example, the output lines of register PC are connected to the data bus, and this is concisely denoted as data bus ← PC). Thus, a bus will have a micro-operation for each set of lines multiplexed to it and for each set of lines demultiplexed (distributed) from it. For example, with reference to Figure 7.10, we can identify several such possible micro-operations as: collector bus ← A, IR ← distributor bus, collector bus ← R(b), and so on. The execution of a micro-operation at a given time is achieved by generating a specific *control signal* (or control signals). A control signal establishes the desired connection by opening the appropriate set of gates (for example, either in a multiplexer, or in a demultiplexer); other uses of control signals have been discussed in Chapter 6, Sec. 6.6, with regard to the selection of the ALU function.

The first step in the design of the control unit is an inventory of the necessary micro-operations. This is done by inspecting the layout of the system (see Figure 7.10) and by referring to the microsequences necessary to implement the instructions in the machine code. Most of these micro-operations are listed in Table 9.1.

We shall now show how a given concurrent step can be implemented by the simultaneous activation of an appropriate subset of micro-operations. This is illustrated in Figure 9.1 with reference to the FETCH-EXECUTE microsequences of the STA instruction. Only a selected subset of the micro-operations given in Table 9.1 is actually shown in Figure 9.1; an "×" denotes that the corresponding micro-operation is being activated. For example, the microstep

```
MAR ← IR[11:0]
```

is executed by (1) feeding IR[11:0] to the data bus (data bus ← IR[11:0]), that is, to the ALU B-lines; (2) setting the ALU to a direct transfer of the data on the

TABLE 9.1 INVENTORY OF MICRO-OPERATIONS

Connections to collector bus

1. Collector bus ← PC
2. Collector bus ← R(b) (b = 1, . . . , 7)
3. Collector bus ← A

Connections to data bus

4. Data bus ← IR[11:0]
5. Data bus ← distributor bus
6. Data bus ← MBR

Connections from distributor bus

7.	A	← distributor bus
8.	R(b)	← distributor bus
9.	IR	← distributor bus
10.	IR[11:0]	← distributor bus
11.	PC	← distributor bus
12.	Address bus	← distributor bus

ALU functions

13. G-lines ← A-lines
14. G-lines ← B-lines
15. G-lines ← A-lines + B-lines
16. G-lines ← A-lines − B-lines
17. G-lines ← A-lines + 1
18. G-lines ← f(A-lines, B-lines) (f:boolean function)
19. G-lines ← LRS, LLS, ARS, ALS of A-lines
20. OF ← $(c_{21} \oplus c_{20}) + ALS(g_{20} \oplus b_{OUT})$

Memory functions

21.	MAR	← address bus
22.	MBR	← data bus
23.	MBR	← M(MAR)
24.	M(MAR)	← MBR
25.	MAR	← 0

Others

26.	IE	← 0
27.	IE	← 1
28.	c-bus	← IR[20:15]

Concurrent step \ Micro-operation	1. Collector bus ← PC	2. Collector bus ← R(b)	3. Collector bus ← A	4. Data bus ← IR[11:0]	5. Data bus ← Distributor bus	6. Data bus ← MBR	9. IR ← Distributor bus	10. IR[11:0] ← Distributor bus	11. PC ← Distributor bus	12. Address bus ← Distributor bus	13. G-lines ← A-lines	14. G-lines ← B-lines	15. G-lines ← A-lines + B-lines	17. G-lines ← A-lines + 1	21. MAR ← Address bus	22. MBR ← Data bus	23. MBR ← M(MAR)	24. M(MAR) ← MBR
MAR ← PC	X									X	X				X			
MBR ← M(MAR), PC ← PC + 1	X								X					X			X	
IR ← MBR						X	X					X						
IR[11:0] ← IR[11:0] + R(b)		X		X				X					X					
MAR ← IR[11:0]				X						X		X			X			
MBR ← A			X		X						X					X		
M(MAR) ← MBR																		X

Figure 9.1 Implementation of the STA instruction.

B-lines; (3) connecting the distributor bus, that is, the ALU G-lines, to the address bus; and (4) feeding the address bus to the MAR.

9.3 THE FUNCTION OF THE CONTROL UNIT

From the detailed descriptions contained in the two preceding chapters, we can now return to the conception, formulated in Chapter 5, of the computing system as a very complex *synchronous sequential network.* The current state of the "network" is given by the values of the variables that are stored in the flip-flops of the computer. We know that these flip-flops are controlled by pulses that are synchronized with the system clock. In particular, let us consider the CPU and assume that the control unit has been carved out (Figure 9.2). The *internal variables* are the contents of registers A, R(1), . . . , R(7), IR, and PC, and of two individual flip-flops: OF and IE (not shown). The *excitation functions* are carried by the input lines to these flip-flops. The output functions are those carried on the outputs to the address bus and the data bus, plus the OP code and some additional "status variables," such as OF, A[20], and so on, indicating some significant conditions of the computing system. The *input variables,* besides the obvious inputs on the data bus, are the *control signals* coming from the control unit. We may view the collection of the control signals as a binary vector (*control vector*), so that a component equal to "1" means an active signal

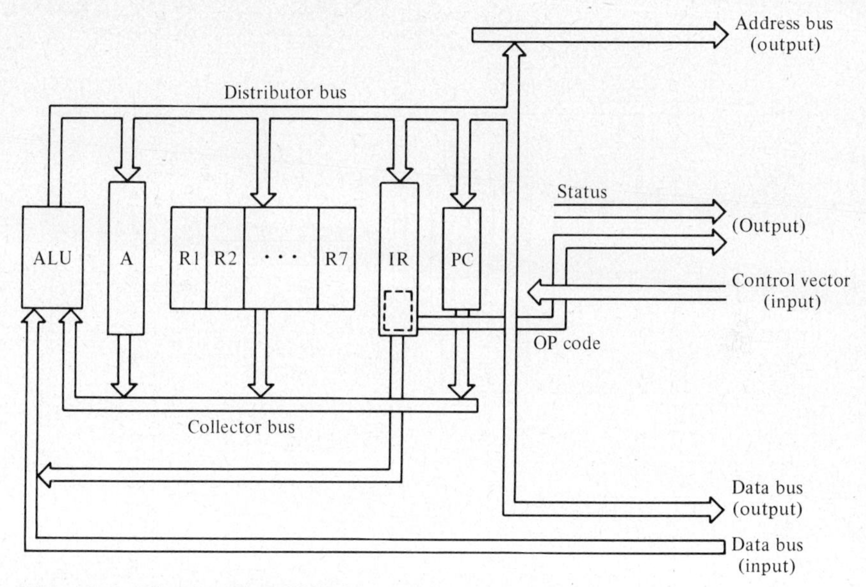

Figure 9.2 The "mutilated" CPU as a sequential network.

and "0" means the opposite (in the chart of Figure 9.2 we have "×" and "blank" for 1 and 0, respectively). Thus we have the following definition:

> An RTL concurrent step describes the set of actions which occur between two consecutive clock pulses; the results of these actions are stored in flip-flops when the clock pulse occurs.

We realize the fundamental role played by the control vector: the values of its components determine the routing of information in a time unit of our synchronous computing system. The overall operation of the computer is realized by sequencing the appropriate control vectors. A control vector in turn can be viewed as the output of another Synchronous Sequential Machine (SSM) called the *control unit* (*timed by the same clock*), whose inputs are the OP code and the status variables from the rest of the system; later we shall discuss the internal variables of this new SSM. In summary we have the effective representation of the computing system as a complex sequential network, consisting of two interacting sequential networks, the "mutilated" CPU and the control unit, as illustrated in Figure 9.3. We shall now concentrate on the control unit.

Before examining specific ways to realize the control unit, let us consider a flowchart describing the various microsequences of our computer. In this flowchart (Figure 9.4) we have described the FETCH microsequence and the EXECUTE microsequences of a subset of the instructions: STA, LDA, ADD, SUB, AND, OR, ENT, JMP, JZA, JPA, and HLT. Each block of the flowchart refers to a "concurrent step" (see Sec. 7.1); the chart can also be easily expanded to include the microsequences for other machine instructions.

An alternative interpretation of the flowchart is as the state diagram of the

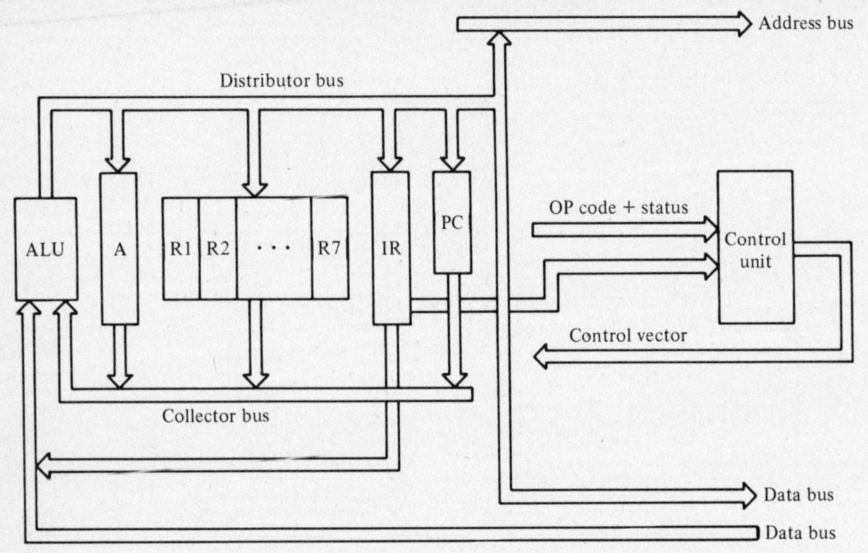

Figure 9.3 CPU and control unit: two interacting sequential networks.

synchronous sequential machine called *control unit*. Therefore we can synthesize our control unit starting from a diagram such as that in Figure 9.4. In Chapter 5, Sec. 5.10, we have recognized that the crucial step in the design of synchronous sequential networks is the state assignment (i.e., the way in which we code the states), and we have outlined three major approaches to the solution of this problem:

Approach 1. Minimal (or near-minimal) number of flip-flops, discrete logic.

Approach 2. One-hot assignment (one flip-flop per state).

Approach 3. Minimal (or near-minimal) number of flip-flops, ROM-based logic.

Approach 1 is certainly the most sophisticated from the designer's standpoint, but it is scarcely attuned to the current situation, due to the very large number of states. Indeed, a minimal set of state variables is quite likely to produce a very complicated combinational logic, which may be difficult to troubleshoot. These considerations basically rule against approach 1, although there are computers whose control units are designed along these lines.

There remains to consider approaches 2 and 3. Approach 2—the sequencer approach—leads to a network that is structurally almost identical to the flow diagram. The flip-flops are connected in cascade as suggested by the flow diagram (compare Figure 9.4 and Figure 9.5), with simple additional combinational circuitry to implement the various branchings and junctions. Since *at any time, there is exactly one flip-flop in the SET state,* the execution of an instruction starts by "injecting" a 1 in the flip-flop corresponding to the first block of the fetch microsequence (s1), and proceeds along the path traversed by

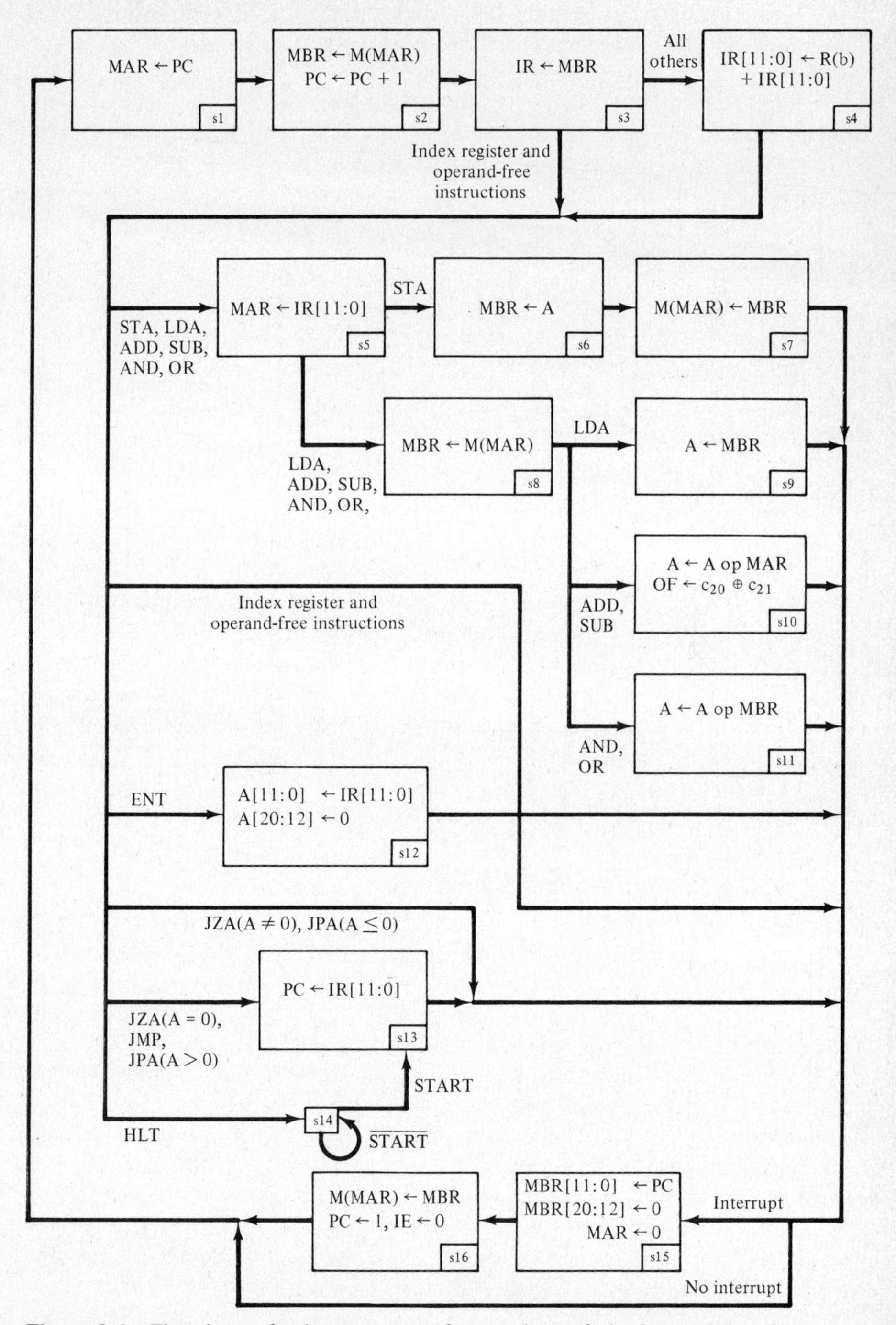

Figure 9.4 Flowchart of microsequences for a subset of the instructions. Steps are labeled s<integer> for future reference.

this "1" in this diagram. After a few steps, this single "1" cycles back to the initial flip-flop (s1), unless the instruction is an HLT, in which case it ends in flip-flop s14.

This realization—which has been adopted in several control unit designs—results in fast operation and is elegant in its simplicity. It is totally of the hardware type, however, and as such leaves no real flexibility for possible

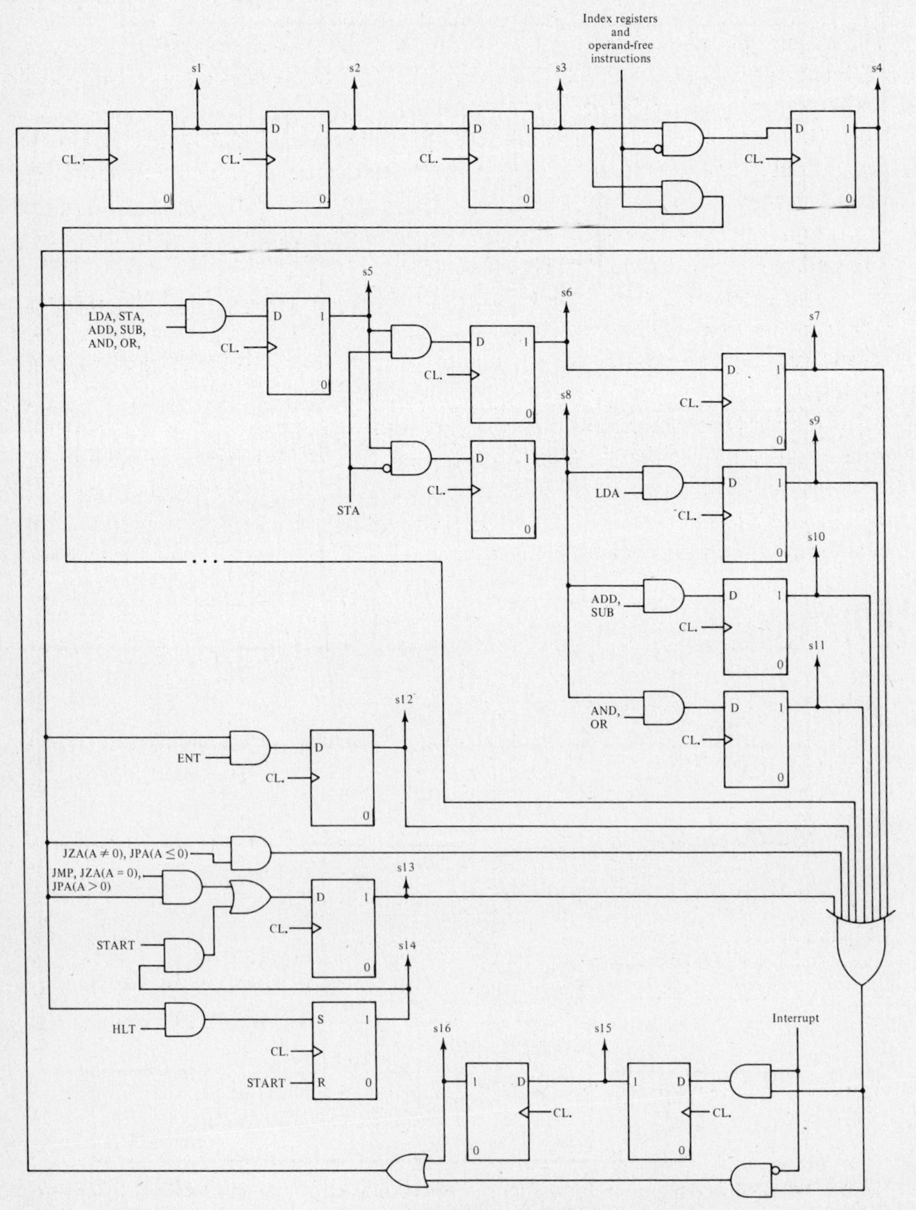

Figure 9.5 (Part of) the control unit for SEC, designed according to the sequencer approach. Each flip-flop is labeled as the step it corresponds to (see Figure 9.4).

modifications or expansions of the instruction repertoire. This drawback disappears in the more modern and versatile realization, corresponding to approach 3, which we are about to describe.

9.4 MICROPROGRAMMED CONTROL UNIT

The general ROM-based design technique of synchronous sequential networks, outlined in Chapter 5, Sec. 5.11, lends itself to a particularly attractive realization of the control unit.

In Figure 9.6(a) we have the conventional scheme of ROM-based SN. If the output signals depend only upon the network state (the content of the register), then we have the scheme of Figure 9.6(b), where the generation of the excitation functions has been assigned—for convenience—to a separate combinational module called ADDRESS (calculation) LOGIC.

The signals labeled OUTPUT form a *control vector,*[1] and the INPUT signals come from the rest of the system. The remaining ROM outputs contain information on the next state: when combined with the INPUT signals in the ADDRESS LOGIC, they produce the next-state vector.

We may now regard the contents of locations of the ROM as "instructions": thus, the control vector becomes the operation code and completely specifies the actions to be carried out, while the other output lines provide information on the "next instruction" (transfer of control). This is the basic idea of *microprogramming:* each location in the ROM contains a *microinstruction*

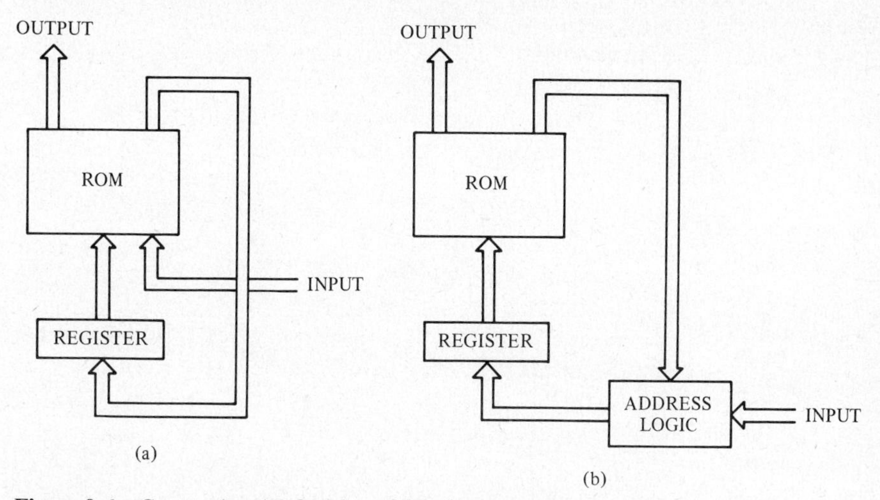

Figure 9.6 Conventional ROM-based SN design, and a convenient variant.

[1]This type of microprogramming is normally referred to as *horizontal,* the reason being that the control signals are produced directly as a ROM-readout, and are thought of as being on a "horizontal line." In another approach, called *vertical microprogramming* (not adopted in the SEC), ROM readouts are more compact strings, which are then transcoded by combinational logic into the control vector. This terminology is an example of dubious jargon.

(implementing a concurrent step), expressed in a very simple language whose "programs" execute the machine language instructions. In a sense, the relation between microinstructions and machine language instructions is analogous to the one between machine language instructions and commands of a high-level language (such as FORTRAN or PASCAL). One difference of degree, if not of substance, is that microprograms are not acccssible to conventional users, and are rarcly tampered with. For this reason, microprograms are referred to as *firmware,* a species between the inflexible hardware and the totally flexible software.

For concreteness we shall now describe a realization of a microprogrammed control unit for the SEC-XR machine, with programmed and interrupt I/O. The ROM (also frequently called *control memory*) has 512 words; besides the bits for the m-bit control vector ($m \simeq 30$–40), the ROM word contains 9 address bits used to select the location of the next microinstruction. The control unit has a 9-bit address register μMAR[8:0] and a detailed layout of it is shown in Figure 9.7. In Figure 9.8 we show the content of portions of the microprogram ROM. For convenience, a control vector will be denoted by the concurrent step it implements rather than by a binary string. (So, a notation MAR ← IR[11:0] will be used instead of the m-bit vector which corresponds to the active micro-operations.) The format of the microinstructions is:

ADDRESS: | NEXT | CONCURRENT STEP |

Our attention here, more than on the action (expressed by CONCURRENT STEP), will be concentrated on the "transfer of control", that is, on where the next microinstruction is to be found.

Note that each bit position of μMAR has an input multiplexer; when the gated logic is blocked (AND gates 1, 2, . . . , 11), the "normal" input to μMAR are the 9 bits of NEXT. So, normally NEXT is the address of the next microinstruction, except in some special "branching" situation to be described below.

The microinstruction (in octal),

```
000:  000, all zeros
```

loops indefinitely to itself, until the content of μMAR is altered by some external intervention as we shall see later. We shall also see that this microinstruction executes the HLT instruction.

The sequence of microinstructions

```
001:  002, MAR ← PC
002:  003, MBR ← M(MAR), PC ← PC + 1
003:  004, IR ← MBR
```

implements the first three concurrent steps of FETCH. For these microinstructions the transfer of control is straightforward.

At the completion of the microinstruction in 003 the operation code is

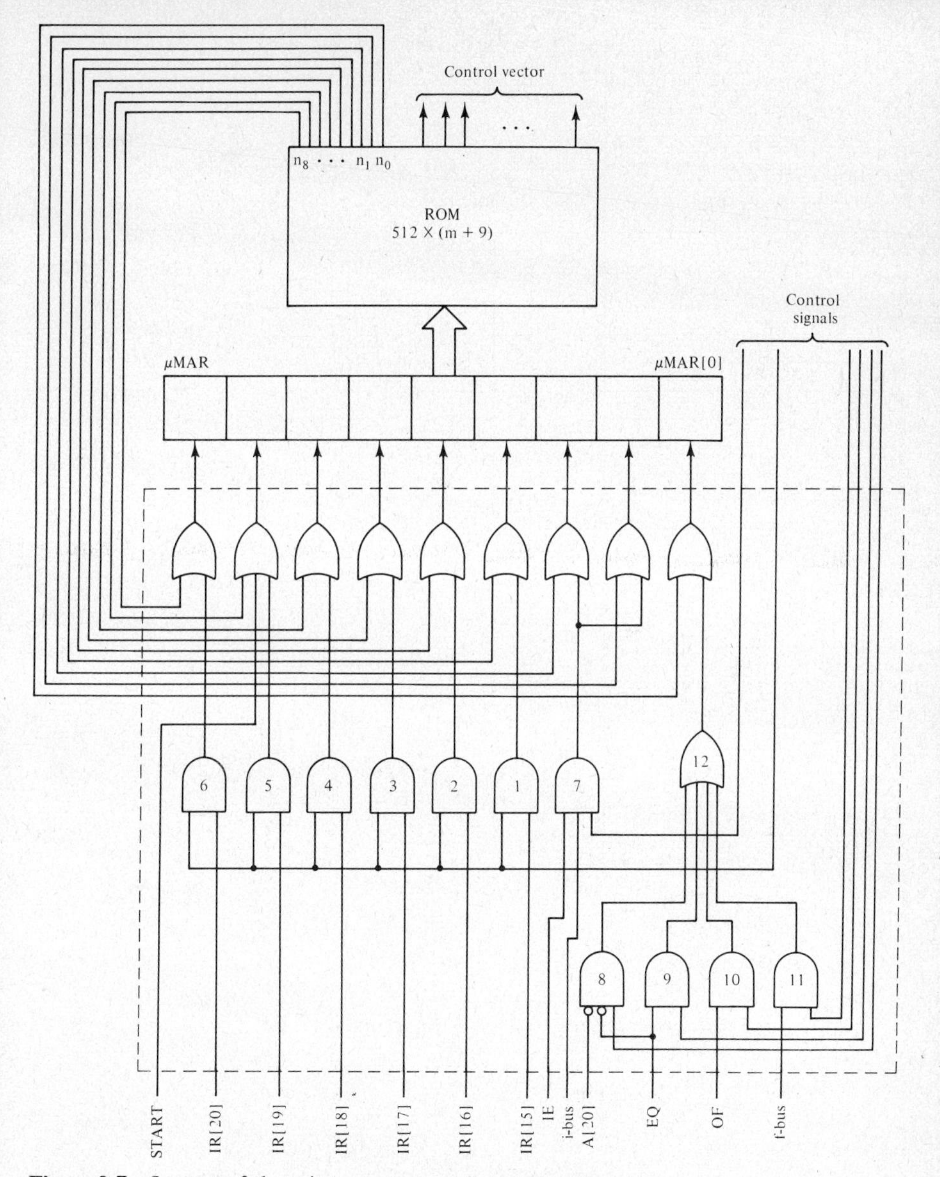

Figure 9.7 Layout of the microprogrammed control unit.

available in the IR register. At this point we proceed to the "address modification step" in 004. With this microinstruction two important actions take place:

1. The first action is the transfer of control to the EXECUTE microprogram of the instruction currently stored in IR. In fact, by opening gates 1–6 (by means of a control signal) the OP code of the instruction is gated into the six most significant positions of μMAR (μMAR[8:3]). Since NEXT = 000, at the completion of this microinstruction the content of μMAR is $(WX0)_8$, where

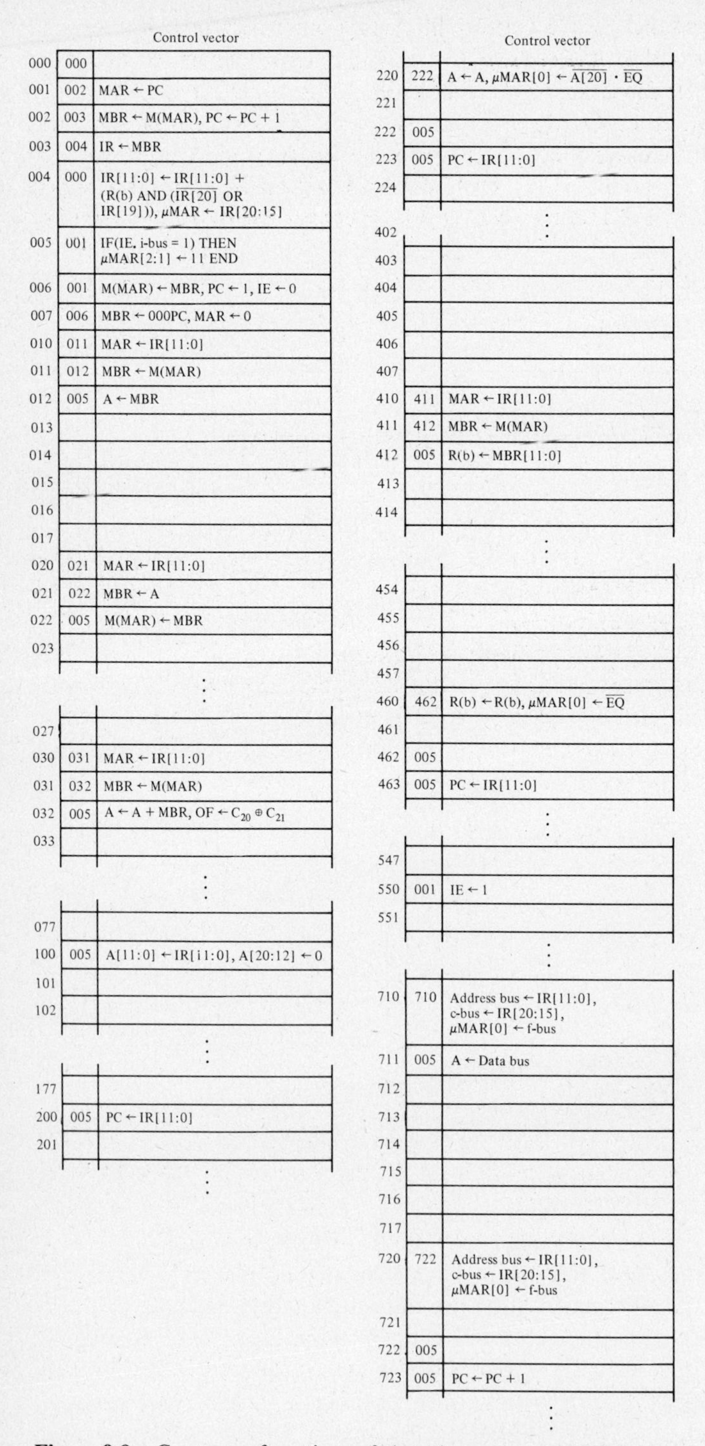

Address	Next	Control vector
000	000	
001	002	MAR ← PC
002	003	MBR ← M(MAR), PC ← PC + 1
003	004	IR ← MBR
004	000	IR[11:0] ← IR[11:0] + (R(b) AND ($\overline{\text{IR[20]}}$ OR IR[19])), μMAR ← IR[20:15]
005	001	IF(IE. i-bus = 1) THEN μMAR[2:1] ← 11 END
006	001	M(MAR) ← MBR, PC ← 1, IE ← 0
007	006	MBR ← 000PC, MAR ← 0
010	011	MAR ← IR[11:0]
011	012	MBR ← M(MAR)
012	005	A ← MBR
013		
014		
015		
016		
017		
020	021	MAR ← IR[11:0]
021	022	MBR ← A
022	005	M(MAR) ← MBR
023		
⋮		
027		
030	031	MAR ← IR[11:0]
031	032	MBR ← M(MAR)
032	005	A ← A + MBR, OF ← $C_{20} \oplus C_{21}$
033		
⋮		
077		
100	005	A[11:0] ← IR[11:0], A[20:12] ← 0
101		
102		
⋮		
177		
200	005	PC ← IR[11:0]
201		
⋮		
220	222	A ← A, μMAR[0] ← $\overline{\text{A[20]}} \cdot \overline{\text{EQ}}$
221		
222	005	
223	005	PC ← IR[11:0]
224		
⋮		
402		
403		
404		
405		
406		
407		
410	411	MAR ← IR[11:0]
411	412	MBR ← M(MAR)
412	005	R(b) ← MBR[11:0]
413		
414		
⋮		
454		
455		
456		
457		
460	462	R(b) ← R(b), μMAR[0] ← $\overline{\text{EQ}}$
461		
462	005	
463	005	PC ← IR[11:0]
⋮		
547		
550	001	IE ← 1
551		
⋮		
710	710	Address bus ← IR[11:0], c-bus ← IR[20:15], μMAR[0] ← f-bus
711	005	A ← Data bus
712		
713		
714		
715		
716		
717		
720	722	Address bus ← IR[11:0], c-bus ← IR[20:15], μMAR[0] ← f-bus
721		
722	005	
723	005	PC ← PC + 1
⋮		

Figure 9.8 Contents of portions of the microprogram ROM.

$(WX)_8$ is used to denote the OP code of the current instruction. It follows that control is transferred to location WX0 in the microprogram ROM. (For example, if the current instruction is 03, b, Y, control is transferred to 030.)

2. The second action is the address modification by means of the index registers. Recall, however, that there are some instructions (*index register instructions,* such as STR, LDR, INR, DER, ENR, JPR, JSR) for which address modification is *forbidden,* and other instructions (*operand-free instructions,* such as NEG, LRS, LLS, ARS, ALS, SIE) for which address modification is *unnecessary.* The choice between the two alternatives (to carry out address modification or not) is simply implemented by either executing

$$\mathrm{IR[11:0]} \leftarrow \mathrm{IR[11:0]} + \mathrm{R(b)}$$

when address modification is needed, or the dummy transfer

$$\mathrm{IR[11:0]} \leftarrow \mathrm{IR[11:0]}$$

when it is not needed. This can be done by a suitable control signal that blocks the output of the index registers onto the collector bus. This signal is readily obtained. Since the operation codes of STR, . . . , JSR are of the form $(4X)_8$ (i.e., IR[20:18] = 4 is characteristic for them) and those of NEG, . . . , SIE are of the form $(5X)_8$, the condition for suppressing address modification is that the most-significant octal digit of the OP code be either 5 or 4, that is, IR[20:18] = $(100)_2$ or $(101)_2$, equivalent to IR[20:19] = $(10)_2$. It is then easy to recognize that the RTL microstep

$$\mathrm{IR[11:0]} \leftarrow \mathrm{IR[11:0]} + (\mathrm{R(b)\ AND\ (}\overline{\mathrm{IR[20]}}\ \mathrm{OR\ IR[19]}))$$

realizes the specified operation.

From 004 we branch to the first microinstruction of the EXECUTE microprograms. Consider, for example, the EXECUTE microprogram of the instruction ADD (OP code = 03):

030: 031, MAR ← IR[11:0]
031: 032, MBR ← M(MAR)
032: 005, A ← A + MBR, OF ← $c_{21} \oplus c_{20}$

The transfers of control are standard; however, the jump to 005 requires some comment. Recall that after the concurrent step A ← A + MBR, OF ← $c_{21} \oplus c_{20}$, we must still execute the following RTL steps:

7. <interrupt test microsequence>
8. GO TO 1.

Therefore in 005 we shall find a microinstruction to test for the presence of interrupts in order to decide whether to branch to the INTERRUPT microprogram or to return to the FETCH microprogram. The condition for an interrupt is IE = 1 and i-bus = 1: if this condition does not occur we must jump back to 001, the first microinstruction of FETCH; otherwise, we shall jump to the first microinstruction of INTERRUPT (in 007). Thus the microinstruction

005: 001, IF(IE.i-bus = 1) THEN μMAR[2:1] ← 11_2END

carries out the prescribed task. Indeed if IE.i-bus = 0, μMAR is loaded with $(001)_2$; otherwise, μMAR is loaded with $(007)_2$ through the action of gate 7 in Figure 9.7.

The INTERRUPT microprogram consists of

007: 006, MBR[11:0] ← PC, MBR[20:12] ← 0, MAR ← 0
006: 001, M(MAR) ← MBR, PC ← 1, IE ← 0

where the transfer of control is standard. After microinstruction 006 has been executed, control is transferred to the FETCH microprogram.

As another example, the EXECUTE microprogram of the JMP instruction (OP code = 20) is given by the single microinstruction

200: 005, PC ← IR[11:0]

We shall now discuss the handling of conditional microsteps in microprogramming. For concreteness, consider IF (OF = 1) THEN PC ← IR[11:0]END. One way to realize it is to gate the signal which controls the micro-operation "PC ← distributor bus" with the signal OF. Another, more interesting, way is to branch to a microinstruction realizing PC ← IR[11:0] or to bypass it, depending upon the value of OF; this branching can be realized by conditionally modifying the content of μMAR. Specifically, for the JOF instruction (code 24) we have the following EXECUTE microprogram:

240: 242, μMAR[0] ← OF
241:
242: 005
243: 005, PC ← IR[11:0]

(OF = 0: 240 → 242; OF = 1: 240 → 243)

which we now analyze. Clearly, if OF = 1 then μMAR[0] will be set to 1, thereby resulting in μMAR = 010 100 011 = 243_8, otherwise μMAR = 242_8; both cells 242_8 and 243_8 contain the usual branch to the interrupt test microinstruction. In an exactly analogous way we shall handle the other conditional branching instructions (JPA, JZA, JPR, SFH). Notice that the relevant status signals are EQ, A[20], OF, f-bus [trace in Figure 9.8 the microprogram of JPA (OP code = 22), where the content of A is circulated through the ALU to generate EQ in the process]; the corresponding ADDRESS LOGIC consists of AND gates 8, 9, 10, and 11 feeding OR gate 12.

The EXECUTE microprograms for all other instructions have analogous format, and consist typically of at most three microinstructions. There is, however, a singular case, SIE (Set Interrupt Enable, Chapter 8, Sec. 8.3), for which the INTERRUPT TEST *must not* take place. This is easily done as follows (recall that the EXECUTE of SIE, OP code 55, begins at 550):

550: 001, IE ← 1

Finally, there are two more points to discuss.

TABLE 9.3 SEC-XR SUMMARY OF INSTRUCTION SET

OP code	Instruction	Operation	Address modification	Memory reference
		Transfer, arithmetic, and logic		
01	LDA, b, Y	A ← M(N)	*	*
02	STA, b, Y	M(N) ← A	*	*
03	ADD, b, Y	A ← A + M(N), OF ← $c_{20} \oplus c_{21}$	*	*
04	SUB, b, Y	A ← A − M(N), OF ← $c_{20} \oplus c_{21}$	*	*
05	AND, b, Y	A ← A AND M(N)	*	*
06	OR, b, Y	A ← A OR M(N)	*	*
10	ENT, b, Y	A[11:0] ← N, A[20:12] ← 0	*	
54	NEG, –, –	A ← −A		
50	LRS, –, –	A[i] ← A[i + 1](i = 0, . . . , 19) A[20] ← 0		
51	LLS, –, –	A[i] ← A[i − 1](i = 1, . . . , 20) A[0] ← 0		
52	ARS, –, –	A[i] ← A[i + 1](i = 0, . . . , 19)		
53	ALS, –, –	A[i] ← A[i − 1](i = 1, . . . , 20) A[0] ← 0, OF ← $g_{20} \oplus b_{OUT}$		
		Control		
00	HLT, –, –	Halt, no operation		
20	JMP, b, Y	PC ← N	*	
21	JZA, b, Y	IF (A = 0) THEN PC ← N	*	
22	JPA, b, Y	IF (A > 0) THEN PC ← N	*	
46	JPR, b, Y	IF (R(b) > 0) THEN PC ← Y		
47	JSR, b, Y	1. R(b) ← PC 2. PC ← Y		
24	JOF, b, Y	IF (OF = 1) THEN PC ← N	*	
23	JS, b, Y	1. M(N) ← PC 2. PC ← N + 1	*	*
		Index register manipulation		
41	LDR, b, Y	R(b) ← M(Y)		*
42	STR, b, Y	M(Y) ← R(b)		*
43	INR, b, Y	R(b) ← R(b) + Y		
44	DER, b, Y	R(b) ← R(b) − Y		
45	ENR, b, Y	R(b) ← Y		
		Input/output		
70	WRT, b, Y	Unit (N) ← A	*	
71	RD, b, Y	A ← Unit(N)	*	
72	SFH, b, Y	IF (Flag(N) = 1) THEN PC ← PC + 1	*	
74	ACT, b, Y	Unit(N) ← START	*	
75	STP, b, Y	Unit(N) ← STOP	*	
55	SIE, –, –	IE ← 1		

Notation: Instruction format

CODE	b	Y
6	3	12

, N = Y + R(b).

What happens when an "illegal" OP code (i.e., one corresponding to no instruction) is used in a machine language program? If we stipulate that all unused locations of the ROM contain 0s, then an illegal OP code, such as 37, will cause a Jump-to-EXECUTE in 370: 000, 0 . . . 0; obviously the latter microinstruction causes a jump to 000, where the computer halts.

How do we restart the computer? This can only be initiated by external intervention, such as pressing a "start" button on the computer panel. Suppose now that this action produces a signal START which is multiplexed into μMAR[7] (Figure 9.7). The content of μMAR becomes $(010 \quad 000 \quad 000)_2 = (200)_8$, that is, it refers to the first microinstruction of the EXECUTE microprogram of JMP, which is

200: 005, PC ← IR[11:0]

This microinstruction loads into PC the current content of IR[11:0] and transfers control to 005 and from here to 001 (no interrupts are present!), the address of the first microinstruction of the FETCH microprogram. Thus, at the SEC machine language level, computation is resumed from the instruction stored in M(IR[11:0]). The typical use of the feature is in combination with the HLT instruction. In fact, suppose that computation stops with the instruction HLT, 0, Y; pressing the start button resumes computation from the instruction stored in location Y.

This concludes the description of the microprogram control unit of the SEC. The realization we have chosen is sufficiently simple to permit the illustration of the basic features of microprogramming. In keeping with the common jargon, the proposed solution has a very high degree of "horizontality," since both the control signals and the next address information are *fully* decoded. However simple, this solution requires more storage in the ROM than more complicated and efficient solutions, which would exhibit some degree of "verticality," and therefore trade storage space with additional logic.

In keeping with the character of this text, we shall not explore this fascinating avenue. The curious reader, however, has at his or her disposal all the tools to successfully continue this journey without the assistance of a guide.

NOTES AND REFERENCES

Although some form of microprogramming is to be found in practically all processors of today, the other approaches to the realization of the control unit (compact sequential machines or one-hot sequencers) have been quite popular in the earlier history of digital computers. In some machines, when using the traditional sequential network approach, particular care has to be exercised in the choice of state codes in order to reduce the complexity of the ensuing combinational logic.

The invention of the microprogram appears very early in the evolution of digital computers, and is credited—like the invention of the index registers—to the Manchester group. The classical paper by Wilkes and Stringer (1953) clearly outlines the concept and its technological requirements. Perhaps the economical constraints of the then available technology accounted for the somewhat slow acceptance of the concept—the emerging flexibility did not fully justify the hardware cost of a microprogrammed control unit—which developed fully when the technological obstacles became less significant.

In this chapter we have described only one typical realization of a microprogrammed control unit. Not only is the alluded to distinction between "horizontal" and "vertical" microprogramming important, but once we accept the notion of a microprogram as an "interpreter" of the machine language, we may immediately conceive of additional deeper interpretative levels, for which the interesting name of nanoprogramming has been proposed. In this view, the conventional microprogram is just at one level in a hierarchy of programming languages, at the top of which are the machine-independent languages (such as FORTRAN and PASCAL).

Practically any text on computer architecture contains a more or less extensive discussion of microprogramming techniques. The reader is again referred to Hayes (1978) and Baer (1980), for further reading on the topic.

PROBLEMS

9.1. Give the description in terms of micro-operations (such as in Figure 9.1) for the EXECUTE microsequences of the following instructions

(a) LDA (b) AND
(c) JMP (d) SUB
(e) JSR (f) DER
(g) ENT (h) INR
(i) OR (j) STR

9.2. Consider the instruction NEG, –, (A ← –A). Referring to the logical diagram of the ALU bit unit (Figure 6.20):

(a) Design the EXECUTE microsequence of NEG and, whenever appropriate, the pattern of the control signals (c_0, S_3, S_2, S_1, S_0, L).

(b) Give the description of this microsequence in terms of micro-operations.

9.3. Describe the microprograms of the EXECUTE phases of the following instructions:

(a) STR (b) LDA
(c) WRT (d) INR
(e) ADD (f) JZA
(g) ENT (h) JS
(i) JSR

9.4. Describe in words the function of the SEC instruction executed by the following microprogram (the instruction is 11, b, Y):

```
110:  111, IF (A > 0) THEN μMAR[4] ← 1 END
111:  112, PC ← PC + 1
112:  005, –
131:  112, A[11:0] ← IR[11:0], A[20:12] ← 0
```

Bibliography

J. L. Baer, *Computer Systems Architecture,* Computer Science Press, Rockville, MD, 1980.

C. G. Bell and A. Newell, *Computer Structures: Readings and Examples,* McGraw-Hill, New York, 1971.

E. R. Berlekamp, *Algebraic Coding Theory,* McGraw-Hill, New York, 1968.

T. L. Booth, *Digital Networks and Computer Systems,* Wiley, New York, 1971.

A. W. Burks, H. H. Goldstine, and J. von Neumann, "Preliminary discussion of the logical design of an electronic computing instrument," *U.S. Army Ordnance Department Report,* 1946.

W. I. Fletcher, *An Engineering Approach to Digital Design,* Prentice-Hall, Englewood Cliffs, NJ, 1980.

I. Flores, *The Logic of Computer Arithmetic,* Prentice-Hall, Englewood Cliffs, NJ, 1963.

R. W. Hamming, "Error Detecting and Error Correcting Codes," *Bell System Tech. Jour.,* pp. 147–160, April 1950.

J. Hartmanis and R. E. Stearns, *Algebraic Structure Theory of Sequential Machines,* Prentice-Hall, Englewood Cliffs, NJ, 1966.

J. P. Hayes, *Computer Architecture and Organization,* McGraw-Hill, New York, 1978.

F. J. Hill and G. R. Peterson, *Introduction to Switching Theory and Logical Design,* Wiley, New York, 1968 (second edition 1974).

F. E. Hohn, *Applied Boolean Algebra,* Macmillan, New York, 1966.

J. E. Hopcroft and J. D. Ullman, *Introduction to Automata Theory, Languages and Computation,* Addison-Wesley, Reading, MA, 1979.

D. A. Huffman, "The synthesis of sequential switching circuits," *J. Franklin Inst.,* Vol. 257, pp. 161–190 and 257–303, 1954.

K. Hwang, *Computer Arithmetic. Principles, Architecture and Design,* Wiley, New York, 1979.

M. Karnaugh, "The map method for synthesis of combinational logic circuits," *Trans. AIEE,* vol. 72, pt. 1, pp. 593–598, 1953.

D. Knuth, *The Art of Computer Programming, Vol. 2: Seminumerical Algorithms,* Addison-Wesley, Reading, MA, 1969.

Z. Kohavi, *Switching and Finite Automata Theory,* McGraw-Hill, New York, 1970 (second edition 1978).

L. A. Leventhal, *Introduction to Microprocessors: Software, Hardware, Programming,* Prentice-Hall, Englewood Cliffs, NJ, 1978.

H. R. Lewis and C. H. Papadimitriou, *Elements of the Theory of Computation,* Prentice-Hall, Englewood Cliffs, NJ, 1981.

S. Lin, *An Introduction to Error-Correcting Codes,* Prentice-Hall, Englewood Cliffs, NJ, 1970.

M. Morris Mano, *Computer System Architecture,* Prentice-Hall, Englewood Cliffs, NJ, 1976.

E. J. McCluskey, Jr., "Minimization of boolean functions," *Bell System Tech. J.,* Vol. 35, pp. 1417–1444, November 1956.

W. S. McCulloch and W. Pitts, "A logical calculus of the ideas immanent in nervous activity," *Bulletin of Mathematical Biophysics,* Vol. 5, pp. 115–133, 1943.

F. J. McWilliams and N. J. A. Sloane, *The Theory of Error-correcting Codes,* North Holland, Amsterdam, 1978.

G. H. Mealy, "A method for synthesizing sequential circuits," *Bell System Tech. J.,* Vol. 34, pp. 1045–1079, September 1955.

E. F. Moore, "Gedanken-experiments on sequential machines," in *Automata Studies* (eds. Shannon-McCarthy) Princeton University Press, Princeton, NJ, 1956.

D. E. Muller, "Boolean algebras in electric circuit design," *Am. Math. Monthly,* vol. 61, no. 7, pp. 27–28, 1954.

D. E. Muller and W. S. Bartky, "A theory of asynchronous circuits," *The Annals of the Computation Laboratory, Harvard,* vol. XXIX, 1957.

S. Muroga, *Logic Design and Switching Theory,* Wiley-Interscience, New York, 1979.

F. P. Preparata and R. T. Yeh, *Introduction to Discrete Structures,* Addison-Wesley, Reading, MA, 1973.

W. W. Quine, "The problem of simplifying truth functions," *Am. Math. Monthly,* vol. 59, pp. 521–531, October 1952.

B. Randell, ed., *The Origins at Digital Computers,* Springer-Verlag, Berlin, 1973.

R. K. Richards, *Arithmetic Operations in Digital Computers,* Van Nostrand, Princeton, NJ, 1955.

C. E. Shannon, *The Mathematical Theory of Communications,* The University of Illinois Press, Urbana, IL, 1949.

C. E. Shannon, "A symbolic analysis of relay and switching circuits," *Trans. AIEE,* vol. 57, pp. 713–723, 1938.

M. Sloan, *Computer Hardware and Organization,* SRA, Chicago, IL, 1976.

A. Turing, "On computable numbers with applications to the Entscheidungsproblem," *Proc. London Math. Soc.,* Ser. 2, Vol. 42, pp. 230–265, Vol. 43, pp. 544–546, 1937.

M. V. Wilkes and J. B. Stringer, "Microprogramming and the design of the control circuits in electronic digital computers," *Proc. Cambridge Phil. Soc.,* Vol. 49, pp. 230–238, 1953.

Appendix

Microprocessors

A.1 INTRODUCTION

The SEC computing system described in the preceding chapters exhibits, in a simplified way, most of the fundamental characteristics of the stored-program machine, or, as it is normally referred to, the von Neumann machine.[1] The von Neumann machine is, in some sense, the model of most computing systems existing today, with the exception of the emerging parallel processing arrays.

In physical size and computing power, the von Neumann machines today span a very wide range. Before the invention of integrated circuits (in the early sixties) and the resulting hardware miniaturization, all computers—even the most unsophisticated ones—were of large physical size, and required complex installations. Today, only the most powerful and complex systems—referred to as "mainframe"—and their peripheral devices require large-size installations; on the other hand, the CPU of very sophisticated computers can today be miniaturized to the size of a single silicon chip whose area is less than 1 cm^2! Such single chip processors are known as *microprocessors* and our SEC could be viewed as an (somewhat clumsy, perhaps) example of a microprocessor.

Microprocessors are today extremely popular digital modules, and have become the symbols of the success of the computer industry. When they first appeared in the early 1970s, microprocessors were rather simple modules. The term "micro" was presumably chosen to suggest both the *single chip packaging* and the small word size (4 bits or, later, 8 bits), which required a diminutive qualifier. A microprocessor of the 1980s, on the other hand, contains in a single chip, the CPU, a sizable memory, and several input/output synchronizers; in

[1]As we mentioned in the Notes at the end of Chapter 2, John von Neumann was a key figure in the group which designed the EDVAC, the first acknowledged stored-program computer.

addition, it rivals large computing systems of the sixties in word size, and it even surpasses them in speed and sophistication. Therefore, thanks to the VLSI revolution, the qualifier "micro" applies nowadays only to the physical size of a very powerful module.

The reader may wonder why in this text we have chosen to describe a hypothetical machine when there are so many "real-world" examples to refer to. As we already suggested in the preface, there are several reasons for this choice.

One reason is that even the simplest "real" microprocessors have a substantially more complex structure than our educational machine. A lot of this structural complexity is the result of clever engineering solutions to specific technological obstacles, or of the painstaking optimization of the use of resources. Although every design effort must strive to achieve the desired performance at the least cost, there is undoubtedly merit in separating the basic principles (the fundamentals of the stored-program computer) from the contingent implementation twists arising from engineering optimization. Frequently, particular solutions are dictated by current technological constraints, which may disappear as technology evolves.

A second reason is that a commercial microprocessor is available and documented as a "finished product," and does not easily permit the pedagogically effective approach (followed in this text) of building the system step by step, so that each added feature enhances the performance in a clearly identifiable way. This approach enabled us to illustrate in a sharp way for the SEC the very important hardware–software trade-off.

A third reason is that, in a rapidly evolving technology, one may doubt the wisdom of a substantial intellectual investment in all the intricacies of a real microprocessor, which may be targeted for early obsolescence (as has happened for so many predecessors of today's modules). By keeping the level of complexity to a minimum (as in the SEC), while highlighting the basic principles, we have developed a framework that greatly facilitates the understanding of the more complex commercial microprocessors as the state-of-the-art evolves.

A fourth, and final, reason is our wish to illustrate in the most elementary detail the structure of a digital computer (the *designer's architecture*). Unfortunately, this objective is not readily achievable with commercial microprocessors, since normally the available literature deals just with that facet of the system that is accessible (and necessary) to the users (the *programmer's architecture*), and the manufacturer is reluctant to disclose details of the circuit realization.

While the preceding four reasons justify the approach taken in this text, it is now appropriate to expand our horizon and to establish a more direct tie with the state of the art by examining a commercially available microprocessor. Obviously, a real microprocessor is considerably more complicated than our educational machine, not only because of the more elaborate implementation of many of the functions existing in the SEC, but also because of the presence of important additional capabilities.

We have clearly indicated in the last three chapters that the SEC should be viewed as an intermediate stage in the development of a full-fledged computer system. Besides some extra hardware functions, a "finished" microprocessor has a collection of programs, which are an integral part of the system. Such programs are the loader (a simple version of which was described in Chapter 8, Sec. 8.5), the assembler, the editor, the I/O routines, and, in general, the so-called "operating system." The discussion of these important software components is well beyond the scope of this text and is the subject of more advanced studies of computer science and engineering. The existence of these programs, however, determines some hardware structural features that are appropriate to consider at this point. For example, the input/output handling routines are programs whose integrity is essential to the functionality of the computer. In other words, one must prevent a user's program from altering any of these "system's" programs. Thus there will be mechanisms for protecting the system's programs, such as the subdivision of memory into two portions, a system's portion and a user's portion. Correspondingly, the CPU will operate in a restricted mode (i.e., without the possibility of modifying the system's program) when executing a user's program (*user state*), while it will operate in an unrestricted mode (*supervisor state*) when full access to the system's programs is warranted. In the user's mode selected portions of memory are protected and only a subset of the instruction repertoire is available for execution.

In this Appendix we shall study the Motorola MC68000, a very interesting microprocessor and one of the most powerful available today (early 1980s).[2] Of course our goal is not a description of all of the details of the systems, for which an excellent user's manual is available. Instead, our objective is to highlight the *main* features of the MC68000, and this is best done by contrasting its sophistication with the simplicity of the SEC, where applicable. Once the design philosophy is on firm ground, the interested reader can easily fill in the (many) additional details by referring to the technical literature on the MC68000.

A.2 DESIGNER'S ARCHITECTURE AND PROGRAMMER'S ARCHITECTURE

By *designer's architecture* of a CPU we mean a detailed description of the equipment consisting of the following items:

1. A list of all registers.
2. A functional specification of all combinational modules.
3. A specification of all CPU data paths used in the execution of the machine language instructions (system layout).
4. The instruction set, where the action of each instruction is described in RTL.
5. The structure of the control unit (e.g., the control unit layout and the content of the microprogram ROM).

[2]Every detail, in text and illustrations, of the MC68000 reported in this appendix is derived from: *MC68000 16-bit Microprocessor. User's Manual*, Prentice-Hall, Englewood Cliffs, NJ, 1982, and is reprinted by permission of Motorola, Inc.

It should be apparent that the level of detail of the description of the SEC-XR achieved in the preceding chapters meets the above specifications, that is, it provides a designer's architecture of the computer. This satisfies our original goal, set forth in Chapter 2, to obtain a most detailed description of the machine. On the other hand, the (machine language) programmer of a given computer does not need to have a complete knowledge of the above items 1–5 to be able to successfully write programs for that computer. Indeed, a little reflection shows that all the programmer needs are items 1 (the list of all registers) and 4 (the instruction set). So we can define the programmer's architecture as follows:

1′. A list of all registers.
2′. The instruction set, where the action of each instruction is described in RTL.

We shall now consider the user's architectures of the SEC and the MC68000.

A.2.1 SEC

The register list (item 1′) is concisely illustrated in Figure A.1. There is one data register A (the accumulator) and seven index registers R(1), . . . , R(7). In addition we have the program counter (PC) and two "status" flip-flops, OF and IE, which identify current special conditions of the system. The description of the SEC programmer's architecture is completed by Table 9.3, which contains the instruction format, the instruction repertoire, and the way in which each instruction identifies its operands (*addressing*).

The addressing modes of the SEC—a fundamental aspect of its architecture—have been examined in detail in Sec. 7.9, to which the reader is referred.

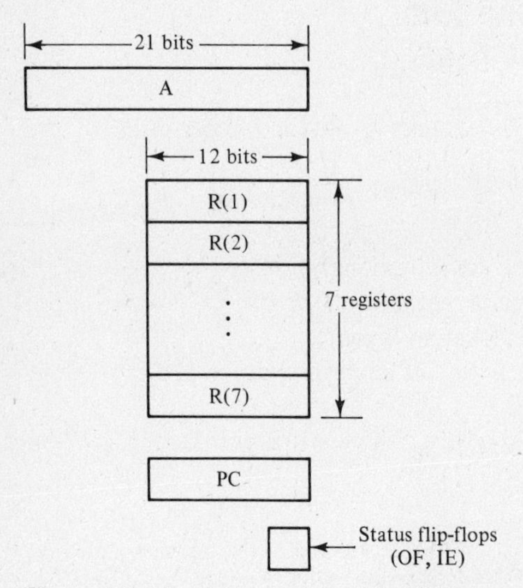

Figure A.1 Register catalog of the SEC.

We simply recall that the following modes are available (in the SEC-XR):

For Data Manipulation Instructions

Accumulator immediate with index
Register immediate
Accumulator direct with index
Register direct
Implicit

For Control Instructions

Direct with index
Direct

A.2.2 MC 68000

The programmer's architecture of the MC68000 is considerably more elaborate and powerful. The register catalog, illustrated in Figure A.2, contains 16 32-bit registers R(0), . . . , R(15) subdivided into two groups: eight data registers D(0), . . . , D(7), and eight address registers A(0), . . . , A(7). [Specifically, denoting by b an octal digit we have R(0b) = D(b) and R(1b) = A(b).]

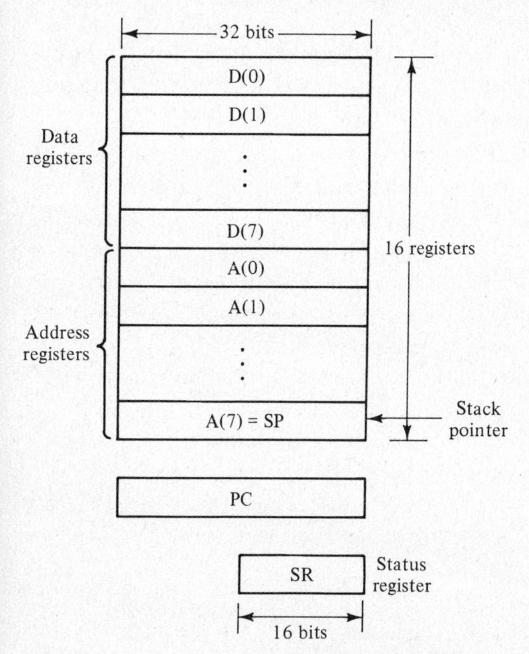

Figure A.2 Register catalog of the MC68000.

Register A(7) also has the specialized function of "stack pointer" and is frequently referred to as SP.[3] In addition, three is a 32-bit program counter PC and a 16-bit status register.

The MC68000 has two operating modes, a supervisor mode and a user mode, whose respective functions implement the philosophy outlined at the end of Sec. A.1. The status register SR is subdivided into two halves, the "supervisor byte" and the "user byte" (as shown below),

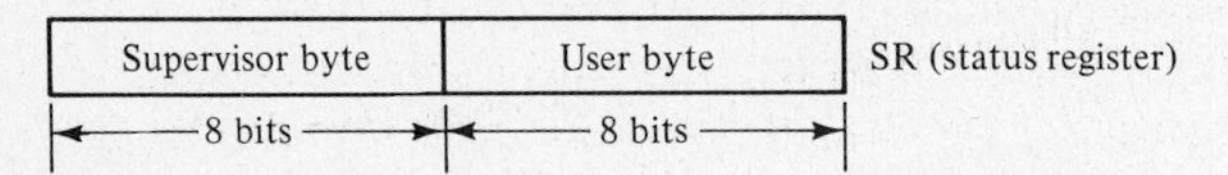

each of which is accessible only to the corresponding mode, and contains relevant parameters. In the following description, however, we shall concentrate almost exclusively on the user mode, except for the discussion of how interrupts are handled by the supervisor mode in Secs. A.4 and A.5.

Whereas in the SEC the registers [A on one side, R(1)–R(7) on the other] have distinct functions, in the MC68000 they all share common capabilities, with additional features selectively exhibited by a subset of them [A(0)–A(7), and in particular A(7), the stack pointer]. In other words, R(0)–R(15) form a small memory, where each register can be used either as an accumulator or as an index [and A(0)–A(7) as stack pointers]. For this reason, they are referred to as general-purpose registers, or, briefly, *general registers*. The capabilities afforded by general registers are particularly useful in computations involving a rather small number of operands with frequent data interchanges (such as in sorting and merging applications), which can be carried out without time-consuming main memory references.

The bidirectional CPU data bus has a bandwidth of 16 bits. The memory is therefore organized in 16-bit words. Since—as we shall see shortly—the address bus has 23 bits, the memory is concisely described as $M(2^{23}-1:0)$ [15:0].

The MC68000 supports several different data types (whereas the SEC supports only one, the 21-bit word). These data types are: bit, byte (8 bits), word (16 bits), long word (32 bits), and BCD (binary coded decimal). A data item of a given type occupies the lowest order bit positions of the register where it is stored. Support of a given data type means that an instruction will exclusively affect the subregister assigned to that specific data type. To avoid long and inessential details, hereafter we shall consider only word-size (16-bit) data.

The instruction has variable length, tailored to the complexity of the command (SEC, which has a fixed length instruction, has several unused bits in instructions as NEG, LRS, etc.). In particular, each instruction has a number of bits which is a multiple of 16, the word length, and consists of a maximum of five words. We see here a typical engineering compromise where the optimization of the use of memory (for storing the program instructions) and of instruction

[3]In reality, there are two distinct 32-bit stack pointers, which are used in a mutually exclusive way, one in the supervisor mode and the other in the user mode.

fetching time (which is proportional to the number of words in the instruction) is achieved at the expenses of a rather elaborate fetch sequence.

As is natural, the instruction code specifies both the number of operands and the addressing mode to be used. For example, an ADD instruction must identify the sources of the addends and the destination for the sum; similarly a transfer instruction (called MOVE) must identify both the source and the destination. Although the MC68000 has 56 distinct instruction codes, they can be conveniently grouped in classes of similarity, so that all instructions in the same class have the same format. Since it is well beyond our scope to account for the entire instruction repertoire (for which again we refer to the user handbook), we shall content ourselves with the illustration of some major example.

According to the formalism of RTL, by M(PC) we denote the first word of the current instruction; the second (if any) is denoted by M(PC + 1); and so on. For ease of reference, we also denote M(PC) with the single word W. Word W—the first one to be fetched—contains the operation code of the instruction and all the necessary information on subsequent necessary fetches. The operation code does not occupy in the MC68000 a fixed set of bit positions; rather, the designers have adopted the approach to assign long codes to the instructions offering few or no choices to the user, while assigning short codes to those where the flexibility to select variants is a desirable feature. The most frequent format for W is shown in Figure A.3. The word is subdivided into five fields, which we have labeled both symbolically (OP, x, m2, m1, and y) and with references that suggest their respective functions. Of these fields, two, the operation code W[15:12] and the operation mode W[8:6], are the most frequently used to specify the instruction command (typically, the operation code specifies the function, such as ADD, and the operation mode specifies the data type and selects the destination of the result). The format of Figure A.3 is well suited for instructions executing binary operations (such as ADD and SUB) or data transfers.

As we saw earlier, instructions of the first type (binary operations) involve two operands and a result, and therefore need to specify (or imply) three locations, two sources and one destination.[4] Referring to Figure A.3, one source operand is specified as D(x) (i.e., the operand register is a data register), while the second operand is identified by means of an elaborate address calculation

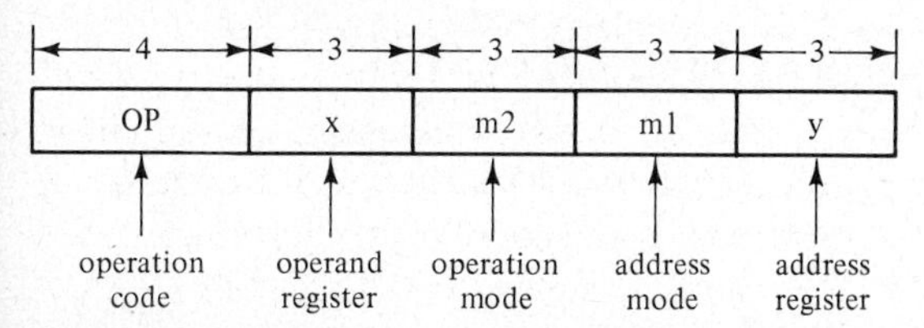

Figure A.3 Typical format of the first instruction word in the MC68000.

[4]The MC68000 has a much richer repertoire of binary operation instructions than the SEC, including multiplication, division, and additional logical operations on strings.

controlled by the two fields m1 and y. This calculation, to be described below, produces an address referred to as EA (Effective Address). The field m2 (operation mode) is then used to select the destination, that is, to determine whether

$$D(x) \leftarrow D(x) \text{ OP } EA$$

is executed, or the alternative

$$EA \leftarrow EA \text{ OP } D(x)$$

We shall now consider in some detail the *address calculation,* because it represents an area where the hardware–software compromise is most clearly illustrated, since many of the features are dictated by the objective to facilitate the programmer's task. The procedure is concisely illustrated in Table A.1 using the RTL notation, but not all of 14 available addressing modes are accounted for.

We now discuss the most salient feature of the addressing modes of Table A.1. The first two modes (m1 = 000 and 001) are quite straightforward and require no further comment. The next five modes, "register indirect," are significant variations of EA = M(A(y)) and afford a very desirable programming flexibility. Beginning with modes m1 = 011 (with postincrement) and 100 (with predecrement), we note that they allow the scanning in ascending or

TABLE A.1 ADDRESSING MODES IN THE MC68000

m1	RTL	Name of addressing mode	
000	EA = D(y)	Data register direct	Register direct
001	EA = A(y)	Address register direct	
010	EA = M(A(y))		Address Register Indirect
011	1. EA = M(A(y)) 2. A(y) ← A(y) + 1	With postincrement	
100	1. A(y) ← A(y) − 1 2. EA = M(A(y))	With predecrement	
101	EA = M(A(y) + M(PC + 1))	With displacement	
110	EA = M(A(y) + M(PC + 1)[7:0] + R(M(PC + 1)[15:12]))	With displacement and index	
111[a]	y = 000: EA = M(PC + 1))	Absolute	Special
	etc.	. . .	

[a]In this mode, the y field is used to specify (up to eight) new addressing modes which do not use a register.

descending order of an array of data items stored in consecutive memory locations. By using A(y) as an index register that is automatically incremented or decremented, we avoid the need of additional instructions for this purpose. But even more interesting is the use of these modes in the handling of stacks (refer to Chapter 7, Sec. 7.7, for the concept of stack). Suppose we have a stack "growing" according to ascending memory addresses. If we store the address following STACK TOP in register A(y) (which acts as the stack pointer), by means of the register indirect with postincrement mode we push data on to the stack, while the register indirect with predecrement mode realizes a stack pop. Since any address register can be used for this purpose, and since there are eight address registers (A(0)–A(7)), the user can concurrently maintain up to eight stacks. Quite useful also is mode m1 = 110 (with displacement and index). This mode needs an extra instruction word (called *extension* word), whose format is

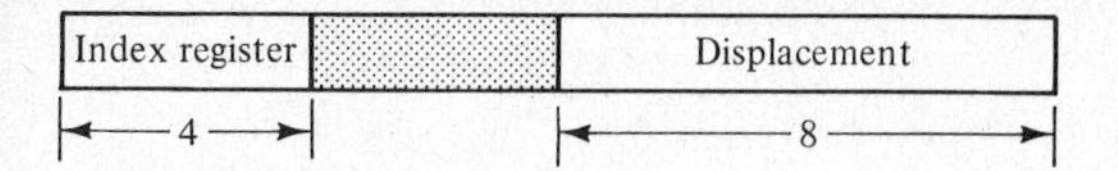

This extension word uses two fields, one for the displacement (as an *immediate* specification) and the other for the index register (as a *direct* specification). This addressing option enables the programmer to relocate a portion of program and use it in a larger program without adjusting the instruction addresses.

The second important class of instructions is represented by the data movement operations, which involve one source and one destination. Again the format of Figure A.3 is well suited, with the difference that the pairs of fields (x, m2) and (m1, y) now play the same role. Specifically, the pair (x, m2) is used to calculate the effective address EA_d of the destination, and (m1, y) is used to calculate EA_s, the effective address of the source. In summary the data transfer (MOVE) instruction carries out the action

$$EA_d \leftarrow EA_s$$

where each of EA_d and EA_s can be either a CPU register or a memory location.

Finally, we note that m1 = 111, y = 000 (absolute) specifies the usual memory reference mode, where the address of the operand is directly specified in the instruction (in an extension word).

The instruction repertoire of the MC68000 is extremely rich and articulate. There are 56 major instruction types, but, by controlling additional fields, such as m2 in Figure A.3, the programmer can deploy as many as about 1000 different commands.

A.3 CONTROL INSTRUCTIONS

A.3.1 SEC

The SEC has a family of instructions to alter the content of the program counter, that is, to transfer control in a manner different from the standard stepping

(PC ← PC + 1). These instructions are HLT, JMP, JPA, JZA, JPR, JOF, SFH, JSR, and JS. The instructions JPA, JZA, JPR, JOF, and SFH are conditional transfers, that is, the alteration of PC effected by any one of them is conditional on the values of some "status" variables which are A[20], EQ, and OF. (Of these only OF is stored in a flip-flop whereas A[20] and EQ are dynamically tested while circulating data through the ALU.) The subroutine linkage may bc accomplished either by JSR or by JS; both the return of control and the stack handling of nested subroutines is done by resorting to standard instructions (see Sec. 7.7).

A.3.2 MC68000

The MC68000 has control transfer capabilities not remarkably different from those of the SEC. However, there are two important features that deserve to be examined in some detail.

The first feature is a single general-purpose instruction for handling conditional transfers of control. As we mentioned earlier, the MC68000 has a 16-bit status register SR, of which only the lower half (the user byte SR[7:0]) is accessible to the user (Figure A.4). Of the 8 bits of the user status byte, only 5 are currently used. The format of the status byte is shown in Figure A.4. The five flip-flops X, N, Z, V, and C are set by the appropriate instructions (arithmetic and logical). Of these bits, N (the result is negative), Z (the result is zero), V (overflow has occurred), and C (a carry has been generated out of the most-significant bit position) represent well-known notions and need no discussion; X (extend) is set equal to C and is used in multiprecision computation (multiword operands), and will not be further considered here. In addition to the status of each of the 4 bits N, Z, V, and C, there are three significant conditions such as "greater than or equal," and so on, which are commonly tested in the program. These three conditions are simple boolean functions of the above variables. In total, these seven conditions and their negations yield 14 testable conditions. To test for any of them, the MC68000 uses a single instruction, called "branch conditionally" (BCC), which consists of two words and has the format

M(PC):	OP	Condition	8-bit displacement
M(PC + 1):	16-bit displacement		

The effected branching is described in RTL as

```
IF <condition> THEN PC ← PC + d
```

where d, the "displacement," is a nonzero integer given either by M(PC)[7:0] or (if M(PC)[7:0] = 0) by M(PC + 1), the content of the extension word.

The second feature to be examined is the mechanism for handling subroutines. Reference addresses (i.e., the current value of PC) are always handled by means of a stack, whose STACK TOP pointer is contained in the SP

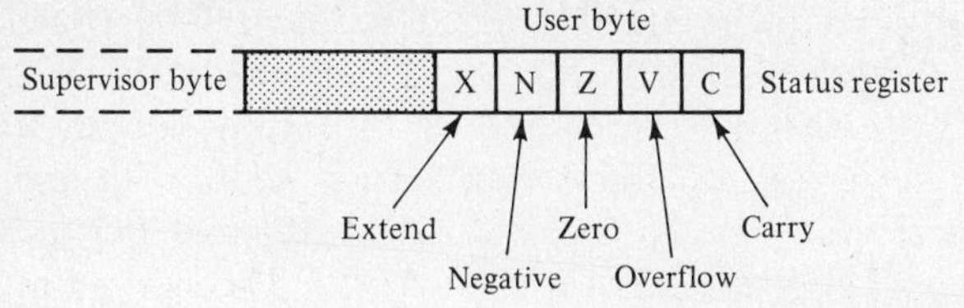

Figure A.4 Structure of the user byte in the status register SR.

register [stack pointer register, or A(7), see Figure A.2]. Thus the JSR (Jump to Subroutine) instruction automatically accomplishes, at the hardware level, the management of the stack (i.e., it pushes PC onto the stack), and is compactly described as follows:

1. M(SP) $\leftarrow$ PC + 1 2. PC $\leftarrow$ EA, SP $\leftarrow$ SP − 1

Recall that EA is the "effective address" and notice that the stack grows toward decreasing addresses. Notice also that the return address will be the location immediately following the one of the current instruction.

Besides this JSR instruction, the MC68000 also has an instruction to perform the reverse linkage, that is, RTS (return from subroutine). This instruction pops the address stored at the top of the subroutine address stacks and transfers control to it, according to the description

1. PC $\leftarrow$ M(SP) 2. SP $\leftarrow$ SP + 1

There are also two variants of the RTS instruction, which we shall not describe here.

A.4 EXTERNAL BUS CONTROL AND INPUT/OUTPUT PROCESSES

The specification of the user's architecture is completed by the way in which the CPU is interfaced with the outside world. Indeed, if we do not limit the notion of "use" to programming, but extend it to include the deployment of the microprocessor in a larger system (such as the use of the microprocessor for communication or process control), then the detailed knowledge of the I/O interface is essential. In this respect there are some substantial differences between the SEC and the MC68000, and they will be pointed out in the following discussion.

A.4.1 SEC

The external bus structure of the SEC-XR is summarized in Figure A.5. This version of SEC does not exhibit the direct memory access feature described in Chapter 8, Sec. 8.4, nor any form of priority (or "vectored") I/O interrupt, as

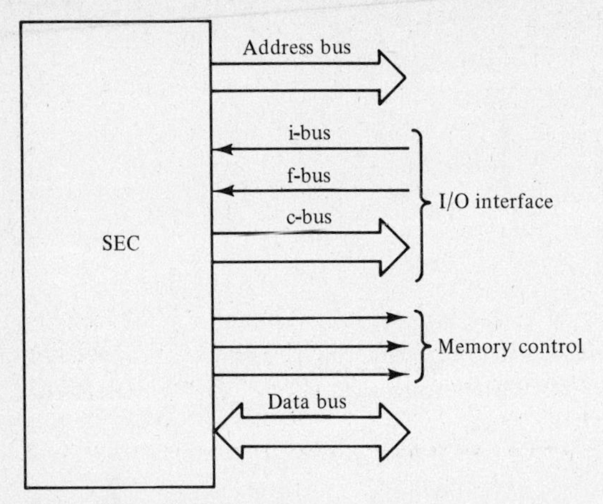

Figure A.5 SEC interfacc.

considered in Problems 8.7 and 8.8. We have a 21-bit bidirectional data bus, a 12-bit address bus, a set of memory control signals, and the I/O interface signals (i-bus, f-bus, and c-bus). Although both the memory and the I/O peripherals share the same address and data buses, the controls for these two types of external units (memory and I/O) are entirely separate. Indeed, there are separate instructions for the two types of units; the RD, WRT, and SFH instructions exclusively concern the peripheral units, which communicate with the CPU only through a rather elaborate protocol. This I/O philosophy is appropriately referred to as *isolated I/O*.

In the diagram of Figure A.5, the CPU has total control of the interface and determines which of the external units is to have access to the data bus. We must remember, however, (see Chapter 8, Sec. 8.4) that, were we to deploy the DMA mode, there would have to be some form of bus arbitration module which would determine which unit (whether the CPU or an external unit) would have control of the address and data buses.

A.4.2 MC68000

The I/O philosophy of the MC68000 represents one of the major alternatives to isolated I/O. Indeed, no distinction is made between memory and other external units, by using the same addressing scheme for memory locations and peripherals. In other words, for the CPU a peripheral unit is just an address in the overall address range from 0 to $2^{23} - 1$, and the same instructions are used to refer to either class of devices. This I/O philosophy is appropriately known as *memory-mapped I/O,* since to the programmer a peripheral port appears just as an address in memory. This of course entails a uniform mechanism for handling the communication between the CPU on one side and memory and the peripherals on the other. The external bus structure is (partially) illustrated in Figure A.6. Besides the 16-bit data bus and the 23-bit address bus and other lines whose discussion is beyond our current scope, there are three sets of control signals, the

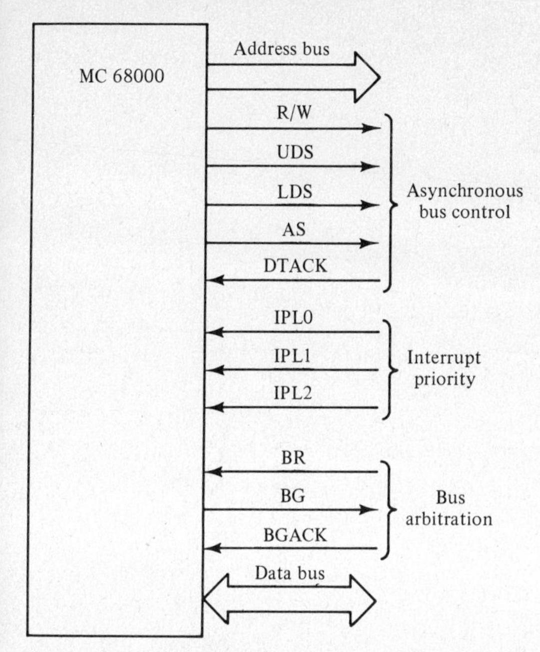

Figure A.6 MC68000 interface.

asynchronous bus control, the interrupt control, and the bus arbitration control, which we shall now consider one by one in detail.

Asynchronous Bus Control The I/O mechanisms must be capable of handling devices with greatly different response times. Thus, it must be based on a request-acknowledge protocol (handshake) in order to correctly exchange data. Such a scheme, which in the isolated I/O applies only to the peripheral units, in the memory-mapped I/O applies to the main memory as well. This protocol is referred to here as *asynchronous bus control.*

We assume for the time being that the CPU has control of the buses, and analyze a read (input) operation, with the aid of Figure A.7.[5] The first action is the issue of a request by the CPU. Referring to the timing diagram of Figure A.7(b), the distinguishing signal is ADDRESS STROBE (abbreviated AS). Signal AS is accompanied by other signals, which specify the type of request, read or write (input or output), and data type (byte or word). The accessed peripheral unit receives the request (by recognizing its address) and, whenever ready to transfer data, it acknowledges receipt of it by activating the signal DATA TRANSFER ACKNOWLEDGE (abbreviated DTACK). Simultaneously, it places data on the data bus. The CPU then latches the data in an input buffer and releases the peripheral unit by deactivating AS, which in turn causes deactivation of DTACK and disconnection of the peripheral from the data bus.

[5]In the physical realization, the MC68000 uses *negative logic* in the I/O interface. To avoid totally unnecessary confusion, we stipulate calling a signal *active* when its action takes place (corresponding to the true conditon), and keep with the convention of representing it as a *high signal* in timing diagrams.

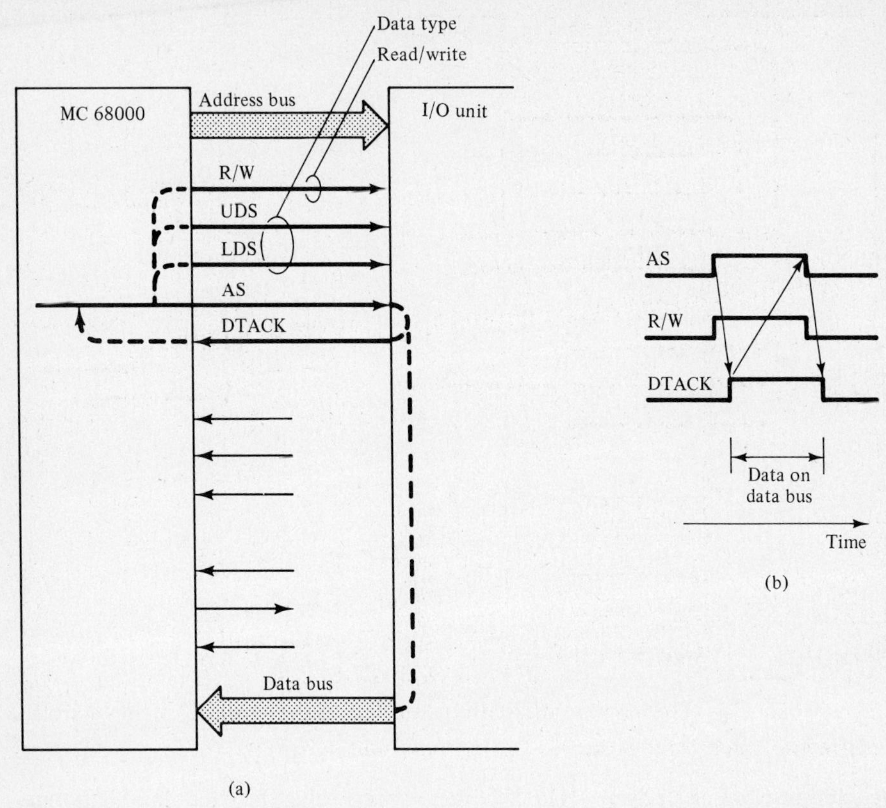

Figure A.7 Asynchronous bus control interface. (a) Signal lines. (b) Timing diagram.

The actions occurring in a write (output) operation are analogous, except for the appropriate modifications. Specifically, upon issuing the data transfer request (AS and the accompanying signals are activated), the CPU also places the output data on the data bus. The peripheral unit recognizes the request and, whenever ready to accept the data, it latches it in its own buffer and returns DTACK to the CPU. The conclusion of the transaction is then straightforward.

Interrupt Handling The above general mechanism is a satisfactory way to handle the data transfers between the CPU and the peripheral units. Of these, the memory is totally "enslaved" to the CPU, since—as long as the CPU retains control of the buses—the memory is constantly available to it. This is not so for the peripheral devices, which require service from CPU when they are ready to exchange data. For these, the MC68000 adopts the interrupt I/O mode, which we described for the SEC in Sec. 8.3. We shall now briefly illustrate the details of interrupt handling in the MC68000.

First of all, there are eight distinct priority levels (0 through 7), 0 denoting the lowest priority and corresponding to the execution of user's program CPU-memory instructions. Interrupts can be nested under the constraint of priority, in the sense that an interrupt is accepted by the CPU if it has higher priority than the one currently being serviced. With reference to Figure A.8(a),

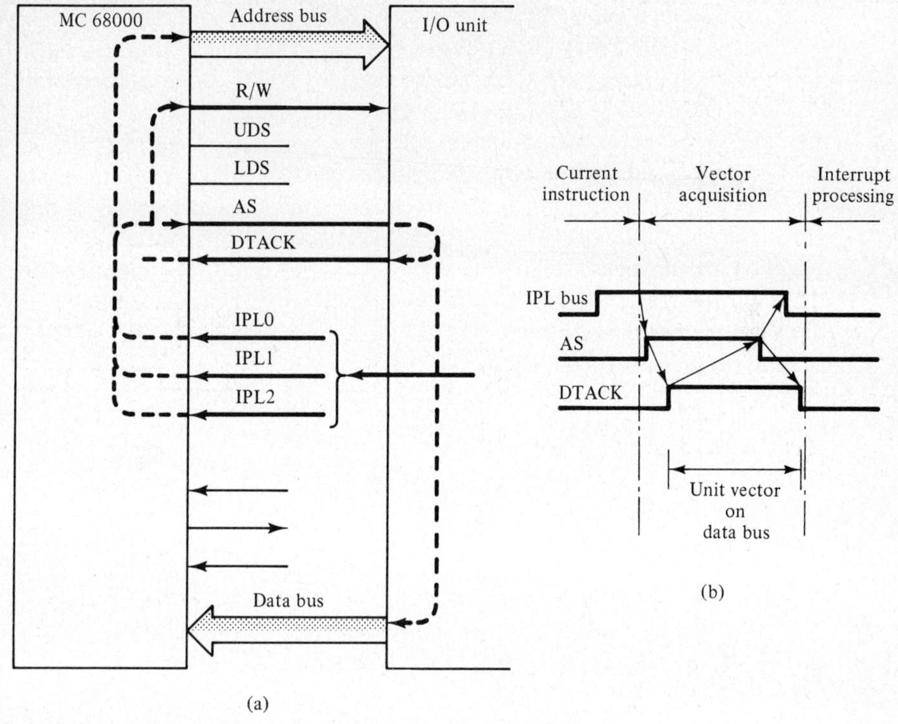

Figure A.8 Illustration of the interrupt handling protocol.

there is a three-line interrupt priority bus (lines IPL2, IPL1, and IPL0). This bus not only carries the interrupt request to the CPU, but also encodes in binary the priority level of it. It is assumed that external logic ensures that, when these lines are in a configuration different from 000, there is a single device originating the interrupt request. The processor stores the current priority level in the system byte of the status register SR as shown in Figure A.9. When an interrupt request is received, the following sequence of actions takes place [refer to the timing diagram in Figure A.8(b)]:

1. At the completion of the current instruction (just as in SEC), the CPU compares the priority level IPL[2:0] of the interrupt with the current interrupt level SR[10:8]. If IPL[2:0] $\leq$ SR[10:8], the interrupt request is put on hold; otherwise it is accepted and acknowledged by issuing in turn a request for the peripheral identification (while SR[10:8] $\leftarrow$ IPL[2:0]). This request is in the form of a "read" opera-

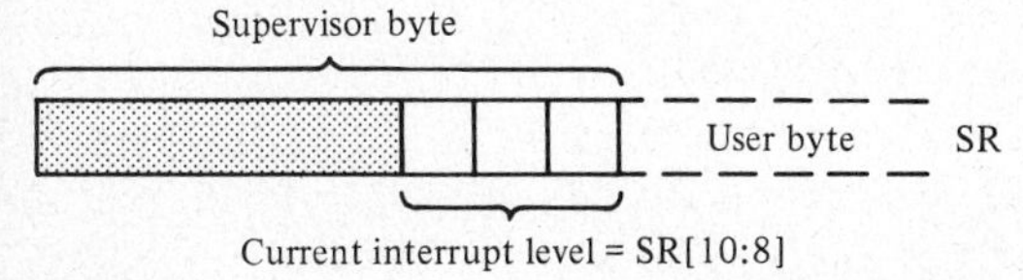

Figure A.9 Structure of the supervisor byte in the status register SR.

tion and is implemented by the following signals (the following description is a simplified version of the actual operation):

a. the new interrupt level is placed on lines A[3:1] of the address bus (this acts as an address for the interrupting unit).

b. R/W is set to read and AS is activated.

2. The peripheral unit receives the request, acknowledges it by activating DTACK, and transmits its identification (conventionally called *vector*) on the data lines D0–D7.
3. The CPU receives the identification vector, stores it in the input buffer, and disactivates the AS.
4. The peripheral unit disactivates DTACK, recognizing that the transaction has been successfully completed.

At this point the CPU initiates the interrupt processing. The acquired peripheral vector is used to identify uniquely the address of the I/O subroutine handling that particular unit (of course, the CPU operates now in the supervisor mode).

Bus Arbitration So far we have assumed that the CPU has control of the bus structure. However, if the system includes direct memory access devices such as a disk drive, then there must be provisions to assign bus control to any of these devices. As we saw in Chapter 8, Sec. 8.4, this function is called *bus arbitration.*

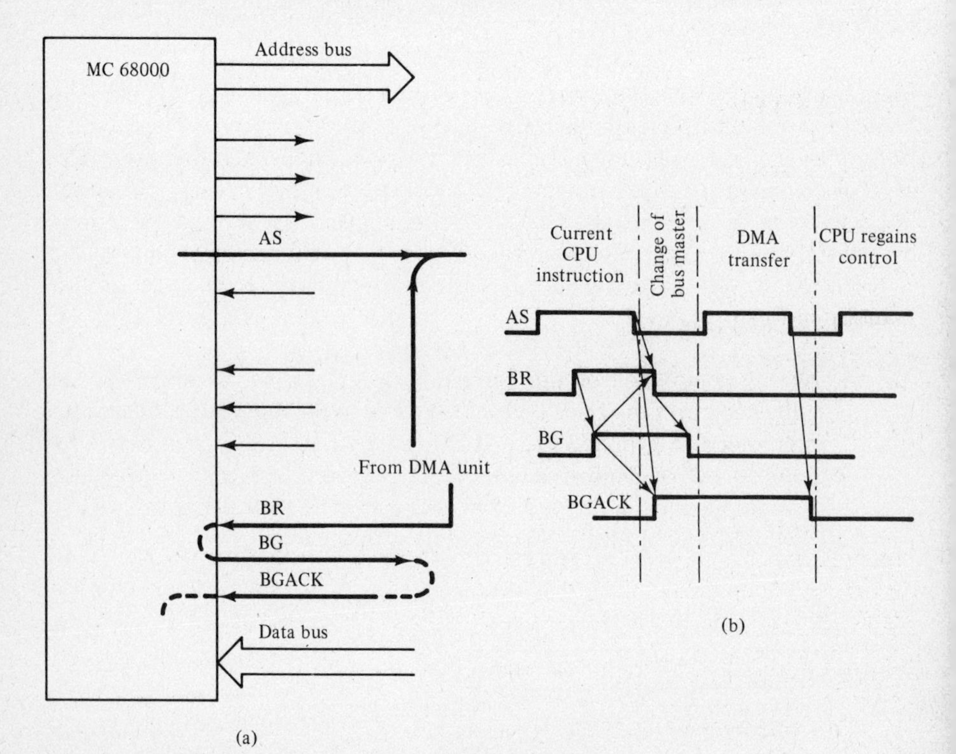

Figure A.10 Bus arbitration interface. (a) Signal lines. (b) Timing diagram.

The unit currently having control of the buses is called the "bus master," and the normal bus master is obviously the CPU. The CPU, however, has the lowest priority as bus master and must relinquish control of the buses (at the completion of the current instruction) to any DMA device requesting to become bus master. Referring to Figures A.10(a) and A.10(b), suppose that the CPU is the current bus master. A DMA device seeking bus mastership sends its request on the BUS REQUEST bus (abbreviated BR). The CPU immediately acknowledges this request, by activating the BUS GRANT (abbreviated BG) signal. The DMA device must now wait for the current bus use to be completed (as evidenced by the active AS) before gaining control; when this occurs, it simultaneously cancels its request (by disactivating its input to the BR bus) and sends an acknowledgment to the CPU by activating the BUS GRANT ACKNOWLEDGE signal (abbreviated BGACK). At this point the CPU relinquishes control of the buses by disactivating BG and waits for future availability of the buses. The DMA device, which has become the new master, performs a memory data transfer according to the same protocol outlined for the CPU. At the completion of this transfer, it returns the bus mastership to the CPU by disactivating its BGACK signal (in other words, a DMA device asserts its current bus master status by means of the BGACK signal).

The preceding discussion outlines the main features of the general I/O organization in the MC68000. Many details (some very significant) have been omitted, in favor of the more fundamental aspects. The interested reader will, again, find a wealth of details in the literature already cited.

A.5 SOFTWARE INTERRUPTS—TRAP INSTRUCTIONS

There is a special type of software interrupts, used in the MC68000 and in other computers, of which there is no counterpart in the SEC (although nothing prevents us from adding this feature to the instruction repertoire of the SEC). This software interrupt is provided by the so-called *trap instructions*. A trap instruction is basically an aid for the programmer in debugging a program. Before discussing how it can be used, it is convenient to describe its action. In the MC68000, the main trap instruction (mnemonic code, TRAP) has the following format:

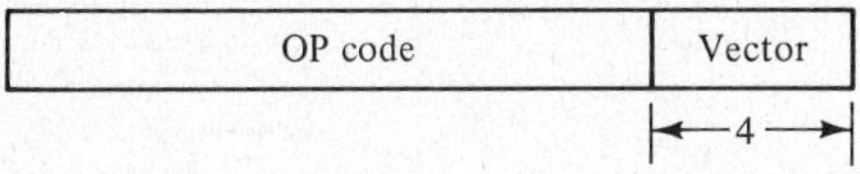

OP code	Vector
	←4→

The rightmost 4 bits of the instruction word, called "vector," are translated by the CPU into an actual memory address, which is the starting address of a trap handling subroutine. In RTL, the action of TRAP may be described as follows:

1. SP ← SP − 1 } stack push
2. M(SP) ← PC }
3. SP ← SP − 1 } stack push
4. M(SP) ← SR }
5. PC ← M (128_{10} + <vector>)

In words, the contents of the registers PC and SR are saved on the system's stack, while control is transferred to the address contained in the memory cell at location 128_{10} + <vector>. This address is the starting address of the subroutine mentioned above. This subroutine will complete its execution by a "return" instruction which will restore the original contents of SR and PC.

Since the trap instruction saves the contents of PC and SR, the programmer has access to the value of these important parameters (particularly SR) at crucial points of the program, by inserting at these points trap instructions and then designing appropriate trap handling subroutines.

Index

Absorption, 56
Abstract terms. *See* Nonterminal terms
Accumulators, 28, 220
ACT instruction, 249, 251
  EXECUTE microsequence of, 254
Adder, 13, 47, 173
  carry-look-ahead, 187–195
  full, 184–186
  parallel, 186–195
  ripple-through, 186
Adder cell, 47–48, 184–186
ADD instruction, 29
  EXECUTE microsequence of, 222
Addition, 12–13
  of integers in two's complement notation, 178–183
Address, 26
  explicit and implicit, 30
Address logic (in microprogrammed control unit), 279, 284
Addressing modes
  in MC68000, 298–299
  in SEC, 239–240
Address modification, 223–225
Adjacent minterms, 79
Algebra
  boolean. *See* Boolean algebra
  switching. *See* Switching algebra
Alphanumeric codes, 16–17
ALS instruction, 223
ALU. *See* Arithmetic-logic unit
American Standard Code for Information Interchange (ASCII), 16–17
Analog signals, 4
AND function, 54–57
AND gate, 50–51
AND instruction, 29, 222
AND matrix (AND plane), 103
AND-to-OR network, 60, 61, 78
Architectural design level, 37
Architecture
  designer's, 293–294
  programmer's, 294
Arithmetic
  binary, 12–13
  fixed-point, 203
  floating point, 203
Arithmetic-logic instructions, 29–30
Arithmetic-logic unit (ALU), 25, 195–202, 220
Arithmetic shift, 200–201
ARS instruction, 223
ASCII. *See* American Standard Code for Information Interchange
<Assignment> (in RTL), 209, 212–213
Associativity, 56

Asynchronous sequential networks
(ASN), 121–130
Automata theory, 165

Backus-Naur Form (BNF), 208
Base (of positional notation), 6
Base conversions, 6–12
of integers, 7–9
of fractions, 9–12
BCD. *See* Binary coded decimal
Binary adder. *See* Adder
Binary arithmetic, 12–13, 20, 173–184
Binary coded decimals (BCD), 15
Binary comparator, 89–92
Binary counters, 146, 147, 163–164
modulo-k, 168
Binary operation, 54
Binary time waveform, 117–118
Block codes, 18
BNF. *See* Backus-Naur Form
Boolean algebra, 54, 68–69
fundamental theorem of, 63
Boolean expressions, 51
normal and canonical forms, 58–63
Bootstrap loader, 269
Borrow (in subtraction), 12
Buffer register, 247
Bus arbitration
in SEC, 264–265
in MC68000, 306–307
Buses, 214–215
Busy wait, 247

Calculating machines, 40
Canonical form (of boolean expressions),
59–62
transformation from normal form to,
59–63
Card, punched, 267
Carry (in addition), 12
end-around, 184
"Carry-look-ahead" parallel adders,
187–195
C-bus. *See* Command bus
Central processing unit (CPU), 220–221
Character (input/output), 245
Clocked flip-flops, 134–138
CLOCK signal, 130–134
Closed subroutines, 229
Code translators, 103
Coding (of information), 5
Coding theory, 20
Cold start, 267–269
Combinational components, 38, 47
Combinational modules, 99–112
Combinational networks, 49, 51–53,
72–113
analysis of, 73–75
design of all-NAND, all-NOR, 97–99
input-assigned, 52
minimization of. *See* Minimization procedure
synthesis of, 75–78
Command bus (c-bus), 253
Commutativity, 55
Complementarity, 55
Complement function, 50, 54–57
Complement notations (for signed integers)
one's complement, 183–184
two's complement, 174–178
Compound statement, 75
Computer word, 26
Concrete terms. *See* Terminal terms
<Concurrent step> (in RTL), 209
<Condition> (in RTL), 208
<Conditional assignment> (in RTL), 209
Consensus, 57
Constants (constant functions), 50
Contact debouncer, 126–128
Control instructions, 30, 299–301
of SEC, 299–300
of MC68000, 300–301
Control memory, 280
Control signals, 35, 195–202, 273–279
Control unit, 25, 275–276
function of, 274–279
microprogrammed, 279–286
Control vector, 274, 279
Conversions. *See* Base conversions
Counter
in loop programs, 31, 225
binary. *See* Binary counters
CPU. *See* Central processing unit

DAR instruction, 266–267
Data break, 264–267
Data Buffer (DB), 247
Data selector/multiplexer, 107
Data transfers, 28–29, 249

DAW instruction, 266
DB. *See* Data Buffer
Decoder, 101–103
Decoder/demultiplexer, 112
Degenerate switching functions, 49–50
Delay diagrams, 190, 193
Delay flip-flop, 135
De Morgan's Law, 56
Demultiplexers, 111–112
  decoder/demultiplexer, 112
DER instruction, 226
Designer's architecture, 293–294
Design levels, sequence of, 40
Device activation, 249
Device design level, 39
Digital signals, 4
Diodes, 39*n*
Direct addressing, 29, 239
Direct Memory Access (DMA), 264–267
Distributivity, 55
DMA. *See* Direct Memory Access
DO LOOP, 31
Don't care entry, 89–92
Duality, principle of, 56

Edge trigger, 128–130
Edge-triggered flip-flop, 132, 134
EDVAC, 41
Electronic design level, 39
Encoder, 100-101
  priority encoder, 101
End-around carry, 184
ENIAC, 41
ENR instruction, 226
ENT instruction, 29
  EXECUTE microsequence of, 222
Equivalent states, 149
Error control, 6
  codes for, 17–19
Errors, 17
Essential k-cube, 84
Essential prime implicant, 93
Excitation functions, 137, 142, 153
EXCLUSIVE OR connective, 65–67
EXECUTE microsequence, 221–223
Execute phase (of instruction) 35, 36
Explicit addresses, 30

Fan-in, 53
Fan-out, 53
F-bus. *See* Flag bus
Feedback, 118–121
FETCH microsequence, 221
Fetch phase (of instruction), 35, 36
Firmware, 280
Fixed-point arithmetic, 203
Flag bus (f-bus), 253
Flip-flop, 38
  clocked, 134–138
  edge-triggered, 132, 134
  delay (D), 135
  gated set-reset (SR), 131–133
  JK, 135, 136
  master-slave, 132–134
  toggle (T), 135, 137
Floating-point arithmetic, 203
Flowchart, 24, 32
Fractions, conversion of, 9–12
Formal Grammar, 207
Full adder, 184–186
Functional design level, 37
Fundamental mode (of asynchronous networks), 124–125
Fundamental product. *See* Minterm
Fundamental sum. *See* Maxterm

Gated latches, 131–133
Gated set-reset flip-flops (SR flip-flops), 131–133
Gates, 38–39
General registers, 296
GO TO (in RTL), 209
Grammar. *See* Formal Grammar
Gray code, 15–16

Handshake, 261, 265, 303
Hardware-software trade-off, 227, 231, 236, 264
Hexadecimal notation, 13–14
High-level languages, 41
HLT instruction, 30

I-bus. *See* Interrupt bus
Idempotency, 55
Identity function, 50
Immediate addressing, 29, 239, 299
Implicant, 92
Implicit addresses, 30, 240
Incompletely specified function, 89
Index register instructions, 226, 283

Index registers, 223–228
Indirect addressing, 237–239, 240
Input-assigned network, 52
Input/output, 25
  interrupt, 256–264
  isolated, 302
  memory-mapped, 302
  programmed, 249–256
Input/output character, 245
Input/output devices
  exogenous, 246
  static, 246
  transport, 245
Input/output instructions, 249
Input/output port, 248
Input/output service routine, 259–261, 263–264
INR instruction, 226
Instruction, 26
Instruction codes. *See* Operation codes
Instruction execute phase. *See* Execute phase
Instruction fetch phase. *See* Fetch phase
Instruction format, 27
  in SEC, 27
  in MC68000, 297
Instruction register (IR), 220
Instruction repertoire, 28
Instruction set, SEC-XR summary of, 285
Integers, conversion of, 7–9
Internal states, 123
Internal variables, 122, 142
Interrupt
  priority, 261
  software, 307–308
  vectored, 261
Interrupt bus (i-bus), 257
Interrupt handling, 304–306
Interrupt input/output, 256–264
Interrupt protocol, 261
INTERRUPT TEST, 259
Inverter, 50–51
Involution, 54
I/O. *See* Input/output *entries*
IR. *See* Instruction register
Isolated input/output, 302

JK flip-flop, 135, 136
JMP instruction, 30
  EXECUTE microsequence of, 222
JOF instruction, 222
JPA instruction, 30
JPR instruction, 226
JS instruction, 231
JSR instruction, 232
JZA instruction, 30
  EXECUTE microsequence of, 222

Karnaugh map, 79
  minimization technique based on, 79–88
K-cube (in Karnaugh map), 83
  largest, 84
  essential, 84
K-map. *See* Karnaugh map

Language symbols (of RTL), 209
Language terms (of RTL), 209
Large-scale integration (LSI), 77
Last-in-first-out (LIFO) policy, 233, 261
Latch, 119–121, 130, 133
  gated, 130
LDA instruction, 29
  EXECUTE microsequence of, 222
LDR instruction, 226
LIFO. *See* Last-in-first-out policy
Linear selection, 218
Linkage (of subroutine), 229
Literals, 58
LLS instruction, 223
Loaders, 268–269
  bootstrap, 269
Logical completeness, 63
Logical design level, 38
Logical shift, 200
Loop program, 31–34
LRS instruction, 223
LSI. *See* Large-scale integration

Machine-independent languages, 41
Machine language programs, 30–31
MAR. *See* Memory Address Register
Master-slave flip-flops, 132–134
Maxterm, 61–62
MBR. *See* Memory Buffer Register
MC68000, 293
  addressing modes in, 298–299
  control instructions of, 300–301
  instruction format, 297
  programmer's architecture of, 295–299
Medium-scale integration (MSI), 77

Memory, 25
control, 280
paged, 236–237
random access (RAM), 216–220
read-only. *See* Read-only memories
Memory Address Register (MAR), 217
Memory Buffer Register (MBR), 217
Memory-mapped input/output, 302
Memory reference (instructions), 29
Micro-operations, 272–274
Microprocessors, 291–308
Microprogrammed control units, 279–286
Microprogramming, 279–280
horizontal, 279
vertical, 279
<Microsequence> (in RTL), 208–209
Microsequence implementation, 212–215
<Microstep> (in RTL), 209
Minimal normal expressions, 84
Minimization procedure
POS (on Karnaugh map), 87–88
SOP (on Karnaugh map), 84–87
tabular, 92–97
Minterm, 59–62
adjacent, 79
Minterm/prime implicant table, 96
Mnemonic codes (of instructions), 29
Modulus-and-sign representation, 28, 173–178
Motorola MC68000. *See* MC68000
MSI. *See* Medium-scale integration
MSI modules, 99–112
Multiplexers, 106–111
data selector/multiplexer, 107

NAND connective, 63–65
NAND gate, 39
Negation,
of boolean entities, 56
of signed integers, 177
Negative logic, 73
Negative numbers (representation of)
complement form, 174–178, 183–184
sign-and-modulus, 28, 173–178
NEG instruction, 223
Nested subroutines, 233–236
Networks
combinational. *See* Combinational networks
sequential. *See* Sequential networks
synchronous sequential, 138, 140–157
Next-state equations, 134
Next-state functions, 144–145, 152
Next-state table, 144
Noncritical race, 125
Nondegenerate switching functions, 49–50
Nonterminal terms (of formal grammar), 207
NOR connective, 63–65
Normal expressions, minimal, 84
Normal form (of boolean expressions), 58–63
product-of-sums (POS), 61
sum-of-products (SOP), 58
transformation to canonical form, 59–63
NOT gate, 64

Octal notation, 14
One-hot assignment, 151
One's complement notation, 183–184
Open-collector output realization, 214–215
Open subroutines, 228–229
Operand-free instructions, 223, 283
Operands, 27
Operating system, 293
Operation codes, 27
OR function, 54–57
OR gate, 50–51
OR instruction, 29
OR matrix (OR plane), 103
OR-to-AND networks, 78
Overflow, 179–182

Paged memory, 236–237
Pages (of memory), 236
Paper-tape reader, 245
Parallel adder, 186–195
Parallel decoder, 103
Parallel register, 138, 139
Parallel-to-serial converter, 109
Parameter exchanges (of subroutine), 230
Parity check codes, 18–19, 155
Parsing (of RTL), 210
Perfect induction (Principle of), 55–56
Peripheral unit (PU), 264–267
PLA. *See* Programmable logic array
Pointer instruction, 34
Pointers (in programs), 31, 225
Polling, 259
POP (of stack), 234

Positional notation, 6, 19
Positive logic, 73
POS (product-of-sums) form, normal, 61–62
POS minimization procedure, 87–88
Prime implicants, 84, 93
  essential, 84, 93
Priority encoder, 101, 261, 271
Priority interrupt, 261
Product code, 18
Productions, 207
Product-of-sums. *See* POS *entries*
Product terms. *See* K-cube
Program counter (PC), 220
Programmable logic array (PLA), 106
  design technique based on, 157–160
Programmed input/output, 249–256
Programmer's architecture
  of SEC, 294–295
  of MC68000, 295–299
Programs, 26
  loop, 31–34
  in machine language, 30–31
  straight line, 31
Propagation delay, 72, 118, 188
Protocol
  for input/output, 261, 265, 303–304
  for subroutine, 230
PU. *See* Peripheral unit
Pulse synchronizer, 125–126
Punched card, 267
PUSH (of stack), 234

Quine-McCluskey minimization technique, 92–97

Race, noncritical, 125
Radix, of positional notation, 6
Random access memory (RAM), 216–220. *See also* Memory
RD instruction, 254, 256
READ-CARD button, 267
Read-only memories (ROMs), 103–106
  design technique based on, 157–160
Redundancy (in coding), 18
Register, 28
  general, 296
  index, 223–228
  parallel, 138, 139
  universal, 160–163
Register-transfer design level, 37
Register Transfer Language (RTL), 207–212
Representation of information, 3–20
Representation of integers
  modulus-and-sign, 28, 173
  one's complement, 183–184
  two's complement, 174–178
Ripple through adder, 186–195
ROMs. *See* Read-only memories
RTL. *See* Register Transfer Language

SEC (Simplistic Educational Computer), 26–34
  addressing modes of, 239–240
  control instructions of, 299–300
  programmer's architecture of, 294–295
SEC-0 computer, 220–221
SEC-XR computer, 223–224
  external bus structure of, 301–302
  summary of instruction set, 285
SEC-XR-PM computer, 238, 240
Self-modifying code, 34
Sequence detector, 146–147
Sequencer, 147–148, 276
Sequential machine (SM), 141–143
  state diagram, 146, 148–150
  theory, 164–165
Sequential network
  asynchronous. *See* Asynchronous sequential networks
  synchronous. *See* Synchronous sequential networks
Set-reset flip-flop (SR), 130
SFH instruction, 251–252
  EXECUTE microsequence of, 253–254
Shift
  logical, 200
  arithmetic, 200
Shifting network, 200
SIE instruction, 259
Simple statements, 75
Single-error detecting code, 18–19
SM. *See* Sequential machine
SN. *See* Sequential network
Software interrupts, 307–308
SOP minimization procedure, 84–87
SOP (sum-of-products) form normal, 58–59

Stable states, 123
Stacks, 233–236, 262–264
STA instruction, 29
  EXECUTE microsequence of, 222
  implementation of, 274
State assignment, 150–152
  one-hot, 151, 276
State diagram
  SM, 146, 148–150
  SN, 145–146, 150–152
Statement
  simple, compound, 75
States (of sequential network)
  equivalent, 149
  internal, 123
  stable, 123
  total, 123
State variables. *See* Internal variables
Static devices, 246
Storage components, 47
Stored-program computer, 41
STP instruction, 249, 251
  EXECUTE microsequence of, 254
Straight-line program, 31
STR instruction, 226
SUB instruction, 29
Subroutine calls, 229
Subroutines, 228–233
  closed, 229–233
  nested, 233–236
  open, 228–229
Subtraction, 12–13
  of integers in two's complement notation, 178–183
Subtractor, 173–174
Sum-of-products. *See* SOP *entries*
Supervisor state, 293
Switching algebra, 54–57, 68
  summary of, 57
Switching functions, 49
  degenerate, 49
  incompletely specified (with don't cares), 88–92
  nondegenerate, 49
Synchronous sequential networks, 138, 140–157
  analysis, 143–147
  model, 141
  synthesis, 147–157

Tabular minimization procedure, 92–97
Terminal terms (of formal grammar), 207
Terms (of formal grammar), 207
Timing diagrams, 117–118
Toggle flip-flop (T), 135, 137
Total states, 123
Transcoder, 103–104
Transfer instructions, 29
Transfer of control instructions, 30
Transistors, 39*n*
Transition table, 123–124, 144
Transport input/output devices, 245–246
Trap instructions, 307–308
Truth table, 49
Two-out-of-five codes, 19
Two's complement notation, 175–178
  addition and subtraction of integers in, 178–183

Unary MINUS, 209
Unary operation, 54
Universal register, 160–163
User state, 293

Variables, 31
Vectored interrupt, 261
Very-large-scale integration (VLSI), 77
von Neumann computer, 25, 41, 291

Wait, busy, 247
Waveform, binary time, 117–118
Window on the input sequence method, 155
Wired logic, 215
WRT instruction, 249, 251
  EXECUTE microsequence of, 254

XOR connective, 65–67